T0226302

Lecture Notes in Computer Science 10781

Commenced Publication in 1973
Founding and Former Series Editors:
Gerhard Goos, Juris Hartmanis, and Jan van Leeuwen

Editorial Board

David Hutchison
  Lancaster University, Lancaster, UK
Takeo Kanade
  Carnegie Mellon University, Pittsburgh, PA, USA
Josef Kittler
  University of Surrey, Guildford, UK
Jon M. Kleinberg
  Cornell University, Ithaca, NY, USA
Friedemann Mattern
  ETH Zurich, Zurich, Switzerland
John C. Mitchell
  Stanford University, Stanford, CA, USA
Moni Naor
  Weizmann Institute of Science, Rehovot, Israel
C. Pandu Rangan
  Indian Institute of Technology, Madras, India
Bernhard Steffen
  TU Dortmund University, Dortmund, Germany
Demetri Terzopoulos
  University of California, Los Angeles, CA, USA
Doug Tygar
  University of California, Berkeley, CA, USA
Gerhard Weikum
  Max Planck Institute for Informatics, Saarbrücken, Germany

More information about this series at http://www.springer.com/series/7407

Mauro Castelli · Lukas Sekanina
Mengjie Zhang · Stefano Cagnoni
Pablo García-Sánchez (Eds.)

Genetic Programming

21st European Conference, EuroGP 2018
Parma, Italy, April 4–6, 2018
Proceedings

 Springer

Editors
Mauro Castelli
Universidade Nova de Lisboa
Lisboa
Portugal

Lukas Sekanina
Brno University of Technology
Brno
Czech Republic

Mengjie Zhang
Victoria University of Wellington
Wellington
New Zealand

Stefano Cagnoni
University of Parma
Parma
Italy

Pablo García-Sánchez
University of Cádiz
Cádiz
Spain

ISSN 0302-9743   ISSN 1611-3349 (electronic)
Lecture Notes in Computer Science
ISBN 978-3-319-77552-4   ISBN 978-3-319-77553-1 (eBook)
https://doi.org/10.1007/978-3-319-77553-1

Library of Congress Control Number: 2018934369

LNCS Sublibrary: SL1 – Theoretical Computer Science and General Issues

© Springer International Publishing AG, part of Springer Nature 2018
This work is subject to copyright. All rights are reserved by the Publisher, whether the whole or part of the material is concerned, specifically the rights of translation, reprinting, reuse of illustrations, recitation, broadcasting, reproduction on microfilms or in any other physical way, and transmission or information storage and retrieval, electronic adaptation, computer software, or by similar or dissimilar methodology now known or hereafter developed.
The use of general descriptive names, registered names, trademarks, service marks, etc. in this publication does not imply, even in the absence of a specific statement, that such names are exempt from the relevant protective laws and regulations and therefore free for general use.
The publisher, the authors and the editors are safe to assume that the advice and information in this book are believed to be true and accurate at the date of publication. Neither the publisher nor the authors or the editors give a warranty, express or implied, with respect to the material contained herein or for any errors or omissions that may have been made. The publisher remains neutral with regard to jurisdictional claims in published maps and institutional affiliations.

Printed on acid-free paper

This Springer imprint is published by the registered company Springer International Publishing AG
part of Springer Nature
The registered company address is: Gewerbestrasse 11, 6330 Cham, Switzerland

Preface

The 21st European Conference on Genetic Programming (EuroGP) took place in the historical central building of the University of Parma, via Universitá 1, Parma, Italy, April 4–6, 2018.

Genetic programming (GP) is a unique field of research. It uses the principles of Darwinian evolution, already well-known in genetic algorithms and other areas of evolutionary computation, to approach problems in the synthesis, improvement, and repair of computer programs. The universality of computer programs, and their importance in so many areas of our lives, means that the automation of these tasks is an exceptionally ambitious challenge with far-reaching implications. It has attracted a very large number of researchers and a vast amount of theoretical and practical contributions are available by consulting the GP bibliography[1].

Since the first EuroGP event in Paris in 1998, EuroGP has been the only conference exclusively devoted to the evolutionary generation of computer programs. Indeed, EuroGP represents the single largest venue at which GP results are published. It plays an important role in the success of the field, by serving as a forum for expressing new ideas, meeting fellow researchers, and initiating collaborations. It attracts scholars from all over the world. In a friendly and welcoming atmosphere authors present the latest advances in the field, also presenting GP-based solutions to complex real-world problems.

EuroGP 2018 received 36 submissions from around world. The papers underwent a rigorous double-blind peer review process, each being reviewed by at least three members of the international Program Committee.

The members of the Program Committee encountered an exceptionally high standard this year, with papers proposing innovative and disruptive ideas. Among the papers presented in this volume, 11 were accepted for full-length oral presentation (30.6% acceptance rate) and eight for short talks (52.8% acceptance rate for both categories of papers combined). Authors of both categories of papers also had the opportunity to present their work in poster sessions.

The wide range of topics in this volume reflects the current state of research in the field. Thus, we see topics and applications including analysis of feature importance for metabolomics, semantic methods, evolution of Boolean networks, generation of redundant features, ensembles of GP models, automatic design of grammatical representations, GP and neuroevolution, visual reinforcement learning, evolution of deep neural networks, evolution of graphs, and scheduling in heterogeneous networks.

Together with three other co-located evolutionary computation conferences (EvoCOP 2018, EvoMusArt 2018, and EvoApplications 2018), EuroGP 2018 was part of the Evo* 2018 event. This meeting could not have taken place without the help of

---

[1] http://liinwww.ira.uka.de/bibliography/Ai/genetic.programming.html.

many people. The EuroGP Organizing Committee is particularly grateful to the following.

- SPECIES, the Society for the Promotion of Evolutionary Computation in Europe and Its Surroundings, aiming to promote evolutionary algorithmic thinking within Europe and beyond, and more generally to promote inspiration of parallel algorithms derived from natural processes.
- The high-quality and diverse EuroGP Program Committee. Each year the members give freely of their time and expertise, in order to maintain high standards in EuroGP and provide constructive feedback to help authors improve their papers.
- Marc Schoenauer of Inria-Saclay, France, for his continued hosting and maintaining of the MyReview conference management system.
- Stefano Cagnoni, Monica Mordonini, and the local organizing team from the University of Parma, Italy.
- Pablo García-Sánchez (University of Cádiz, Spain) for the Evo* 2018 publicity and website.
- Our invited speakers, Una May O'Reilly and Penousal Machado, who gave inspiring, enlightening, and entertaining keynote talks.
- The Evo* coordinators: Anna I Esparcia-Alcázar, from Universitat Politècnica de València, Spain, and Jennifer Willies.

April 2018

Mauro Castelli
Lukas Sekanina
Mengjie Zhang
Stefano Cagnoni
Pablo García-Sánchez

Organization

Organizing Committee

Program Co-chairs

Mauro Castelli — Universidade Nova de Lisboa, Portugal
Lukas Sekanina — Brno University of Technology, Czech Republic

Publication Chair

Mengjie Zhang — Victoria University of Wellington, New Zealand

Local Chair

Stefano Cagnoni — University of Parma, Italy

Publicity Chair

Pablo García-Sánchez — University of Cádiz, Spain

Conference Administration

Jennifer Willies — EvoStar Coordinator
Anna Isabel — EvoStar Coordinator
  Esparcia-Alcázar

Program Committee

Alexandros Agapitos	University College Dublin, Ireland
Lee Altenberg	University of Hawaii at Manoa, USA
Ignacio Arnaldo	MIT, USA
Douglas Augusto	LNCC/UFJF, Brazil
R. Muhammad Atif Azad	University of Limerick, Ireland
Wolfgang Banzhaf	Michigan State University, USA
Helio Barbosa	LNCC/UFJF, Brazil
Heder Bernardino	LNCC/UFJF, Brazil
Anthony Brabazon	University College Dublin, Ireland
Stefano Cagnoni	University of Parma, Italy
Mauro Castelli	Universidade Nova de Lisboa, Portugal
Ernesto Costa	University of Coimbra, Portugal
Luis Da Costa	Université Paris-Sud XI, France
Antonio Della Cioppa	University of Salerno, Italy
Grant Dick	University of Otago, New Zealand
Federico Divina	Pablo de Olavide University, Spain
Marc Ebner	Ernst-Moritz-Arndt Universität Greifswald, Germany

Aniko Ekart	Aston University, UK
Anna Isabel Esparcia-Alcázar	Universitat Politècnica de València, Spain
Francisco Fernandez de Vega	Universidad de Extremadura, Spain
Gianluigi Folino	ICAR-CNR, Italy
James A. Foster	University of Idaho, USA
Christian Gagné	Université Laval, Québec, Canada
Jin-Kao Hao	LERIA, University of Angers, France
Inman Harvey	University of Sussex, UK
Erik Hemberg	MIT, USA
Malcolm I. Heywood	Dalhousie University, Canada
Ting Hu	Dartmouth College, USA
David Jackson	University of Liverpool, UK
Colin Johnson	University of Kent, UK
Ahmed Kattan	Um Al Qura University, Saudi Arabia
Graham Kendall	University of Nottingham, UK
Michael Korns	Korns Associates, USA
Jan Koutník	IDSIA, Switzerland
Krzysztof Krawiec	Poznan University of Technology, Poland
Jiří Kubalík	Czech Technical University in Prague, Czech Republic
William B. Langdon	University College London, UK
Kwong Sak Leung	The Chinese University of Hong Kong, SAR China
John Levine	University of Strathclyde, UK
Evelyne Lutton	Inria, France
Penousal Machado	University of Coimbra, Portugal
Radek Matoušek	Brno University of Technology, Czech Republic
James McDermott	University College Dublin, Ireland
Andrew McIntyre	Dalhousie University, Canada
Bob McKay	Seoul National University, South Korea
Eric Medvet	University of Trieste, Italy
Jorn Mehnen	Cranfield University, UK
Julian Miller	University of York, UK
Alberto Moraglio	University of Exeter, UK
Xuan Hoai Nguyen	Hanoi University, Vietnam
Quang Uy Nguyen	Military Technical Academy, Vietnam
Miguel Nicolau	University College Dublin, Ireland
Julio Cesar Nievola	Pontificia Universidade Catolica do Parana, Brazil
Michael O'Neill	University College Dublin, Ireland
Una-May O'Reilly	MIT, USA
Fernando Otero	University of Kent, UK
Ender Ozcan	University of Nottingham, UK
Gisele Pappa	Federal University of Minas Gerais, Brazil
Andrew J. Parkes	University of Nottingham, UK
Tomasz Pawlak	Poznan University of Technology, Poland

Clara Pizzuti	Institute for High Performance Computing and Networking, Italy
Thomas Ray	University of Oklahoma, USA
Denis Robilliard	Université Lille Nord de France, France
Peter Rockett	University of Sheffield, UK
Álvaro Rubio-Largo	University of Extremadura, Spain
Conor Ryan	University of Limerick, Ireland
Marc Schoenauer	Inria, France
Lukas Sekanina	Brno University of Technology, Czech Republic
Yin Shan	Medicare, Australia
Sara Silva	INESC-ID Lisboa, Portugal
Moshe Sipper	Ben-Gurion University, Israel
Alexei N. Skurikhin	Los Alamos National Lab, USA
Terence Soule	University of Idaho, USA
Lee Spector	Hampshire College, USA
Jerry Swan	University of York, UK
Ivan Tanev	Doshisha University, Japan
Ernesto Tarantino	ICAR-CNR, Italy
Jorge Tavares	Microsoft, Germany
Leonardo Trujillo	Instituto Tecnológico de Tijuana, Mexico
Ali Vahdat	Dalhousie University, Canada
Leonardo Vanneschi	Universidade Nova de Lisboa, Portugal
David White	UCL, UK
Bartosz Wieloch	Poznan University of Technology, Poland
Man Leung Wong	Lingnan University, Hong Kong, SAR China
Bing Xue	Victoria University of Wellington, New Zealand
Mengjie Zhang	Victoria University of Wellington, New Zealand

Contents

Short Presentations

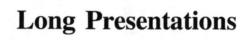

Long Presentations

Using GP Is NEAT: Evolving Compositional Pattern Production Functions

Filipe Assunção[✉], Nuno Lourenço,
Penousal Machado, and Bernardete Ribeiro

CISUC, Department of Informatics Engineering,
University of Coimbra, Coimbra, Portugal
{fga,naml,machado,bribeiro}@dei.uc.pt

Abstract. The success of Artificial Neural Networks (ANNs) highly depends on their architecture and on how they are trained. However, making decisions regarding such domain specific issues is not an easy task, and is usually performed by hand, through an exhaustive trial-and-error process. Over the years, researches have developed and proposed methods to automatically train ANNs. One example is the Hyper-NEAT algorithm, which relies on NeuroEvolution of Augmenting Topologies (NEAT) to create Compositional Pattern Production Networks (CPPNs). CPPNs are networks that encode the mapping between neuron positions and the synaptic weight of the ANN connection between those neurons. Although this approach has obtained some success, it requires meticulous parameterisation to work properly. In this article we present a comparison of different Evolutionary Computation methods to evolve Compositional Pattern Production Functions: structures that have the same goal as CPPNs, but that are encoded as functions instead of networks. In addition to NEAT three methods are used to evolve such functions: Genetic Programming (GP), Grammatical Evolution, and Dynamic Structured Grammatical Evolution. The results show that GP is able to obtain competitive performance, often surpassing the other methods, without requiring the fine tuning of the parameters.

Keywords: Compositional Pattern Production Functions
NeuroEvolution of Augmenting Topologies · Genetic Programming
Grammatical Evolution · Dynamic Structured Grammatical Evolution

1 Introduction

Artificial Neural Networks (ANNs) are hard to train. A lot of algorithmic design choices are involved, and learning is often tuned using an iterative, and cumbersome, trial-and-error process. One of the major decisions is selecting the learning algorithm to use, along with all the needed parameters, which, in turn, have to be optimised. The majority of the learning algorithms are gradient-descent and, as such, have a high probability of becoming trapped in local optima.

© Springer International Publishing AG, part of Springer Nature 2018
M. Castelli et al. (Eds.): EuroGP 2018, LNCS 10781, pp. 3–18, 2018.
https://doi.org/10.1007/978-3-319-77553-1_1

One of the goals of NeuroEvolution (NE) is the automatisation of learning in ANNs. To that end, NE applies Evolutionary Computation (EC) applies to search and tune the best solutions for the values of the weights and bias of ANNs. There are many approaches to that end: evolve the learning algorithm parameters, the actual learning rules, or directly tune the weights and bias values of the networks. The approach adopted in this paper is based on the premise that the network's weights follow some pattern that can be learned. Assuming that this pattern exists, it may be easier to evolve a function that outputs the weights of the connections between nodes than evolving the weights directly.

HyperNEAT [19] is based on the use of NeuroEvolution of Augmenting Topologies (NEAT) [20] to evolve Compositional Pattern Production Networks (CPPNs), which are structurally similar to ANNs, and are used as a means to encode the weights of a network. At a high level, CPPNs are a function that, given the position of two neurons, outputs the synaptic weight of the connection between the two given neurons. As such, and since a CPPN can be seen as a function mapping the coordinates of a pair of nodes into a weight, instead of evolving CPPNs one can use conventional EC techniques, such as GP, to evolve functions to the same task. Such functions are known as Compositional Pattern Production Functions (CPPFs) [7]. Since, in essence, CPPNs are CPPFs that use the representation adopted by NEAT, from here on we will refer to both as CPPFs except when it is necessary to make a distinction. Thus, a CPPF evolved by NEAT is actually a CPPN.

In the current work we apply different methods to the optimisation of CPPFs. More precisely, we optimise CPPFs using NEAT, Genetic Programming (GP), Grammatical Evolution (GE), and Dynamic Structured Grammatical Evolution (DSGE). We test these approaches in two different problems: a visual discrimination, and a line following task. Each problem has two setups of different complexity. The results show that GP consistently obtains the best results.

The remainder of the paper is organised as follows. In Sect. 2 we introduce works related with the evolutionary training of ANNs. Then, in Sect. 3, we detail the specifications of CPPFs and of each approach that is going to be used to evolve them. In Sect. 4 experiments are conducted and the results analysed. To end, in Sect. 5, conclusions are drawn, and future work is addressed.

2   Related Work

The use of Evolutionary Algorithms (EAs) to promote learning in ANNs not only avoids the need for manual tuning, but also reduces the risk of getting stuck in local optima: instead of a having a single solution, we have a population of solutions that is evolved. Each candidate solution encodes a trained network and their quality is determined according to the performance on a specific task. There are many approaches to the evolutionary learning of ANNs; they can be divided into three groups: optimisation of the (i) learning algorithm parameters [12,17]; (ii) learning rules [5,18]; or (iii) weights and bias values [3,8,10,21].

Despite the high number of works on the automatic training of ANNs, in the current paper our focus is on approaches that evolve the learning rules.

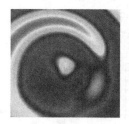

Fig. 1. Example of an image generated by HyperNEAT (left), and NEvAr (right).

More particularly, we are interested in the methods that explore the spatial regularity of the networks. HyperNEAT [19] aims at evolving CPPNs, which can be described as similar to ANNs (in structural terms). CPPNs are able to capture and encode patterns and correlations among network weights. To promote the evolution of the CPPNs it is possible to use NEAT [20]. NEAT was primarily designed as a NE approach for the optimisation of the topology and weights of ANNs; but, since CPPNs are structurally similar to ANNs they can also be evolved by NEAT.

Looking closely at HyperNEAT, what is actually evolved is a function that given spatial data provides an output. For example, one of the first applications of HyperNEAT was Picbreeder [1] – a web platform for evolving images collaboratively. In Picbreeder, the evolved CPPNs take as input two arguments: x, y, which are the coordinates of a pixel in the image; the output is the intensity. Before Picbreeder there were similar systems that aimed at evolving images; one of such methods is NEvAr [15]. The main difference between the two approaches relies on the mechanism used to evolve the function that generates the pixel value: while in Picbreeder the authors rely NEAT, in NEvAr tree-based GP is used to evolve a function, as in common symbolic regression approaches.

The fact that NEAT and GP have been applied to the evolution of functions to generate images is not within the scope of the current work; however it illustrates how both approaches can produce similar results in terms of the images that are generated (check Fig. 1). Additionally, both approaches have already been used in the evolution of functions that map the position of any pair of two nodes into a weight value, i.e., a rule that defines what are the weights of a target network. Stanley et al. trained multiple neuronal controllers using NEAT [20]; Buk et al. followed the same rationale but using GP [7]. The results of [7] show that using GP or NEAT it is possible to find adequate solutions; however, convergence seems to be faster with GP.

3   Generation of CPPFs

The main goal of the current work is to compare the performance of different approaches in the evolution of CPPFs for the training of ANNs. Figure 2 depicts the interaction between evolution, the generated CPPFs and the substrate, which is the network that is trained to solve a specific problem. Four

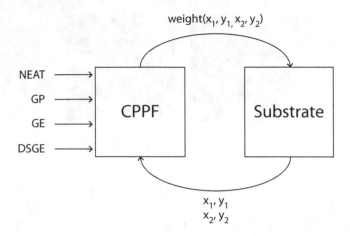

Fig. 2. Overview of the current work, evidencing the interaction between the CPPFs and the substrate.

different approaches are tested: NEAT, GP, GE, and DSGE. These are briefly described in Sects. 3.1, 3.2, 3.3, and 3.4, respectively. CPPFs are functions with 4 inputs: x_1, y_1, x_2, y_2, which similarly to HyperNEAT, are the positions of the neurons in the substrate. The substrate is the ANN that is used to solve the problem at hand. It is a grid of neurons, with a fixed number of inputs and outputs that varies according to the problem to be solved. To know the weights of the connections between the neurons in the substrate we need to query the evolved CPPF; if it returns a value above a defined threshold then the connection between the input neurons $((x_1, y_1)$ and $(x_2, y_2))$ exists and the synaptic weight is the one returned by the CPPF; conversely, if the value is bellow the threshold, the connection does not exist.

3.1 NeuroEvolution of Augmenting Topologies

NeuroEvolution of Augmenting Topologies (NEAT) [20] is a NE approach, where each candidate solution encodes an entire network as a list of neurons along with the connections between them. Whilst the majority of other NE approaches initialise the population at random, in NEAT evolution starts from a minimal structure, i.e., initial networks are composed of no hidden-nodes, and thus the input nodes are directly connected to the output nodes. Other novel aspects of NEAT include innovation protection techniques and speciation. In its vanilla form the only genetic operator that is applied to generate offspring is mutation, which aims at changing any node or structural property of a network.

CPPNs are structurally similar to ANNs, and thus it is possible to evolve them using NEAT and its principles. As such, the target of evolution is a network structure, with four inputs and one output, where the activation function of each node is selected from a defined set, and the connections are evolved and have a weight associated to it, as in ANNs. This approach is known as HyperNEAT [19].

3.2 Tree-Based Genetic Programming

In its standard form Genetic Programming (GP) [13] encodes the solutions as trees, where the inner nodes represent functions and leaves represent terminals. The crossover operator exchanges sub-trees between parents; mutation acts similarly to the crossover operator: a random sub-tree of the individual that is to be mutated is replaced by another valid one.

There are several approaches where GP is used to evolve learning rules of ANNs (e.g. [5,18]). GP is well-known for its results in symbolic regression tasks. Typically this assumes evaluating the evolved individuals for different input values. Likewise, the evolution of CPPFs (or CPPNs using HyperNEAT) involves the evaluation of the individuals for all pairs of neurons. The difference is that while in symbolic regression one typically compares the output values with the desired ones, in the evolution of CPPFs one does not know the desired value and, therefore, fitness is assigned according to the performance of the network on a given task. Due to the similarities between symbolic regression and CPPF evolution we consider GP to be a natural choice for this task, and we expect it to be able of discovering effective functions that encode the substrate weights.

3.3 Grammatical Evolution

Grammatical Evolution (GE) [16] uses an indirect encoding to represent solutions as derivations of a user-defined grammar. One of the advantages of this type of approaches is that they are easily generalisable to deal with different domains, just requiring the change of the grammar production rules. Candidate solutions are encoded as a linear sequence of integers, where each integer represents the possibility to further expand a given non-terminal symbol. To that end, the mathematical modulus operation is used, and the expansion possibility is equal to the integer modulus the number of possibilities for expanding the current non-terminal symbol. In GE both mutation and crossover genetic operators are used; mutation randomly changes an integer to another one; one-point crossover is used.

Like in GP, the majority of the works concern the evolution of the weights of the networks (e.g. [2]). Due to its similarities with GP we will focus on the generation of CPPFs.

3.4 Dynamic Structured Grammatical Evolution

Dynamic Structured Grammatical Evolution (DSGE) [4,14] is another grammar-based GP approach. The main difference between GE and DSGE relies on the way the genotype is encoded. While in GE a single list of integers is used, in DSGE there is a list of integers for each non-terminal symbol. This encapsulation of the genetic material promotes locality, in the sense that there is a direct association between the non-terminal symbols and the integers used for their mapping. In addition, the modulus is no longer needed, thus avoiding the redundancy issue commonly pointed out as a disadvantage of GE. In DSGE only the non-terminal symbols that are used are encoded, and consequently the variation operators are applied to integers that are effectively used in the genotype to phenotype mapping.

To promote evolution crossover and mutation are applied; bit-mask crossover is used, where codons (i.e., the set of integers associated to a specific non-terminal symbol) are changed between two parents; the mutation operator is a standard per gene point mutation. For the same reason than in GP and GE we will tackle the evolution of CPPFs using DSGE.

4 Experiments

We conduct experiments with the four evolutionary methodologies described above: NEAT, GP, GE, and DSGE. The objective of the the experiments is the evolution of CPPFs for the training of neuronal controllers for solving two specific tasks, which are described next.

4.1 Problems Description

The experiments are conducted in two different tasks: (i) visual discrimination; and (ii) line following. We selected these problems because they are common benchmarks used by HyperNEAT. Moreover, its performance has been thoroughly studied, and it possible to compare the approaches without introducing any problem dependency bias. In the upcoming sub-sections we briefly describe the tasks, and the structure of the substrates that are used to solve them.

Visual Discrimination

This problem was one of the first to be used to demonstrate the effectiveness of HyperNEAT [19]. The objective of this visual computation task is to distinguish between two different objects – a target and a distractor – independently of their positions in a field. We test the evolution of CPPFs for two different setups. The two setups have the same input dimensions: a 11×11 image, but the targets and distractors differ. In the big-little setup the target is a 3×3 square, and the distractor is a 1×1 square:

$$\text{target}_{\text{square}} = \begin{bmatrix} 1 & 1 & 1 \\ 1 & 1 & 1 \\ 1 & 1 & 1 \end{bmatrix}, \text{distractor}_{\text{square}} = \begin{bmatrix} 1 \end{bmatrix}.$$

To increase the complexity of the task, we then experiment with a triangular target and distractor, which have the same dimensions, but are mirrored (triup-down setup):

$$\text{target}_{\text{triangle}} = \begin{bmatrix} 1 & 0 & 0 \\ 1 & 1 & 0 \\ 1 & 1 & 1 \end{bmatrix}, \text{distractor}_{\text{triangle}} = \begin{bmatrix} 1 & 1 & 1 \\ 0 & 1 & 1 \\ 0 & 0 & 1 \end{bmatrix}.$$

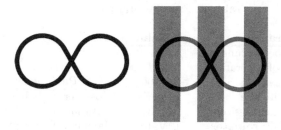

Fig. 3. Line following task. On the left the road that the agent must follow. On the right the same road path with different friction areas.

The identification of the target shape is performed by a neuronal controller, which is trained using a CPPF. To identify the shape, the trained network must generate the highest activation at the center of the target shape. Therefore the goal of evolution is the minimisation of the distance between the response provided by the network and the target center (underlined 1 in the target matrices). To further complexify the problem, as in [7], the positions of the distractor are set at random. This makes the fitness function non-deterministic, which means that two consecutive evaluations can provide two different fitness values.

For this task, the CPPFs receive 6 inputs: additionally to the usual coordinates of the nodes x_1, y_1, x_2, y_2, there are also two delta values $x_1 - x_2$, and $y_1 - y_2$ (as in [6]). The substrate consists of a sandwich network [19], i.e., the input layer is directly connected to the output layer, and both have the same size (in this case 121 neurons, one for each pixel of the image).

Line Following

In the line following task [9] the goal is to evolve the controllers of an agent so that it can effectively navigate a road, i.e., follow a line at maximum speed. The map is made of regions with different friction rates, and thus the agent should strive to steer in those with the lowest resistance. Two setups are tested: in the first one all regions except the road have the same friction rate (Fig. 3, left); on the second there are regions of different friction (marked as the darker stripes on the right side of Fig. 3). The regions outside the road have a friction that is 5 times higher than on the road.

The agent is a robot with two wheels and 5 sensors, each with a range of 3 pixels. The sensors are placed to the front of the agent, and provide a read of the characteristics of the field within range. The substrate is a feedforward network with 15 inputs (which are fed with the sensors data), a hidden-layer, and 2 outputs (that control each of the wheels).

The evolved CPPFs receive the standard 4 inputs, but have 3 outputs, each responsible for encoding a function that represents the weights of a specific part of the substrate: input to output, hidden-connections, and output layer bias. When evolving CPPNs there is no problem in having three different functions, each represented by a different output neuron. However, to accomplish the same with CPPFs authors typically evolve three different functions simultaneously

Table 1. Experimental parameters.

Parameter	Value
Number of runs	30
Number of generations	250 / 100
Population size	100
Elite size	1%
Tournament size	3

NEAT Parameter	Value
Weight range	(-3, 3)
Minimum weight	0.3
Add node probability	0.03
Add connection probability	0.1
Mutate weight probability	0.8
Reset weight probability	0.1
Reenable connection probability	0.01
Disable connection probability	0.01
Mutate bias probability	0.2
Mutate node type probability	0.2
Std weight mutation	0.2
Std bias mutation	0.5

GP Parameter	Value
Crossover probability	0.9
Mutation probability	0.1
Maximum tree-depth	17

GE Parameter	Value
Codon size	127
Wrapping	2
Crossover probability	0.9
Mutation probability	0.05

DSGE Parameter	Value
Crossover probability	0.9
Mutation probability	$1/\text{codon\_size}$
Maximum recursivity	17

(as in [7]). We followed a different direction, and evolve a single function that outputs a vector of size 3. This can be easily accomplished by using a function set that operates on vectors and scalars (the multiplication of two vectors performs the dot product; the same happens for division; the trigonometric operations are applied to each one of the components of the vector); and random constants which are vectors of size 3. This approach has also been used in NEvAr [15].

4.2 Experimental Setup

To promote a fair comparison, we adopted and adapted the vanilla implementations of each of the evolutionary engines, which are easily found in public repositories[1]. Table 1 details the parameterisation of the different algorithms. For the two benchmarks and respective setups the parameters are the same, except for the number of generations of each run, which is 250 for the visual discrimination task setups, and 100 for line following.

For the grammar-based approaches (GE and DSGE) we use the grammar of Fig. 4. This grammar is capable of generating CPPFs that provide the weights for the substrate based on multiple inputs; for the visual discrimination task x_1, y_1, x_2, y_2, d_1, and d_2: the first 4 are the positions of the neurons in the substrate

---

[1] NEAT – https://github.com/noio/peas [6].
  GP – https://github.com/DEAP/deap.
  GE – https://github.com/jmmcd/ponyge.
  DSGE – https://github.com/nunolourenco/dsge.

$$<\text{expr}> ::= <\text{expr}> <\text{op}> <\text{expr}>$$
$$| <\text{var}>$$
$$| <\text{preop}> (<\text{expr}>)$$
$$<\text{var}> ::= x1 \mid y1 \mid x2 \mid y2$$
$$| d1 \mid d2$$
$$| <\text{float}>$$
$$<\text{preop}> ::= + \mid - \mid * \mid /$$
$$<\text{preop}> ::= sin \mid -$$
$$<\text{float}> ::= - <\text{first}>.<\text{number}><\text{number}>$$
$$| <\text{first}>.<\text{number}><\text{number}>$$
$$<\text{first}> ::= 0 \mid 1 \mid 2$$
$$<\text{number}> ::= 0 \mid 1 \mid 2 \mid 3 \mid 4$$
$$| 5 \mid 6 \mid 7 \mid 8 \mid 9$$

Fig. 4. Grammar used for evolving CPPFs with GE and DSGE. For the line following task in the <var> rule the terminal symbols d1 and d2 are removed from the grammar, and the last expansion possibility of <var> is replaced by [<float>, <float>, <float>].

and the last are the deltas, $x_1 - x_2$ and $y_1 - y_2$, respectively. In the line following task the deltas are not considered.

We decided to use function sets that are commonly used in conjunction with each one of the approaches. As such, when evolving CPPFs – with tree or grammar-based GP – for the visual discrimination tasks we consider a simple function set: sin, addition, subtraction, multiplication, and division. The terminals are the ones used by the grammar of Fig. 4, i.e., the inputs of the CPPF, and random float values that may range between -3 to 3. In NEAT the nodes of the network can use the following activation functions: sin, bound, linear, gaussian, sigmoid, and absolute value. When evolving CPPFs for the line following tasks, the function set is expanded in order to handle vectors and the random constants become random vectors of size 3. No changes are required to the function set of NEAT.

4.3 Experimental Analysis

In each experiment we perform 30 independent evolutionary runs so that we can understand the behaviour of the methods in each of the tested problems and setups. We analyse fitness evolution and convergence speed. In addition, a statistical analysis is performed to assess if any of the approaches is statistically superior to the others in terms of the quality of the evolved solutions.

Visual Discrimination

In the visual discrimination task the goal is to reduce the distance to the center of the target shape, and consequently fitness is to be minimised; recall that the

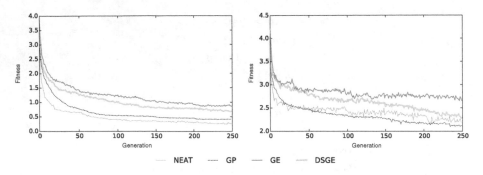

Fig. 5. Evolution of the best individuals across generations in the visual discrimination task for the big-little (left) and triup-down (right) setups. Results are averages of 30 independent runs.

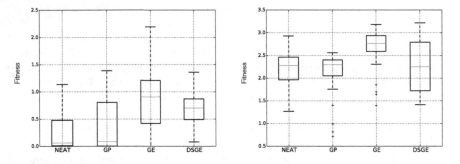

Fig. 6. Analysis of the fitness of the best solutions of the visual discrimination task using box plots. On the left the big-little setup, and on the right the triup-down.

fitness function is not deterministic, and thus the fitness of the same individual evaluated multiple times can vary. Figure 5 depicts, for both setups, the evolution of fitness across generations. The results are averages of the 30 best solutions, one from each of the evolutionary runs. These charts show that with any of the evolutionary engines evolution is promoted and there is convergence. The difference in the setups complexity is noticeable by the analysis of the difference in the fitness scales: in both setups the average fitness of the best solutions starts approximately from the same point, but in the big-little setup it is capable of reaching much lower values than in the triup-down setup. Having the target and distractor shapes with the same size but mirrored makes the problem too challenging for an appropriate function capable of encoding the weights of the substrate to be found in the given number of generations.

Nonetheless, for both setups, in terms of fitness, the results reported by NEAT and GP are superior to those of the grammar-based methods. To better analyse the quality of the generated solutions we use box plots, focusing on the distribution of the quality of the best individuals from each evolutionary run (see Fig. 6). From the box plots it is possible to see that GE has the worse performance. Focusing on each of the setups individually, in the case of the

Table 2. Graphical overview of the statistical results of the visual discrimination experiments with effect sizes for the big-little (left) and triup-down (right) setups.

	NEAT	GP	GE	DSGE
NEAT		∼	+++	+++
GP	∼		+++	++
GE	∼	∼		∼
DSGE	∼	∼	∼	

	NEAT	GP	GE	DSGE
NEAT		∼	+++	∼
GP	∼		+++	∼
GE	∼	∼		∼
DSGE	∼	∼	++	

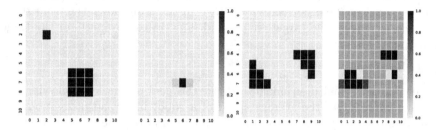

Fig. 7. The left and right figures represent the activations generated by one of the best solutions (discovered using GP), for each of the setups: big-little and triup-down, respectively.

big-little setup NEAT and GP have a close median, with GP having a slightly superior dispersion. For the triup-down setup NEAT and GP also have roughly the same median, but the dispersion is lower in GP. GP has more outliers, but these outliers correspond to runs that achieve better solutions. Looking at the DSGE performance, in the big-little setup it performs worse than NEAT or GP, and in the triup-down setup it has a similar performance, but larger dispersion.

The analysis of the generated solutions reveals that, for the big-little setup, NEAT, GP, GE, and DSGE generate perfect solutions in 5, 8, 3, and 0 out of the 30 runs, respectively. So, despite GP having a slightly higher median and dispersion in the fitness values, it is the approach that finds solutions with a perfect performance most often. In the triup-down setup, no approach is capable of finding a perfect solution.

To better understand if any of the approaches is superior to the others we conduct a statistical study. To check if the samples follow a Normal Distribution we use the Kolmogorov-Smirnov and Shapiro-Wilk tests, with a significance level of $\alpha = 0.05$. The tests reveal that the data does not follow a normal distribution and, as such, a non-parametric test (Mann-Whitney U, $\alpha = 0.05$) is used to perform the pairwise comparison of the approaches (with Bonferroni correction). Table 2 presents a graphical overview of the results of the statistical analysis: ∼ indicates no statistical difference, and + that the approach in the row is statistically better than the one in the column. The effect size is a measure that quantifies the strength of a phenomenon (larger values mean a stronger effect), and is denoted by the number of + signals, where +, ++ and +++

correspond, respectively, to low $(0.1 \leq r < 0.3)$, medium $(0.3 \leq r < 0.5)$ and large $(r \geq 0.5)$ effect sizes. The statistical results show that it is not possible to point out a single approach as the best one; there are no differences between NEAT and GP. GE is surpassed by all other approaches, and DSGE is outperformed in the triup-down setup, but has the same performance as NEAT and GP in the big-little.

Figure 7 presents examples of one of the best solutions (for each setup) regarding the activations that are generated for the identification of the target shapes: darker colours mean higher activation values. As perceptible, in the big-little setup the target shape is correctly identified; in the triup-down the trained network fails to identify the target shape, and is activated by parts of the target and distractor shapes.

Line Following

Contrary to the visual discrimination task, the goal of the line following task is the maximisation of the average speed of a robot in a road navigation task, i.e., maximise the distance travelled in a fixed amount of time (3000 time steps). As before, we start by analysing the evolution of fitness across generations (see Fig. 8). GE is the approach that takes the largest number of generations to converge, reaching the lowest results. Conversely, NEAT, GP, and DSGE are the methods that generate the best solutions, with GP outperforming the other two in the easy and hard setups. The box plots allow stronger conclusions than before (see Fig. 9); while in the visual discrimination task NEAT or GP were not superior in the two setups, in the line following task it is perceptible that GP performs better than the remaining approaches in the easy and hard setups, i.e., despite having a few outliers GP has a median that is higher than those of the remaining approaches, and the interquartile range is smaller, meaning that the results are more consistent.

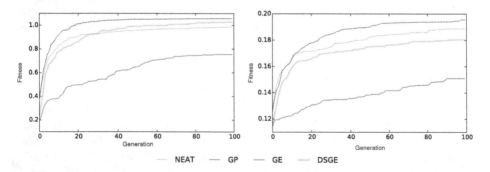

Fig. 8. Evolution of the best individuals across generations in the line following task for the easy (left) and hard (right) setups. Results are averages of 30 independent runs.

To strengthen our analysis, we perform a statistical analysis of the results. The results are reported in Table 3, using the same graphical representation of

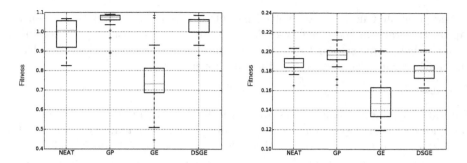

Fig. 9. Analysis of the fitness of the best solutions of the line following task using box plots. On the left the easy setup, and on the right the hard.

Table 3. Graphical overview of the statistical results of the line following experiments with effect sizes for the easy (left) and hard (right) setups.

	NEAT	GP	GE	DSGE
NEAT		~	+++	~
GP	+++		+++	++
GE	~	~		~
DSGE	+++	~	+++	

	NEAT	GP	GE	DSGE
NEAT		~	+++	+++
GP	+++		+++	+++
GE	~	~		~
DSGE	~	~	+++	

the statistical analysis of the visual discrimination task. A brief perusal of the results shows that GP clearly outperforms the other approaches in the easy and hard setups. In fact, the effect size is always large, except in what concerns the comparison between GP and DSGE for the easy setup.

An example of the best models navigating in each of the setups is depicted in Fig. 10. The line marks the path followed by the robot. In the easy setup it is clear that the robot is capable of travelling through the road without any difficulty, never leaving the predefined path. The same cannot be stated for the hard setup, where the regions of different friction make learning more challenging

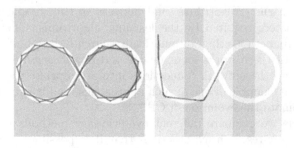

Fig. 10. The left and right figures represent an example of one of the best solutions (found using GP) for the easy and hard setups, respectively.

to the point that none of the evolved models is capable of completing an entire lap on the track in the given execution time; in fact the vast majority of them depict a behaviour similar to the one presented in the example, i.e., they leave the marked path and are unable of getting back to it.

Discussion

Summing up the previous results, we conclude that in the visual discrimination task no approach is superior to the remaining ones. The analysis of the results present in both the evolution and box plots shows that GP and NEAT have a similar performance. The statistical tests also confirm that there are no meaningful differences between the two approaches. Nevertheless, it is possible to say that GP is more effective, since it discovers perfect solutions most often, i.e., solutions with a distance of 0 to the target shape. On the big-little GP discovers perfect solutions 8 times out of 30 runs, and NEAT, GE, and DSGE discover perfect solutions in 5, 2, and 0 runs, respectively.

In the line following task GP outperforms all the other approaches considered in the comparison, which is clearly perceptible by the analysis of the box plots, where the fitness of GP is superior and the quality of the results consistent.

The performance of the grammar-based approaches is fairly disappointing in comparison with the remaining methods, which was an unexpected result. A possible explanation pertains the way float constants are created, which makes the search space larger. While in NEAT and GP the floats are just one terminal, in the grammar-based methods there is the need to associate an expansion possibility to each of the integers of the floats (which have a fixed precision). This might be mitigated by allowing both grammar-based approaches to perform a larger number of evaluations.

Based on the experiments conducted, it is possible to state that GP has better overall performance than NEAT. Additionally, GP requires far less parameterisation, without compromising the end results.

5 Conclusions and Future Work

ANNs are difficult to train, mainly due to the gradient-descent nature of the majority of the learning algorithms, and because of the complexity in setting the hyper-parameters required by the learning algorithms. Therefore, practitioners investigate methods that automatically search for the best weights of the networks. Some of these works focus on the automatic generation of learning rules that can map the network connections into appropriate weights and bias. HyperNEAT is an example of one of such approaches, and it is based on the use of NEAT to promote the evolution of CPPNs.

In this article we compare different evolutionary methods for the generation of CPPFs. Our research hypothesis is that it is possible to replace NEAT in HyperNEAT by a simpler method, that requires far less parameters, without compromising the overall quality of the obtained results. To validate our hypothesis, we apply NEAT, GP, GE and DSGE to the evolution of CPPFs in two benchmarks commonly used in HyperNEAT experiments: visual discrimination and line following, each having two setups that vary in complexity.

The experimental results, supported by statistical analysis, confirm our research hypothesis, i.e., they show that using tree-based GP it is possible to evolve effective CPPFs that, for the considered tasks, outperform the CPPFs discovered by the other methods, including NEAT. This result is somewhat surprising, specially taking into consideration that we used a vanilla implementation of GP and resorted to a basic and generic function set. As such we consider that these results pave the way for the application of GP approaches to the evolution of CPPFs, creating opportunities for the application of more sophisticated GP approaches to this type of tasks.

Future work will focus in four different research directions: (i) applying these approaches to a wider set of problems in order to assess the generality of the conclusions and identify problem specific limitations and strengths of different approaches; (ii) study the impact of the function and terminal sets in evolution; (iii) test how the compared approaches behave in terms of scalability; and (iv) expand the comparison to non-vanilla implementations and graph-based evolutionary methods, such as Cartesian Genetic Programming, which may be suitable for the evolution of CPPFs [11].

Acknowledgments. This work is partially funded by: Fundação para a Ciência e Tecnologia (FCT), Portugal, under the grant SFRH/BD/114865/2016, and is based upon work from COST Action CA15140: ImAppNIO, supported by COST (European Cooperation in Science and Technology): www.cost.eu.

References

1. Secretan, J., et al.: Picbreeder: evolving pictures collaboratively online. In: Proceedings of the SIGCHI Conference on Human Factors in Computing Systems, pp. 1759–1768. ACM (2008)
2. Ahmadizar, F., Soltanian, K., AkhlaghianTab, F., Tsoulos, I.: Artificial neural network development by means of a novel combination of grammatical evolution and genetic algorithm. Eng. Appl. Artif. Intell. **39**, 1–13 (2015)
3. Assunção, F., Lourenço, N., Machado, P., Ribeiro, B.: Automatic generation of neural networks with structured grammatical evolution. In: 2017 IEEE Congress on Evolutionary Computation (CEC), pp. 1557–1564, June 2017
4. Assunção, F., Lourenço, N., Machado, P., Ribeiro, B.: Towards the evolution of multi-layered neural networks: a dynamic structured grammatical evolution approach. In: Proceedings of the Genetic and Evolutionary Computation Conference, GECCO 2017, pp. 393–400. ACM, New York (2017). http://doi.acm.org/10.1145/3071178.3071286
5. Bengio, S., Bengio, Y., Cloutier, J.: Use of genetic programming for the search of a new learning rule for neural networks. In: 1994 Proceedings of the First IEEE Conference on Evolutionary Computation, IEEE World Congress on Computational Intelligence, pp. 324–327. IEEE (1994)
6. van den Berg, T.G., Whiteson, S.: Critical factors in the performance of Hyper-NEAT. In: Proceedings of the 15th Annual Conference on Genetic and Evolutionary Computation, pp. 759–766. ACM (2013)

7. Buk, Z., Koutník, J., Šnorek, M.: NEAT in HyperNEAT substituted with genetic programming. In: Kolehmainen, M., Toivanen, P., Beliczynski, B. (eds.) ICANNGA 2009. LNCS, vol. 5495, pp. 243–252. Springer, Heidelberg (2009). https://doi.org/10.1007/978-3-642-04921-7_25

8. David, O.E., Greental, I.: Genetic algorithms for evolving deep neural networks. In: Proceedings of the Companion Publication of the 2014 Annual Conference on Genetic and Evolutionary Computation, pp. 1451–1452. ACM (2014)

9. Drchal, J., Koutník, J., Snorek, M.: HyperNEAT controlled robots learn how to drive on roads in simulated environment. In: 2009 IEEE Congress on Evolutionary Computation, CEC 2009, pp. 1087–1092. IEEE (2009)

10. Gomez, F., Schmidhuber, J., Miikkulainen, R.: Accelerated neural evolution through cooperatively coevolved synapses. J. Mach. Learn. Res. 9(May), 937–965 (2008)

11. Khan, M.M., Khan, G.M., Miller, J.F.: Evolution of neural networks using Cartesian genetic programming. In: 2010 IEEE Congress on Evolutionary Computation (CEC), pp. 1–8. IEEE (2010)

12. Kim, H.B., Jung, S.H., Kim, T.G., Park, K.H.: Fast learning method for back-propagation neural network by evolutionary adaptation of learning rates. Neurocomputing 11(1), 101–106 (1996)

13. Koza, J.R.: Genetic Programming: On the Programming of Computers by Means of Natural Selection, vol. 1. MIT Press, Cambridge (1992)

14. Lourenço, N., Pereira, F.B., Costa, E.: Unveiling the properties of structured grammatical evolution. Genet. Program Evolvable Mach. 17(3), 251–289 (2016)

15. Machado, P., Cardoso, A.: All the truth about NEvAr. Appl. Intell. 16(2), 101–118 (2002)

16. O'Neil, M., Ryan, C.: Grammatical evolution. In: O'Neil, M., Ryan, C. (eds.) Grammatical Evolution, pp. 33–47. Springer, New York (2003). https://doi.org/10.1007/978-1-4615-0447-4

17. Parra, J., Trujillo, L., Melin, P.: Hybrid back-propagation training with evolutionary strategies. Soft. Comput. 18(8), 1603–1614 (2014)

18. Radi, A., Poli, R.: Discovering efficient learning rules for feedforward neural networks using genetic programming. In: Abraham, A., Jain, L.C., Kacprzyk, J. (eds.) Recent Advances in Intelligent Paradigms and Applications, pp. 133–159. Springer, Heidelberg (2003). https://doi.org/10.1007/978-3-7908-1770-6_7

19. Stanley, K.O., D'Ambrosio, D.B., Gauci, J.: A hypercube-based encoding for evolving large-scale neural networks. Artif. Life 15(2), 185–212 (2009)

20. Stanley, K.O., Miikkulainen, R.: Evolving neural networks through augmenting topologies. Evol. Comput. 10(2), 99–127 (2002)

21. Whitley, D., Starkweather, T., Bogart, C.: Genetic algorithms and neural networks: optimizing connections and connectivity. Parallel Comput. 14(3), 347–361 (1990)

Evolving the Topology of Large Scale Deep Neural Networks

Filipe Assunção[(⊠)](ID), Nuno Lourenço(ID), Penousal Machado(ID),
and Bernardete Ribeiro(ID)

CISUC, Department of Informatics Engineering,
University of Coimbra, Coimbra, Portugal
{fga,naml,machado,bribeiro}@dei.uc.pt

Abstract. In the recent years Deep Learning has attracted a lot of attention due to its success in difficult tasks such as image recognition and computer vision. Most of the success in these tasks is merit of Convolutional Neural Networks (CNNs), which allow the automatic construction of features. However, designing such networks is not an easy task, which requires expertise and insight. In this paper we introduce DENSER, a novel representation for the evolution of deep neural networks. In concrete we adapt ideas from Genetic Algorithms (GAs) and Grammatical Evolution (GE) to enable the evolution of sequences of layers and their parameters. We test our approach in the well-known image classification CIFAR-10 dataset. The results show that our method: (i) outperforms previous evolutionary approaches to the generations of CNNs; (ii) is able to create CNNs that have state-of-the-art performance while using less prior knowledge (iii) evolves CNNs with novel topologies, unlikely to be designed by hand. For instance, the best performing CNN obtained during evolution has an unexpected structure using six consecutive dense layers. On the CIFAR-10 the best model reports an average error of 5.87% on test data.

Keywords: Convolutional Neural Networks · Deep Neural Networks
Genetic Algorithm · Dynamic Structured Grammatical Evolution

1 Introduction

Machine Learning (ML) enables machines to learn from large volumes of data, where often there is the need to pre-process the data in order to extract features. To do that, it is required expert knowledge about the problem domain, and then we have to manually design a model that can learn the data patterns. Deep Learning (DL) avoids this by building models that are aimed at learning a representation of the data, thus reducing the amount of required domain knowledge. But, DL models tend to require deep Artificial Neural Networks (ANNs) so that the learning of the problem features is effective. Nonetheless, DL has been successfully applied in many domains, such as, computer vision [7,18,23], speech recognition [6,11], or machine translation [33].

© Springer International Publishing AG, part of Springer Nature 2018
M. Castelli et al. (Eds.): EuroGP 2018, LNCS 10781, pp. 19–34, 2018.
https://doi.org/10.1007/978-3-319-77553-1_2

An example of a deep network, which is often used for object recognition is VGG, introduced by Simonyan and Zisserman in [25]. VGG is a 16 to 19 deep Convolutional Neural Network (CNN), which has pushed the boundaries on the ImageNet Challenge 2014. CNNs, as other DL models, involve a large number of design choices. For instance, one needs to decide on the number of layers, the type of layers, and the parameterisation of the multiple receptive fields that compose it such as the number of filters, stride, or filter sizes. Everyday these models get more and more complex and optimising all the involved parameters is becoming an increasingly arduous task. For that reason, researchers have focused their efforts on automating the design of deep networks. The current work is a step forward in this line of research.

In this article in propose DENSER, a novel representation that is capable of searching for adequate topologies (and the learning hyper-parameters) of CNNs. Although in the current paper we only apply it to the evolution of CNNs, we argue that the proposed method can be applied to different network structures. By combining the principles of a standard Genetic Algorithm (GA) with Grammatical Evolution (GE) [20] we allow the direct evolution of a sequential list of layers, where the parameter values of each layer are encapsulated in a position of the GA genotype, facilitating the application of the genetic operators. In this way we can reuse the method for different network structures and domains, as we only need to change the underlying grammar. We test the proposed approach on an image classification dataset, namely the CIFAR-10 dataset. The results reveal that our method is able to find competitive CNNs, often superior to others reported in the literature. In concrete, the CNN that obtains the best performance on training data, has an accuracy of 94.13%, i.e., an error of 5.87% on test data.

The remainder of the document is organised as follows. In Sect. 2 we survey NeuroEvolution (NE) works, with special focus on those targeting the evolution of deep structures. Next, in Sect. 3, we introduce our approach, detailing how we combine the principles of GAs with GE. Experimental results are reported in Sect. 4. To end, in Sect. 5 conclusions are drawn, and future work and open questions are addressed.

2   State of the Art

When designing learning models, an exhaustive trial-and-error process is often followed in an attempt to discover which is the configuration that performs best. In particular, and focusing our attention on ANNs (shallow or deep), decisions have to be made considering the topology and weights/learning parameters of the networks. For that reason, the approaches that try to automatically tune the networks are grouped according to the aspects of the network they try to optimise: (i) learning; (ii) structure; (iii) learning and topology.

Several iterative, non-evolutionary approaches, have already been successfully applied to the optimisation of ANNs (e.g., [8]). In the vast majority of these methods only a solution is being optimised, and consequently it is likely that the search procedure will become trapped in local optima. In addition, the aim is often to find the simplest solution; however, the simplest solution is not necessarily the one that performs best, or is even the one that is easiest to train [2].

The use of Evolutionary Computation (EC) techniques to optimise ANNs defines NeuroEvolution (NE). In NE, the population of candidate solutions that is evolved throughout generations represents ANNs for solving a specific task. The quality of the candidate solutions is measured on how well the encoded networks perform when solving the problem.

The application of Evolutionary Algorithms (EAs) to the optimisation of the learning of fixed network topologies can happen at different levels. The most simple strategy consists of the use of EAs to optimise the hyper-parameters. Examples of such approaches are described by Kim et al. [13], and Parra et al. [22], focusing on the optimisation of the multiple parameters of the Back-Propagation (BP) algorithm (and its variants). An alternative to evolving the hyper-parameters consists on evolving the actual learning rules that are used for updating the synaptic weights [24]. A particular example of such approaches are those based on the evolution of composition pattern producing functions, i.e., functions that given the position of two neurons in a grid are able to generate the weight associated with that connection [3, 28].

If on the one hand, optimising the parameters of the learning algorithms is a difficult task, it is also true that the majority of the learning algorithms have a gradient-descent nature, and as such, are susceptible to become trapped in a local optimum. Using a population-based search heuristic (such as EAs) is a way to minimise the impact of this issue; to that end, the weights and bias values of ANNs can be directly evolved.

There are various NE works on the search for the appropriate weights and bias values. Usually, the values of the weights are encoded linearly, i.e., a linear sequence of bits [31] or real-values [4], each representing a specific connection; or using a matrix representation [12]. It is also important to mention approaches specifically designed for tuning the weights of ANNs, such as Cooperative Synapse NeuroEvolution (CoSyNE) [9], as well as those that more generally aim at optimising real-values, e.g., G3PCX [5].

When focusing on the development of NE approaches for training ANNs the topology of the networks is often fixed. Notwithstanding, as previously stated, defining the topology is also a laborious process, which requires domain expertise and multiple attempts. The NE methods that tune the structure of networks can be partitioned into three groups, according to how they address the optimisation task: (i) connection; (ii) node; or (iii) layer-based.

In connection-based encodings, the majority of the approaches optimise the connections that are used in a large, a-priori, defined network [15, 21]. However, this limits the search space, disabling the exploration of alternative network structures that are not considered in the pre-defined network. On the other hand, node-based approaches have as base-unit of evolution each single neuron and the connections from and to that neuron. Consequently, they are the most flexible type of approach in what respects the exploration of the search space, as they allow the creation of any sort of structure or node network. Examples of well-known node-based approaches are: EPNET [32], Symbiotic, Adaptive Neuro-Evolution (SANE) [19], or NeuroEvolution of Augmenting Topologies (NEAT) [29].

Although node-based approaches make the search less restricted and unbiased, they also make it more difficult to search for deep networks, which can be made of thousands or even millions of nodes. That is the reason why the majority of the works focusing on the generation of deep structures use the layers as base unit of evolution. Coevolution DeepNEAT (CoDeepNEAT) [16] combines the ideas behind SANE and NEAT for the evolution of deep networks, where two populations of modules and blueprints are evolved simultaneously. Following the same line of research, in CGP-NN [30] Cartesian Genetic Programming is used in the evolution of the architecture of CNNs. However, instead of promoting the automatic discovery of the most appropriate modules, they are defined a-priori, and only their combination and placement is evolved.

3   Proposed Approach

To promote the evolution of the structure and parameters of the ANNs we propose a novel representation, called DENSER (Deep Evolutionary Network StructurEd Representation), that combines the basic principles of GAs with Dynamic Structured Grammatical Evolution (DSGE) [1]. The sequence of layers is encoded using the GA, and the parameterisation of each layer using DSGE. By doing this, we are able to evolve networks where the genetic material of each layer is kept together, and therefore the manipulation of the solutions is easier, since there is a one-to-one mapping between the layers and their parameters.

In the upcoming sub-sections we further detail the representation used, the genetic operators, and how the generated networks are evaluated, respectively in Sects. 3.1, 3.2 and 3.3.

3.1   Representation

Each candidate solution encodes the structure of a single ANN by means of an ordered linear structure, where each position is a functional unit of the network, i.e., a layer. It is also possible to evolve the learning algorithm that should be used to train the network and its hyper-parameters. The motivation to promote a layer-based evolution rather than node or connection-based is related with the desire to tackle challenging problems, which often require deep networks. Such structures have a large number of neurons and connections, which makes their optimisation using a low level representation hard to accomplish.

To facilitate the application of the approach to different network structures and layer types we encode each layer similarly to DSGE, meaning that evolution acts on grammatical derivations. As a results the genotype of each position of the GA (which represents a layer) is encoded as a list of genes, each of them responsible for keeping the expansion possibilities for specific non-terminal symbols of the grammar. In addition to the standard DSGE genotype we introduce a special coding block to deal with integer and float values. This block is represented in the grammar in the form of [variable name, variable type, number of values, minimum value, maximum value]. An example of their use in a grammar can be found in Fig. 1. At the genotypic level, the block values are kept together with the integers encoding the non-terminal expansion possibilities.

<pre>
        <features> ::= <convolution>
                     | <pooling>
     <convolution> ::= layer:conv [num-filters,int,1,32,256] [filter-shape,int,1,1,5]
                       [stride,int,1,1,3] <padding> <activation> <bias>
                       <batch-normalisation> <merge-input>
<batch-normalisation> ::= batch-normalisation:True
                     | batch-normalisation:False
     <merge-input> ::= merge-input:True
                     | merge-input:False
         <pooling> ::= <pool-type> [kernel-size,int,1,1,5] [stride,int,1,1,3] <padding>
       <pool-type> ::= layer:pool-avg
                     | layer:pool-max
         <padding> ::= padding:same
                     | padding:valid
  <classification> ::= <fully-connected>
 <fully-connected> ::= layer:fc <activation> [num-units,int,1,128,2048 <bias>
      <activation> ::= act:linear
                     | act:relu
                     | act:sigmoid
            <bias> ::= bias:True
                     | bias:False
         <softmax> ::= layer:fc act:softmax num-units:10 bias:True
        <learning> ::= learning:gradient-descent [lr,float,1,0.0001,0.1]
</pre>

Fig. 1. Example grammar for the encoding of CNNs.

The combination of a GA with DSGE not only makes the approach easily generalisable, but also enables the incorporation of domain knowledge. To define the allowed structure of the networks (i.e., the allowed sequence of layers) the method requires the definition of a list of tuples, where each index of the list indicates the valid grammar starting symbols (for that layer) along with the minimum and maximum number of layers of that type. For example, for searching CNNs the following structure can be specified: [(features, 1, 10), (classification, 1, 2), (softmax, 1, 1), (learning, 1, 1)]. Using the previous example and the grammar of Fig. 1, the search space encompasses networks that are formed by at least one and up to ten convolution or pooling layers (that can be placed in any order). These first convolution and pooling layers are followed by one or two fully-connected ones, and then by an output layer. The output layer is usually encoded as a fully-connected layer with a specific number of neurons corresponding to the number of classes of the problem. On top of the definition of the network topology we also allow the learning parameters to be optimised.

Figure 2 depicts an example of the genotype of a candidate solution, based on the grammar of Fig. 1 and on the GA structure introduced above: [(features, 1, 10), (classification, 1, 2), (softmax, 1, 1), (learning, 1, 1)]. As previously explained, the candidate solution has two genotypic levels: (i) the GA level which defines the structure, and points out to the grammar non-terminal symbol that is to be used as the start symbol; and (ii) the DSGE level that stores the ordered sequence of integers encoding the expansion possibilities for each specific non-terminal, and the real-values needed by the networks. Figure 3 presents the phenotype corresponding to the layer which has the DSGE genotype detailed in Fig. 2.

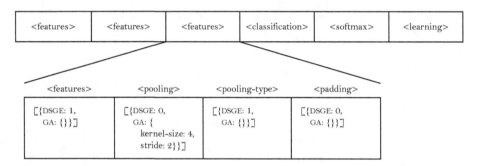

Fig. 2. Example of the genotype of a candidate solution that encodes a CNN.

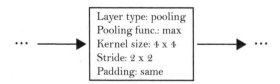

Fig. 3. Phenotype corresponding to the layer specified in Fig. 2.

3.2   Genetic Operators

To promote the evolution of the candidate solutions we rely on crossover and mutation operators specifically designed for the manipulation of ANNs.

Crossover

One of the advantages of having two genotypic levels is that the outer level encodes each layer separately. Since the genetic material is encapsulated, devising efficient crossover operators becomes easier. Based on the nature of the genotype, we developed two crossover operators, which are applied probabilistically, each having the same likelihood (i.e., 50%).

Before describing the crossover operators we need to define the notion of module. In this context, the term module does not refer to a set of layers that can be replicated multiple times, but is rather the set of layers that belongs to the same GA structural index. For example, in the above example of a GA structure, the module (features, 1, 10) is composed by all those layers that have their derivation starting with the same non-terminal symbol *features*. The two crossover operators are applied at different levels: one changes layers within a specific layer module, while the other swaps entire modules between individuals.

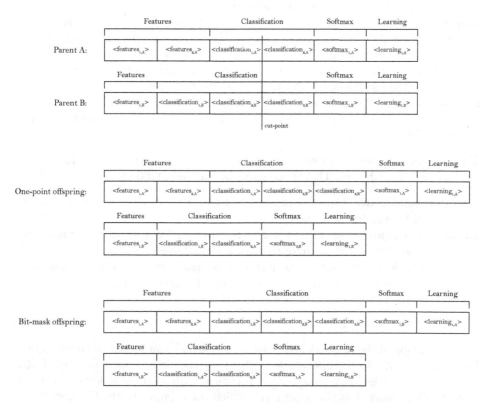

Fig. 4. Example of the introduced crossover operators. The example focuses on the GA level of the genotype. For the bit-mask crossover the mask is 1001, which is associated to the features, classification, softmax and learning modules, respectively.

The first operator is based on the principle that each layer has its genetic material encapsulated. Hence, we designed a crossover operator, that generates two offspring by crossing the layers that belong to the same module of two parents (chosen by tournament selection). The same module in different parents can have a distinct number of layers; to deal with that the cutting point is randomly generated considering the individual that has less layers in the module.

The other crossover operator is loosely based on the uniform operator for binary representations, and acts upon the modules swapping them between individuals. Figure 4 shows an example of the application of the crossover operators.

Mutation

The operators that work at the GA level aim at manipulating the layers and their parameters. For this purpose we developed the following operators:

Add layer – a new layer is generated at random with the initial symbol for the grammatical derivation being the one of the module where the layer will be placed. This operator can only be applied in modules where the maximum number of layers has not been reached yet;

Replicate layer – similar to the previous mutation operator, but instead of generating a new random layer uses one that is already in the genotype and copies it into another position of the module. This copy is done by reference, which means that if at any given time the layer or some of its parameters are changed, the modifications are propagated to their replicas;

Remove layer – deletes a random layer from a given module. It is only possible to remove a layer if after removal the number of layers in that module is still above the minimum threshold.

The previous operators act only at a macro level, and thus do not change the parameters of the layers. This is accomplished at the DSGE level:

Grammatical mutation – as in standard DSGE, an expansion possibility is replaced by another valid one;

Integer mutation – an integer block is replaced by a new one, where the integers are generated at random, within the allowed range;

Float mutation – similar to the integer mutation, but where instead of randomly generating new values, a Gaussian perturbation is applied.

3.3   Evaluation

The evaluation of the network is divided into two different steps: (i) the mapping from the genotype to the phenotype; and (ii) the training of the generated ANN.

To decode the genotype, the outer level of each candidate solution is traversed linearly. Remember that the outer level (which corresponds to the GA genotype) is where the initial start symbol for expanding the grammatical derivations of the layers is stored. The grammatical genotype is decoded similarly to DSGE: the integers encoding the expansion possibilities of each non-terminal symbol are used only once sequentially. The main difference is that when the expansion of the non-terminal symbol hits a block that represents an integer or float value the corresponding integer or float value is read from the grammatical GA genotype.

To train the evolved networks we used Keras, running on top of TensorFlow. The dataset used to validate our approach is partitioned into three disjoint sets:

Train – used to train the network using the defined or evolved learning parameters. The parameters vary depending on the used learning algorithm;

Validation – used to evaluate the performance of the network during evolution;

Test – kept aside from the evolutionary process, and used to evaluate the performance of the best models on unseen data, so we can better understand their generalisation ability.

Each network is trained during 10 epochs, and the fitness is the best performance on the validation set on the 10 epochs. Data augmentation is used, namely, padding, horizontal flips, and random crops. A more detailed explanation of the data augmentation approach followed can be found in [30].

4   Experimentation

To test the approach we conducted experiments on the evolution of CNNs for the classification of the CIFAR-10 dataset (further detailed in Sect. 4.1). The experimental setup used is described in Sect. 4.2, and the analysis of the experimental results is carried out in Sect. 4.3.

4.1   Problem Description

The CIFAR-10 dataset [14] is composed of images of 10 disjoint classes, namely: airplane, automobile, bird, cat, deer, dog, frog, horse, ship and truck. For each class there are 6000 cases, making a total of 60000 instances. Each instance is a 32×32 RGB colour image. The goal is to train a CNN that can correctly identify the class of each sample, maximising the accuracy of the object recognition task.

4.2   Experimental Setup

Table 1 shows the parameters used in the experiments. As discussed before, each network is trained during 10 epochs, using the backpropagation learning algorithm, and a learning rate of 0.01. Fitness is measured using the validation set. After the evolutionary cycle, and to further tune the best generated models, we merge the train and validation sets, so that more data is available for training the best topologies found; the networks are trained during 400 epochs with the same learning rate policy. The test data is not changed, and is used to measure the final performance of the best networks found during evolution.

The topology of the evolved networks is constrained to the following GA structure: [(features, 1, 30), (classification, 1, 10), (softmax, 1, 1)], and the experiments are conducted with a grammar similar to the one presented in Fig. 1. This way, we allow the evolution of networks that can have up to 40 hidden-layers: up to 30 convolution or pooling layers followed by at most 10 fully-connected layers. We use the same data augmentation strategy of [30]: each training instance is applied a padding of 4; then we randomly crop the padded image to 32×32, followed by random horizontal flipping.

Table 1. Experimental parameters.

Evolutionary engine parameter	Value
Number of runs	10
Number of generations	100
Population size	100
Crossover rate	70%
Mutation rate	30%
Tournament size	3
Elite size	1%
Dataset parameter	**Value**
Train set	42500 instances
Validation set	7500 instances
Test set	10000 instances
Data augmentation parameter	**Value**
Padding	4
Random crop	4
Horizontal flipping	50%

4.3   Experimental Analysis

Figure 5 depicts the evolution of the average fitness and number of layers of the best CNNs across generations. A brief perusal of the results indicates that evolution is occurring, and solutions tend to converge around the 80th generation. Two different and contradictory behaviours are observable. From the start of evolution and until approximately the 60th generation an increase in performance is accompanied by a decrease in the number of layers; this changes from the 60th generation until the last generation where an increase in performance is followed by an increase in the number of hidden-layers of the best networks. To support this analysis we compute the correlation between the average fitness values of the best individuals and the average number of layers, per generation.

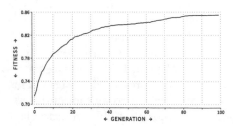 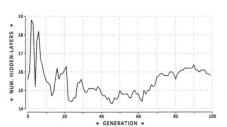

Fig. 5. Evolution of the fitness (left) and number of layers (right) of the best individuals across generations. Results are averages of 10 independent runs.

The Pearson correlation reports a coefficient of -0.7166 (moderate negative correlation) for the correlation between the two metrics before the 60th generation; after the 60th generation the coefficient is 0.9204 (strong positive correlation).

This analysis reveals an apparent contradiction, that is explained after the fact that in the first generation the randomly generated solutions have a large number of layers (approximately 15.6), which correspond to very deep networks. However, since the numeric parameters of each layer are set at random, they would hardly provide any meaningful results. As evolution proceeds and optimises the numeric values, the best solutions can steadily increase the number of layers to improve their performance. This indicates that it may be advantageous to start the evolutionary process with shallower networks.

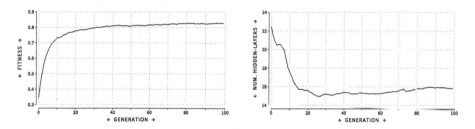

Fig. 6. Evolution of the fitness (left) and number of layers (right) of the overall population across generations. Results are averages of 10 independent runs.

In addition to analysing the best evolved solutions we also inspect the overall quality of the population. Figure 6 shows the evolution of the fitness, and number of layers, across generations at the population level. The conclusions are in line with those reported for the analysis of the best solutions, however the change in behaviour occurs earlier, around the 25th generation. Before the 25th generation the Pearson correlation reports a coefficient of -0.89 (strong negative correlation), and after a coefficient of 0.8801 (strong positive correlation). The change in behaviour happens earlier than when considering only the best solutions because in the first generations the population has many low performing solutions that are quickly discarded.

The fittest network found during evolution (in terms of validation accuracy) is represented in Fig. 7. As it can be observed, several lambda layers exit. This is due to the fact that the employed grammar allows merging the output of the convolution layers (Conv2D) with the input, using the Add layer. When the number of channels to be merged is different, we pad the one that has less channels, using the Keras Lambda layer. When the signals do not have the same width and height we down-sample the largest one, by applying max pooling.

The most puzzling characteristic of the evolved network is the importance and number of the fully-connected (i.e., dense) layers that are used at the end of the topology. Other approaches on the evolution of CNNs tend to disregard fully-connected layers, and focus only on convolution and pooling layers. We tried to

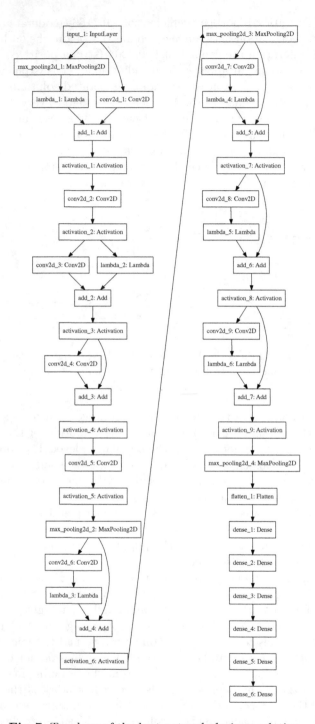

Fig. 7. Topology of the best network during evolution.

remove some of the fully-connected layers, and preliminary results show that the performance of the network degenerated. Moreover, to the best of our knowledge, the sequential use of a such large number of dense layers is unprecedented, and it is fair to say that a human would never think of such topology, which makes this evolutionary outcome remarkable.

Once the evolutionary process is completed, the best network found in each run, i.e. the one obtaining the highest fitness value, is re-trained 5 times, with different initial weights. These networks are selected according to their accuracy on the validation set, to ensure that we have an unbiased selection. The results regarding accuracy reported bellow are averages of these 5 trains for each network.

First, we train the networks with the same learning rate policy used in evolution, but during 400 epochs. With this setup we obtain, on average, a classification accuracy of 88.41% (error of 11.59%) on the test set. To further enhance the accuracy of the networks we adopt the strategy described by Snoek et al. in [26], i.e., for each instance of the test set we generate 100 augmented images. The label assigned by the model is the class that has the maximum average confidence value on the 100 generated augmented images. Following this validation approach the average accuracy on the test set of the best evolved networks increases to 89.93% (10.07% of error).

Although the average accuracy of the fittest models seems low when compared with state of the art approaches, the accuracy of the fittest network is slightly higher: 92.70% (an error of 7.30%). To investigate if it is possible to increase the performance of the fittest networks we re-train them using the same strategy of CGP-CNN [30]. We use a varying learning rate: it starts at 0.01; on the 5th epoch it is increased to 0.1; by the 250th epoch it is decreased to 0.01; and finally at the 375th it is reduced to 0.001. With the previous training policy the average accuracy of the fittest network increases to 93.38%. Finally, this accuracy is further improved if we follow the guidelines from [26], and perform data augmentation on the test data: 94.13%, i.e., an error of 5.87%, which is a highly competitive result.

Obviously, in an ideal scenario, all the training strategies described above would be used during evolution, however it is unfeasible to do so, since it would require immense computational power. Thus, the experimental results indicate that it is possible to obtain competitive results using evolutionary means and that it is possible to do so with limited computational resources, using a low number of training epochs (10) during evolution.

Table 2 shows a comparison with the best results reported by other methods. An analysis of the results shows that DENSER (i.e., our approach) is the one that reports the lowest error. The number of trainable parameters is much higher in our methodology because we allow the placement of fully-connected layers in the evolved CNNs. In addition to the increase in performance, our approach attains these results without any prior knowledge about the domain. Whilst in CGP-CNN the authors have to define fixed modules of layers that are placed and connected by the evolutionary algorithm to form a CNNs, we do not require

Table 2. Comparison of the best results obtained by different methods on the CIFAR-10 dataset. The error rate is measured on the test set. The number of parameters is the number of values that need to be tuned during training.

Method	Error rate	Number of parameters
CoDeepNEAT [29]	7.3%	–
Snoek et al. [26]	6.37%	–
CGP-CNN (ConvSet) [30]	6.75%	1.52×10^6
CGP-CNN (ResSet) [30]	5.98%	1.68×10^6
DENSER	**5.87%**	10.81×10^6

any definition of modules, which implies that the algorithm must discover the appropriate sequence of layers to construct effective networks.

5   Conclusions and Future Work

The definition of the structure and parameterisation of learning models is a hard and time consuming task. This problem is even more pressing when dealing with deep architectures, where the high number of layers makes the tuning task more difficult to accomplish by hand. To this end several evolutionary methods that seek to automatically solve this issue have been proposed.

In this article we combine two evolutionary methods: GAs and GE. With this combination we are able to evolve linear sequences of layers, where each layer is encoded using a grammar-based approach. Consequently the genetic material associated with each layer is encapsulated, making it easier to apply genetic operators to the candidate solutions. The use of a grammar to specify how the layers are encoded makes the approach easily adaptable to other network structures, layer types and domains.

The experimental results confirm the effectiveness of the approach, which outperforms CGP-CNN [30], CoDeepNEAT [29] and the work by Snoek et al. [26], without resorting to prior knowledge. As such, DENSER is, currently, the most successful method for automatic construction of networks in CIFAR-10 dataset. Moreover, its performance is only surpassed by [10,17,27], which resort to prior knowledge.

As future work we intend to further test our approach, performing more experiments to ensure the quality and consistency of the results. Moreover, we plan to evaluate the performance on other classification tasks, to assess the generality of the proposed method.

Acknowledgments. This work is partially funded by: Fundação para a Ciência e Tecnologia (FCT), Portugal, under the grant SFRH/BD/114865/2016, and is based upon work from COST Action CA15140: ImAppNIO, supported by COST (European Cooperation in Science and Technology): www.cost.eu. We would also like to thank NVIDIA for providing us Titan X GPUs.

References

1. Assunção, F., Lourenço, N., Machado, P., Ribeiro, B.: Towards the evolution of multi-layered neural networks: a dynamic structured grammatical evolution approach. In: Proceedings of the Genetic and Evolutionary Computation Conference, GECCO 2017, pp. 393–400. ACM, New York (2017). http://doi.acm.org/10.1145/3071178.3071286
2. Blum, A., Rivest, R.L.: Training a 3-node neural network is NP-complete. In: Proceedings of the 1st International Conference on Neural Information Processing Systems, pp. 494–501. MIT Press (1988)
3. Buk, Z., Koutník, J., Šnorek, M.: NEAT in HyperNEAT substituted with genetic programming. In: Kolehmainen, M., Toivanen, P., Beliczynski, B. (eds.) ICANNGA 2009. LNCS, vol. 5495, pp. 243–252. Springer, Heidelberg (2009). https://doi.org/10.1007/978-3-642-04921-7_25
4. David, O.E., Greental, I.: Genetic algorithms for evolving deep neural networks. In: Proceedings of the Companion Publication of the 2014 Annual Conference on Genetic and Evolutionary Computation, pp. 1451–1452. ACM (2014)
5. Deb, K., Anand, A., Joshi, D.: A computationally efficient evolutionary algorithm for real-parameter optimization. Evol. Comput. **10**(4), 371–395 (2002)
6. Deng, L., Hinton, G., Kingsbury, B.: New types of deep neural network learning for speech recognition and related applications: an overview. In: 2013 IEEE International Conference on Acoustics, Speech and Signal Processing (ICASSP), pp. 8599–8603. IEEE (2013)
7. Farfade, S.S., Saberian, M.J., Li, L.J.: Multi-view face detection using deep convolutional neural networks. In: Proceedings of the 5th ACM on International Conference on Multimedia Retrieval, ICMR 2015, pp. 643–650. ACM, New York (2015). http://doi.acm.org/10.1145/2671188.2749408
8. Franco, L., Jerez, J.M.: Constructive Neural Networks, vol. 258. Springer, Heidelberg (2009). https://doi.org/10.1007/978-3-642-04512-7
9. Gomez, F., Schmidhuber, J., Miikkulainen, R.: Accelerated neural evolution through cooperatively coevolved synapses. J. Mach. Learn. Res. **9**(May), 937–965 (2008)
10. Graham, B.: Fractional max-pooling. arXiv preprint arXiv:1412.6071 (2014)
11. Graves, A., Mohamed, A.R., Hinton, G.: Speech recognition with deep recurrent neural networks. In: 2013 IEEE International Conference on Acoustics, Speech and Signal Processing, pp. 6645–6649, May 2013
12. Junyou, B.: Stock price forecasting using PSO-trained neural networks. In: 2007 IEEE Congress on Evolutionary Computation, CEC 2007, pp. 2879–2885. IEEE (2007)
13. Kim, H.B., Jung, S.H., Kim, T.G., Park, K.H.: Fast learning method for backpropagation neural network by evolutionary adaptation of learning rates. Neurocomputing **11**(1), 101–106 (1996)
14. Krizhevsky, A., Hinton, G.: Learning multiple layers of features from tiny images (2009)
15. Leung, F.H.F., Lam, H.K., Ling, S.H., Tam, P.K.S.: Tuning of the structure and parameters of a neural network using an improved genetic algorithm. IEEE Trans. Neural Netw. **14**(1), 79–88 (2003)
16. Miikkulainen, R., Liang, J., Meyerson, E., Rawal, A., Fink, D., Francon, O., Raju, B., Navruzyan, A., Duffy, N., Hodjat, B.: Evolving deep neural networks. arXiv preprint arXiv:1703.00548 (2017)

17. Mishkin, D., Matas, J.: All you need is a good init. arXiv preprint arXiv:1511.06422 (2015)
18. Mnih, V., Kavukcuoglu, K., Silver, D., Rusu, A.A., Veness, J., Bellemare, M.G., Graves, A., Riedmiller, M., Fidjeland, A.K., Ostrovski, G., et al.: Human-level control through deep reinforcement learning. Nature **518**(7540), 529–533 (2015)
19. Moriarty, D.E., Miikkulainen, R.: Forming neural networks through efficient and adaptive coevolution. Evol. Comput. **5**(4), 373–399 (1997)
20. O'Neil, M., Ryan, C.: Grammatical evolution. In: O'Neil, M., Ryan, C. (eds.) Grammatical Evolution, pp. 33–47. Springer, Boston (2003). https://doi.org/10. 1007/978-1-4615-0447-4_4
21. Palmes, P.P., Hayasaka, T., Usui, S.: Evolution and adaptation of neural networks. In: 2003 Proceedings of the International Joint Conference on Neural Networks, vol. 1, pp. 478–483. IEEE (2003)
22. Parra, J., Trujillo, L., Melin, P.: Hybrid back-propagation training with evolutionary strategies. Soft. Comput. **18**(8), 1603–1614 (2014)
23. Plis, S.M., Hjelm, D.R., Salakhutdinov, R., Allen, E.A., Bockholt, H.J., Long, J.D., Johnson, H.J., Paulsen, J.S., Turner, J.A., Calhoun, V.D.: Deep learning for neuroimaging: a validation study. Front. Neurosci. **8**, 229 (2014)
24. Radi, A., Poli, R.: Discovering efficient learning rules for feedforward neural networks using genetic programming. In: Abraham, A., Jain, L.C., Kacprzyk, J. (eds.) Recent Advances in Intelligent Paradigms and Applications, pp. 133–159. Springer, Heidelberg (2003). https://doi.org/10.1007/978-3-7908-1770-6_7
25. Simonyan, K., Zisserman, A.: Very deep convolutional networks for large-scale image recognition. arXiv preprint arXiv:1409.1556 (2014)
26. Snoek, J., Rippel, O., Swersky, K., Kiros, R., Satish, N., Sundaram, N., Patwary, M., Prabhat, M., Adams, R.: Scalable Bayesian optimization using deep neural networks. In: International Conference on Machine Learning, pp. 2171–2180 (2015)
27. Springenberg, J.T., Dosovitskiy, A., Brox, T., Riedmiller, M.: Striving for simplicity: the all convolutional net. arXiv preprint arXiv:1412.6806 (2014)
28. Stanley, K.O., D'Ambrosio, D.B., Gauci, J.: A hypercube-based encoding for evolving large-scale neural networks. Artif. Life **15**(2), 185–212 (2009)
29. Stanley, K.O., Miikkulainen, R.: Evolving neural networks through augmenting topologies. Evol. Comput. **10**(2), 99–127 (2002)
30. Suganuma, M., Shirakawa, S., Nagao, T.: A genetic programming approach to designing convolutional neural network architectures. In: Proceedings of the Genetic and Evolutionary Computation Conference, GECCO 2017, pp. 497–504. ACM, New York (2017). http://doi.acm.org/10.1145/3071178.3071229
31. Whitley, D., Starkweather, T., Bogart, C.: Genetic algorithms and neural networks: optimizing connections and connectivity. Parallel Comput. **14**(3), 347–361 (1990)
32. Yao, X., Liu, Y.: Evolutionary artificial neural networks that learn and generalise well. In: 1996 IEEE International Conference on Neural Networks, Washington, DC, USA, Volume on Plenary, Panel and Special Sessions, pp. 159–164 (1996)
33. Zhang, J., Zong, C.: Deep neural networks in machine translation: an overview. IEEE Intell. Syst. **30**(5), 16–25 (2015)

Evolving Graphs by Graph Programming

Timothy Atkinson[(✉)] [iD], Detlef Plump[iD], and Susan Stepney[iD]

Department of Computer Science, University of York, York, UK
{tja511,detlef.plump,susan.stepney}@york.ac.uk

Abstract. Rule-based graph programming is a deep and rich topic. We present an approach to exploiting the power of graph programming as a representation and as an execution medium in an evolutionary algorithm (EGGP). We demonstrate this power in comparison with Cartesian Genetic Programming (CGP), showing that it is significantly more efficient in terms of fitness evaluations on some classic benchmark problems. We hypothesise that this is due to its ability to exploit the full graph structure, leading to a richer mutation set, and outline future work to test this hypothesis, and to exploit further the power of graph programming within an EA.

1 Introduction

Representation is crucial in computer science, and an important specific representation is the graph. Graphs are used in a wide range of applications and algorithms, see for example [4,5,22]. In evolutionary algorithms (EAs), graphs are used in some applications, but are usually encoded in a linear genome, with the genome undergoing mutation and crossover, and a later "genotype to phenotype mapping" used to decode the linear genome into a graph structure. For example in Cartesian Genetic Programming (CGP) [12,14], the connections of feed forward networks are encoded in a linear genome. NEAT [23,24] provides a linear encoding of ANNs which are seen as graph structures. Trees (a subset of more general graphs) are also used in EAs. Grammatical Evolution [15,21] uses a linear genome of integers to indirectly encode programs. Genetic Programming [6,7] is unusual for an EA: rather than using a linear genome, it typically directly manipulates abstract syntax trees. Poli [19,20] uses a 'graph on a grid' representation: the underlying structure is a graph, but the nodes are constrained to lie on discrete grid points. MOIST [8] proposes using trees with multiple output nodes and sharing to extend traditional genetic programming to domains where problems have multiple, related outputs. Pereira et al. [16] represent Turing machines as graphs encoded in a linear genome, and develop a crossover operator based on the structure of the underlying graph.

There are arguments for and against linear genomes representing graphs. Linear genomes are standard in EAs, and they can exploit the knowledge about

T. Atkinson—Supported by a Doctoral Training Grant from the Engineering and Physical Sciences Research Council (EPSRC) in the UK.

© Springer International Publishing AG, part of Springer Nature 2018
M. Castelli et al. (Eds.): EuroGP 2018, LNCS 10781, pp. 35–51, 2018.
https://doi.org/10.1007/978-3-319-77553-1_3

evolutionary operators. However, they can hide the underlying structure of the problem, and can have biases in the effect of the evolutionary operators. There may be advantages in evolving graphs directly, rather than via linear genome encodings or 2D grid encodings, and defining mutation operators that respect the graph structure. Direct graph transformation has deep theoretical under-pinnings, and has become increasingly accessible through efficient graph pro-gramming languages such as GP 2 [2,17]. GP 2 enables high-level problem solv-ing in the domain of graphs, freeing programmers from handling low-level data structures. It has a simple syntax whose basic computational units are graph transformation rules which can be graphically edited. Also, GP 2 comes with a concise operational semantics to facilitate formal reasoning on programs.

Here we exploit an extension to GP 2 [1] that has probabilistic elements to support EA applications. We perform experiments of evolving graphs directly, and compare the results with experiments previously done with CGP. Using graph transformations, we write evolutionary operators as graph transformation rules, and we calculate fitness in the same context. Our results indicate that direct evolution can be significantly more efficient (significantly fewer fitness function evaluations) than basic CGP, due to the increased number of mutations available, allowing more effective exploration of the search landscape.

The paper is organised as follows. In Sect. 2 we overview Graph Programming. In Sect. 3 we describe how we have incorporated an EA into graph programming (EGGP). In Sect. 4 we compare our EGGP setup with Cartesian Genetic Pro-gramming (CGP). In Sect. 5 we describe benchmark experiments, and in Sect. 6 provide the results, demonstrating that EGGP is significantly more efficient, in terms of fitness evaluations, than vanilla CGP. In Sect. 7 we draw conclusions and outline future work in examining the reasons for this improvement.

2   Graph Programming

This section is a (very) brief introduction to the graph programming language GP 2; see [18] for a detailed account of the syntax and semantics of the language. A graph program consists of declarations of *graph transformation rules* and a main command sequence controlling the application of the rules. Graphs are directed and may contain loops and parallel edges. The rules operate on *host graphs* whose nodes and edges are labelled with integers, character strings or lists of integers and strings. Variables in rules are of type int, char, string, atom or list, where atom is the union of int and string. Atoms are considered as lists of length one, hence integers and strings are also lists. For example, in Fig. 1, the list variables a, c and e are used as edge labels while b and d serve as node labels. The small numbers attached to nodes are identifiers that specify the correspondence between the nodes in the left and the right graph of the rule.

Besides carrying list expressions, nodes and edges can be *marked*. For exam-ple, in the program of Fig. 3, blue and red node marks are used to prevent the rule mutateEdge from creating a cycle.

The principal programming constructs in GP 2 are conditional graph-trans-formation rules labelled with expressions. The program in Fig. 1 applies the single rule link *as long as possible* to a host graph. In general, any subprogram can be iterated with the postfix operator "!". Applying link amounts to non-deterministically selecting a subgraph of the host graph that matches link's left graph, and adding to it an edge from node 1 to node 3 provided there is no such edge (with any label). The application condition where not edge(1,3) ensures that the program terminates and extends the host graph with a minimal number of edges. Rule matching is injective and involves instantiating variables with concrete values. We remark that GP 2's inherent non-determinism is useful as many graph problems are naturally multi-valued, for example the computation of a shortest path or a minimum spanning tree.

```
Main := link!

link(a,b,c,d,e:list)
```

```
where not edge(1,3)
```

Fig. 1. A GP 2 program computing the transitive closure of a graph.

Given any graph G, the program in Fig. 1 produces the smallest transitive graph that results from adding unlabelled edges to G (A graph is *transitive* if for each directed path from a node v_1 to another node v_2, there is an edge from v_1 to v_2.). In general, the execution of a program on a host graph may result in different graphs, fail, or diverge. The *semantics* of a program P maps each host graph to the set of all possible outcomes [17]. GP 2 is computationally complete in that every computable function on graphs can be programmed [18]. Commands not used in this paper are the non-deterministic application of a set of rules and various branching commands.

While rule matching in GP 2 is non-deterministic, the refined language P-GP 2 (for *Probabilistic GP 2*) selects a match for a rule uniformly at random [1]. This language has been used to obtain the results described in the rest of this paper.

3   Evolving Graphs by Graph Programming (EGGP)

3.1   Representation

Our approach uses the following representation of individual solutions. An individual I over function set $F = \{f_1, f_2, ...f_n\}$ is a directed graph containing a set V_i of input nodes which have no outgoing edges and a set V_o of output nodes which have one outgoing edge and no incoming edges. Each non-input and non-output node is associated with some function in F. For simplicity we assume that all functions in F (and the fitness function) operate on a single domain. EGGP individuals are defined in Definition 1.

Further, for each function node v labelled with a function f of arity n, v has outgoing edges $e_1, ..., e_n$ such that for $i = 1, ..., n$, $a(e_i) = i$. Then a provides the order in which to pass v's inputs to f, resolving ambiguity for asymmetric functions. We assume acyclic graphs in this work, and hence an individual I represents a solution as a network, where each node computes a function on its inputs (which are given by its outgoing edges). We refer to acyclic EGGP individuals as being "feed-forward".

Definition 1 (EGGP Individual). *An EGGP Individual over function set F is a directed graph $I = \{V, E, s, t, l, a, V_i, V_o\}$ where V is a finite set of nodes and E is a finite set of edges. $s: E \rightarrow V$ is a function associating each edge with its source. $t: E \rightarrow V$ is a function associating each edge with its target. $V_i \subseteq V$ is a set of input nodes. Each node in V_i has no outgoing edges and is not associated with a function. $V_o \subseteq V$ is a set of output nodes. Each node in V_o has one outgoing edge, no incoming edges and is not associated with a function. $l: V \rightarrow F$ labels every "function node" that is not in $V_i \cup V_o$ with a function in F. $a: E \rightarrow \mathbb{Z}$ labels every edge with a positive integer.*

Such a representation may contain neutral material; nodes to which there is no path from any output and therefore do not contribute to the functionality of the solution. This is a direct encoding, and the conversion from genotype to phenotype is given by simply removing material which does not contribute to any output. We present an example individual in Fig. 2. This individual is both feed-forward and satisfies the arities of its associated function set

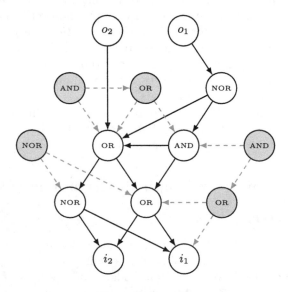

Fig. 2. An example EGGP Individual for a binary logic problem.

F = {OR, AND, NOR} This individual has two input nodes, labeled i_1 and i_2, and two output nodes, labeled o_1 and o_2. Neutral material, which does not contribute to the phenotype of the individual, is coloured gray. Edge labels are omitted for visual clarity, and is unambiguous for this example as all of the functions in F are symmetrical.

3.2 Atomic Mutations

We describe two point mutations for an EGGP individual that appear maximally simplistic; changing the function associated with a node and changing a single input to a node.

Main := pickEdge; markOutput!; mutateEdge; unmark!

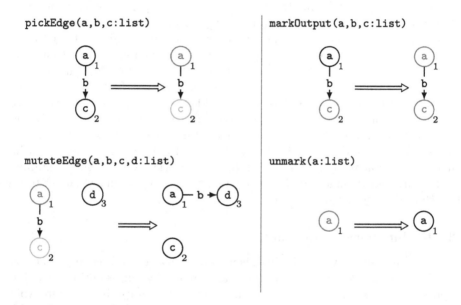

Fig. 3. Mutating an edge of an EGGP individual while preserving feed-forwardness using a graph program.

As we assume that individuals are feed-forward in this work we require a mutation that respects this constraint. A point mutation of a node's input edge while maintaining feed-forwardness is shown in Fig. 3. Firstly an edge to mutate is chosen and marked, with uniform probability, with pickEdge and then all nodes which have a path to the source of the chosen edge are marked using markOutput!. The edge is then mutated to target some unmarked node, chosen with uniform probability, using mutateEdge, and clearly there cannot be a cycle introduced as the earlier steps of the mutation have not marked that node and therefore that node does not have a path to the source of the chosen edge. Finally,

mutateFunction-f_y(f_x:string)

Fig. 4. Mutating the function of an EGGP individual's node to some function f_y. f_x is the existing function of the node being mutated, and an equivalent rule can be constructed for each function in the function set.

we unmark all marked nodes using the **unmark** rule. Using the given mutation, we are able to mutate an individual while respecting feed-forwardness without applying any restraints to the individual or the mutation, by transforming the individual using a graph program.

A point mutation of a node's function is shown in Fig. 4. Here node 1 has its function updated to some function $f_y \in F$. This operator clearly preserves feed-forwardness as it introduces no new edges. In this work we deal with function sets with fixed, common arities. However, this may not always be desirable, for example when attempting symbolic regression over a function set containing both addition and sin operators. In the C-based library for Cartesian Genetic Programming this is overcome by simply using the first few inputs for lower arity functions [26], but in EGGP this would prevent some feed-forward preserving mutations when a node appears to contribute to the input of another but in truth does not. Although we do not address this issue in this work, we propose that it would be possible to add and delete input edges when mutating a node's function to maintain correct function arities, while also maintaining feed-forwardness when adding those new input edges in a similar manner to the algorithm for edge mutations given in Fig. 3.

3.3   Evolutionary Algorithm

Crossover between EGGP individuals is not obvious as there is no apparent relationship between the nodes and edges of any two individual solutions. It might be possible to use historical markers, as used in the graph-based neuroevolution algorithm NEAT [24], or some other approach, but this is not attempted in this work. Without a crossover operator, it is natural to consider single-survivor evolutionary algorithms. As we intend to benchmark against Cartesian Genetic Programming in Sect. 5, we propose the use of the evolutionary algorithm most commonly used with it, the $1 + \lambda$ evolutionary algorithm shown in Fig. 5. This algorithm is an extended form of Random Hill Climbing, where in each generation λ new individuals are generated by mutating the sole surviving parent from the previous generation. Additionally, we allow a new individual with equal fitness to its parent to replace its parent in the next generation, facilitating the phenomena of "neutral drift". Propagating changes in the genotype which result in neutral changes in the phenotype is known to positively influence the performance of CGP [13] and we see no obvious reason why this would not also be the case in EGGP.

```
1: procedure 1 + λ(maxGenerations, λ)
2:     parent ← generateindividual
3:     parentScore ← evaluate(parent)
4:     generation ← 0
5:     while solution not found and generation ≤ maxGenerations do
6:         newParent ← parent
7:         for i = 0 to λ do
8:             child ← mutate(parent)
9:             childScore ← evaluate(child)
10:            if childScore ≤ parentScore then
11:                newParent ← child
12:                parentScore ← childScore
13:        parent ← newParent
14:        generation ← generation + 1
```

Fig. 5. The $1 + \lambda$ evolutionary algorithm with neutral drift enabled.

3.4 Parameters

Use of the EGGP representation and the $1 + \lambda$ algorithm is parameterised by the following items:

- m_r: the mutation rate. Nodes and edges have no particular order in EGGP, and the order in which feed-forward edges are mutated may influence the availability of future mutations. We therefore opt to generate the number of node mutations and edge mutations to apply to the individual using binomial distributions (simulating one probabilistic mutation for each node or edge) and then distribute these mutations across the individual at random.
- λ: the number of individuals to generate in each generation of the $1 + \lambda$ algorithm.
- F: A function set. The maximum arity of functions in F is used to specify the number of input edges associated with each function node.
- n: the number of function nodes to use in each individual.
- p: the number of inputs that each individual should have to interface with the fitness function ($|V_i| = p$).
- q: the number of outputs that each individual should have to interface with the fitness function ($|V_o| = q$).
- A fitness function used to evaluate each individual.

4 Relation to Cartesian Genetic Programming

4.1 Cartesian Genetic Programming

Cartesian Genetic Programming (CGP) is a type of evolutionary algorithm in which individuals are represented as linear sequences of genes corresponding to a directed acyclic graph. Each gene is an integer representing either (1) where a node gets its inputs from or (2) the function of a node. These nodes are ordered so that all input connections must respect that ordering, preventing cycles. When evolving over a function set where each function takes 2 inputs, there are 3 genes for each node in the individual; 2 representing each of the node's inputs, and 1 representing the node's function. Outputs are represented as single genes describing the node in the individual which corresponds to that output. These connection genes (nodes' input genes and the singular output genes) point to other nodes based on their index in the ordering.

An example genotype-phenotype mapping is given in Fig. 6. Here an individual consisting of 3 nodes over a function set of arity 2, 1 input and 1 output is represented by 10 genes. These genes decode into the shown directed acyclic graph. In CGP individuals may be seen as a grid of n_r rows and n_c columns; a node in a certain column may use any node from any row in an earlier column as an input. Hence the total $n = n_r \times n_c$ nodes are ordered under a $\leq$ operator. The example shown in Fig. 6 is a single row instance of CGP.

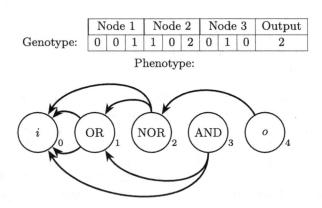

	Node 1			Node 2			Node 3		Output
Genotype:	0	0	1	1	0	2	0	1	0	2

Phenotype:

Fig. 6. The genotype-phenotype mapping of a simple CGP individual consisting of 1 input, 3 nodes and 1 output and arity 2. Each node is represented by 3 genes; the first 2 describe the indices of the node's inputs (starting at index 0 for the individual's input i) and the third describing the node's function. Function indices 0, 1 and 2 correspond to AND, OR and NOR respectively. The final gene describes the index of the node used by the individual's output o.

4.2   Comparison to EGGP

Here we demonstrate that EGGP provides a richer representation than CGP:

- For a fixed number of nodes n and function set F, any CGP individual can be represented as an EGGP individual, whereas the converse may not always hold when the number of rows in a CGP individual is greater than one.
- Any order-preserving CGP mutation can be represented as a feed-forward preserving mutation in EGGP, whereas some feed-forward preserving mutations may not be order-preserving nor valid in the CGP framework.

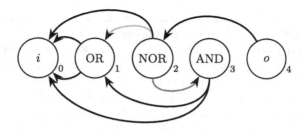

Fig. 7. A feed-forward preserving edge mutation. An edge (red) directed from node 2 to node 1 is replaced with an edge (blue) directed to node 3. This mutation produces a valid circuit but is impossible in CGP as it does not preserve order. (Color figure online)

Firstly, consider the genotype-phenotype decoding of a CGP individual. Here we have clearly defined sets of input, output and function nodes. Additionally each function node is associated with some function from the function set, and there are ordered input connections (edges) from each function node to its inputs. Clearly this decoded individual can be treated as an EGGP individual fitting Definition 1. Conversely, consider the case where $n_r > 1$. Then there is the trivial counter example of an EGGP individual with a solution depth greater than n_c (as $n > n_c$) which clearly cannot be expressed as a CGP individual limited to depth n_c.

We now consider mutations available over a CGP individual in comparison to those for an EGGP individual where feed-forward preserving mutations are used. Clearly, as each order preserving mutation is feed-forward preserving, any valid mutation for the CGP individual is available for its EGGP equivalent. However, consider the example shown in Fig. 7. Here a feed-forward mutation, connecting node 3 to node 4 is available in the EGGP setting but is not order preserving so is impossible in the CGP setting. Additionally, the semantic change that has occurred here, where an active node has been inserted between two adjacent active nodes is a type of phenotype growth that is impossible in CGP. Hence every mutation available in CGP is available in EGGP for an equivalent individual but the converse may not be true.

Therefore the landscape described by EGGP over the same function set and number of nodes is a generalisation of that described by CGP, with all individual solutions and viable mutations available, alongside further individual solutions and mutations that were previously unavailable.

4.3   Ordered EGGP (O-EGGP)

To demonstrate whether any differences in performance between EGGP and CGP arise from the freedom of mutation using feed-forward preserving, rather than order-preserving, input mutations, we compare performance against an ordered variant of EGGP, called Ordered EGGP.

Each node in an O-EGGP individual is associated with an order in an analogous manner to CGP. Node function mutations from EGGP are used, but input mutations are order-preserving rather than feed-forward preserving. Hence the same set of mutation operators, with the same probability distribution over their outcomes, is available for equivalent O-EGGP and CGP individuals. This approach simulates the landscape and search process of CGP under identical conditions, so should produce identical results to an equivalent CGP implementation. By also benchmarking O-EGGP we demonstrate that it is EGGP's free graphical representation and the associated more general ability to mutate input connections with respect to preserving feed-forwardness that yields higher quality results.

5   Benchmarking

To benchmark EGGP we compare against basic CGP and O-EGGP for a set of Digital Circuit problems taken from [28], which are used in comparisons between CGP and its Embedded variant ECGP. We produce our own CGP benchmark results, which are roughly in line with those available in [28], by using the C-based CGP library [26]. The set of Digital Circuit problems studied is given in Table 1; we study bit adders, bit multipliers and even-parity circuits of various sizes alongside a 3:8-Bit de-multiplexer and a 4×1-Bit comparator (see [28] for details). As many of these circuits are typically constructed manually using XOR gates, we use the function set $\{AND, OR, NAND, NOR\}$ to artificially increase the difficulty of these problems. We use the number of incorrect bits produced by a candidate solution in comparison to the full truth table of the given problem as the fitness function.

Table 1. Digital Circuit benchmark problems.

Digital Circuit	Number of inputs	Number of outputs
1-bit adder (1-Add)	3	2
2-bit adder (2-Add)	5	3
3-bit adder (3-Add)	7	4
2-bit multiplier (2-Mul)	4	4
3-bit multiplier (3-Mul)	6	6
3:8-bit de-multiplexer (DeMux)	6	6
4×1-bit comparator (Comp)	4	18
3-bit Even Parity (3-EP)	3	1
4-bit Even Parity (4-EP)	4	1
5-bit Even Parity (5-EP)	5	1
6-bit Even Parity (6-EP)	6	1
7-bit Even Parity (7-EP)	7	1

Each algorithm is run 100 times, with a maximum generation cap of 20,000,000; every run in each case successfully produced a result with the exception of the 3-Mul for CGP, which produced a correct solution in 99% of cases. In all 3 benchmarks, 100 nodes are used for each individual. Following conventional wisdom for CGP, we use a mutation rate of 4% for CGP and O-EGGP benchmarks. Additionally, a single row of nodes is used in each of these cases ($n_r = 1$). However, from our observations EGGP works better with a lower mutation rate, so for EGGP benchmarks we use 1%. An investigation of how mutation rate influences the performance in EGGP is left for future work. The $1 + \lambda$ algorithm is used in all 3 cases, with $\lambda = 4$. Due to time constraints, O-EGGP is only benchmarked on easier problems; 1-Add, 2-Add, 2-Mul, DeMux, 3-EP, 4-EP and 5-EP. We argue that if the results from these benchmarks are in line with the CGP benchmark results we may extrapolate that O-EGGP is indeed simulating CGP. In this case the two distinguishing factors between the EGGP and CGP benchmarks are the use of mutation operator (feed-forward preserving vs. order preserving) and mutation rate (1% vs. 4%).

To provide comparisons, we use the following metrics; median number of evaluations (ME), median absolute deviation (MAD; median of the absolute deviation from the evaluation median ME), and interquartile range (IQR). The number of evaluations taken for each run is calculated as the number of generations used multiplied by the total population size ($1 + \lambda = 5$). The hypotheses we investigate for these benchmarks are:

1. EGGP performs significantly better than CGP on the same problems under similar conditions. This hypothesis, if validated, would demonstrate the value of our approach.
2. O-EGGP does not perform significantly better or worse than CGP on the same problems under identical conditions. This hypothesis, if validated, would indicate that the possible factors influencing EGGP's greater performance for the first hypothesis would be reduced to the use of the feed-forward mutation operator and the mutation rate.

6   Results

Here we present results from our benchmarking experiments. Digital circuit results for EGGP and CGP are given in Table 2; results for O-EGGP on a smaller benchmark suite are given in Table 3.

To test for statistical significance we use the two-tailed Mann-Whitney U test [10], which (essentially) tests the null hypothesis that two distributions have the same medians (the non-parametric analogue of the t-test applicable only to normally distributed data). In the case where we get a statistically significant result ($p < 0.05$), we also calculate the effect size, using the non-parametric Vargha-Delaney A Test [27].

Comparing EGGP to CGP in Table 2, we find no significant improvement of EGGP over CGP for small problems (1-Add, 2-Mul). Indeed, for 2-Mul CGP

Table 2. Results from Digital Circuit benchmarks for CGP and EGGP. The p value is from the two-tailed Mann-Whitney U test. Where $p < 0.05$, the effect size from the Vargha-Delaney A test is shown; large effect sizes ($A > 0.71$) are shown in **bold**. The values for CGP on the 3-Mul problem include the single failed run.

Problem	EGGP			CGP			p	A
	ME	MAD	IQR	ME	MAD	IQR		
1-Add	5,723	3,020	7,123	6,018	3,312	7,768	0.62	–
2-Add	74,633	32,863	66,018	180,760	88,558	198,595	10^{-15}	**0.82**
3-Add	275,180	114,838	298,250	2,161,378	957,035	1,837,942	10^{-31}	**0.97**
2-Mul	14,118	5,553	12,955	10,178	5,258	14,459	0.018	0.60
3-Mul	1,241,880	437,210	829,223	15,816,940	7,948,870	19,987,744	10^{-34}	**0.99**
DeMux	16,763	4,710	9,210	20,890	6,845	14,063	0.013	0.60
Comp	262,660	84,248	174,185	1,148,823	425,758	1,012,149	10^{-31}	**0.97**
3-EP	2,755	1,558	4,836	4,365	2,530	5,345	0.038	0.58
4-EP	13,920	5,803	11,629	22,690	11,835	24,340	10^{-6}	0.69
5-EP	34,368	15,190	30,054	106,735	55,615	126,063	10^{-18}	**0.86**
6-EP	83,053	33,273	66,611	485,920	248,150	535,793	10^{-3}	**0.97**
7-EP	197,575	61,405	131,215	1,828,495	843,655	1,860,773	10^{-33}	**0.99**

significantly outperforms EGGP ($p < 0.05$), albeit with a small effect size ($0.56 < A < 0.64$). As the problems get larger and harder we find significant ($p < 0.05$) improvement of EGGP over CGP in all cases. The effect size is small ($0.56 < A < 0.64$) for the 3:8-Bit De-Mux and 3-Bit Even Parity, and medium ($0.64 < A < 0.71$) for 4-EP. We find highly significant ($p < 0.001$) improvements along with large effect sizes ($0.71 < A$) on all other problems, including the most difficult problems: 3-Add, 3-Mul, 4×1-Bit Comparator and 7-Bit Even Parity. So there is a clear progression of increasing improvement with problem difficulty.

We visualise some highly significant results as box-plots, with raw data overlayed and jittered, in Fig. 8. For each of the named problems, it can be clearly seen that EGGP's interquartile range shares no overlap with CGP's, highlighting the significance of the improvement made. Overall, we see these results to validate our first hypothesis that EGGP performs significantly better than CGP when addressing the same harder problems, although we note that no significant improvement is made for simpler problems.

When comparing O-EGGP to CGP in Table 3, we find no significant difference between either approach on any of the problems in the smaller benchmark set. The results show similar numbers of median evaluations (ME) in each case, and produce p values indicating no significant difference between the samples. We believe that these findings support our hypothesis that O-EGGP does not perform significantly better or worse than CGP on identical problems under identical conditions. As O-EGGP theoretically simulates CGP, this indicates that we can consider the differences between the runs of EGGP and O-EGGP, namely

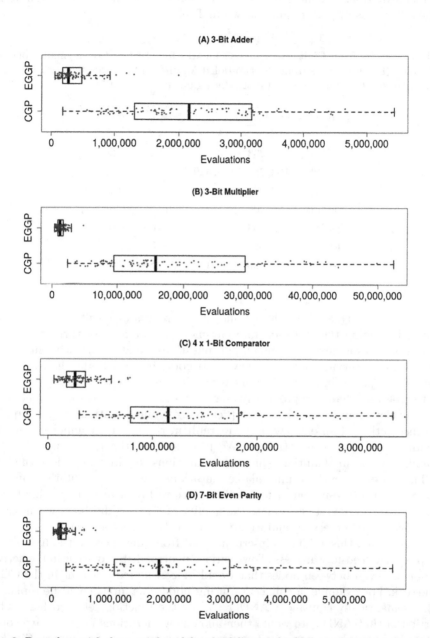

Fig. 8. Box-plots with data overlayed for the following highly significant results; (A) 3-Bit Adder, (B) 3-Bit Multiplier, (C) 4 × 1-Bit Comparator and (D) 7-Bit Even Parity. Overlayed data is jittered for visual clarity.

feed-forward preserving mutations and mutation rate, as the major contributors to the differences in performance shown in Table 2.

Table 3. Results from Digital Circuit benchmarks for O-EGGP on a smaller benchmark suite. The p value is from the two-tailed Mann-Whitney significance test comparing against CGP; no result is statistically significant ($\alpha = 0.05$).

Problem	O-EGGP			p
	ME	MAD	IQR	
1-Add	6,253	3,610	9036	0.66
2-Add	193,753	109,420	239,133	0.95
2-Mul	13,930	7,905	19,104	0.12
DeMux	21,406	5,115	10,065	0.66
3-EP	3,903	2,315	4,831	0.64
4-EP	23,360	11,893	21,865	0.84
5-EP	121,820	51,150	107,868	0.56

Further, we suggest that the significant differences in results would not be resolved by tuning the mutation rate parameter. Therefore we turn our attention to the feed-forward preserving input mutation operator. As shown in Fig. 7, feed-forward preserving mutations may insert nodes between nodes that would be considered adjacent in the CGP framework. This allows a subgraph of the solution to grow and change in previously unavailable manners. Performing functionally equivalent mutations with order preserving input mutations might require the construction of an entirely new subgraph in the neutral component of the individual which is then activated. We propose that the former mutation is more likely to occur than the sequence of mutations required to achieve the latter. Therefore where those unavailable mutations are "good" mutations in the sense of the fitness function, better performance will be achieved by using them directly. A future investigation into the quality of the neighborhood when using the feed-forward preserving mutation would clarify this hypothesis.

Additionally, this ability to insert material from anywhere in the individual that preserves feed-forwardness allows various neutral drifts to occur in the active component, even between nodes that would be considered adjacent in the CGP framework. For example, a connection using node x as input could be replaced by the semantically equivalent $AND(x, x)$, for the function set used here. The insertion of that AND gate would then allow new mutations in the active component; for example changing its function, or mutating one of its inputs. Similar neutral mutations exist in this domain, such as the insertion of double negations using NAND gates. Additionally, the reverses of these transformations are also possible, freeing up genetic material to be used elsewhere. How useful these neutral mutations in the active component are is left for future work.

7 Conclusion and Future Work

We have proposed graphs as a fundamental representation for evolutionary algorithms and in particular the use of rule-based graph programming as a means to perform mutations. We have developed an algorithm, Evolving Graphs by Graph Programming, and demonstrated significantly improved performance on a suite of classic benchmark problems in comparison to CGP. We have demonstrated an ordered variant of EGGP, O-EGGP, that simulates and produces similar results to CGP, to support our hypothesis that the feed-forward preserving input mutation leads to improved performance. We believe this sets a clear precedent for future work on evolutionary algorithms using graphs as a fundamental representation and graph programming as a mechanism for transforming them.

There are a number of directions in which this work may be built upon. Further investigation into the value of the feed-forward preserving mutation operator and how the phenotype is able to change under it is necessary. If our hypothesis that neutral mutations in the active component are useful is confirmed, it may then be possible to force similar neutral mutations by encoding equivalence laws for a given domain as graph programming mutations, such as logical equivalence laws for circuits [11] or the ZX-calculus's equivalence rules for quantum graphs [3]. Additionally, a study of whether strict adherence to function arities for function sets with varying arities is helpful, as discussed in Sect. 3.2, may be worthwhile. In the present work we have avoided crossover operators, but a thorough investigation into how graphs can be usefully recombined would be of interest. Existing ideas such as history-based crossover [23] and subgraph swapping [9,16] offer potential inspiration. We note the possibility of transferring the active component of one individual into the neutral component of another to be reabsorbed via future mutations (in a manner analogous to horizontal gene transfer in bacteria [25]), a mechanism made possible by the lack of constraint on our representation.

References

1. Atkinson, T., Plump, D., Stepney, S.: Probabilistic graph programming. In: Proceedings of the International Workshop on Graph Computation Models (GCM 2017) (2017)
2. Bak, C., Plump, D.: Compiling graph programs to C. In: Echahed, R., Minas, M. (eds.) ICGT 2016. LNCS, vol. 9761, pp. 102–117. Springer, Cham (2016). https://doi.org/10.1007/978-3-319-40530-8_7
3. Coecke, B., Duncan, R.: Interacting quantum observables: categorical algebra and diagrammatics. New J. Phys. **13**(4) (2011). 86 p
4. Cormen, T.H., Leiserson, C.E., Rivest, R.L., Stein, C.: Introduction to Algorithms, 3rd edn. MIT Press, Cambridge (2009)
5. Jungnickel, D.: Graphs, Networks and Algorithms, 4th edn. Springer, Heidelberg (2013). https://doi.org/10.1007/978-3-642-32278-5
6. Koza, J.R.: Genetic Programming: On the Programming of Computers by Means of Natural Selection. MIT Press, Cambridge (1993)

7. Koza, J.R., Bennett, F.H., Stiffelman, O.: Genetic programming as a Darwinian invention machine. In: Poli, R., Nordin, P., Langdon, W.B., Fogarty, T.C. (eds.) EuroGP 1999. LNCS, vol. 1598, pp. 93–108. Springer, Heidelberg (1999). https://doi.org/10.1007/3-540-48885-5_8

8. Galván-López, E., Rodríguez-Vázquez, K.: Multiple interactive outputs in a single tree: an empirical investigation. In: Ebner, M., O'Neill, M., Ekárt, A., Vanneschi, L., Esparcia-Alcázar, A.I. (eds.) EuroGP 2007. LNCS, vol. 4445, pp. 341–350. Springer, Heidelberg (2007). https://doi.org/10.1007/978-3-540-71605-1_32

9. Machado, P., Correia, J., Assunção, F.: Graph-based evolutionary art. In: Gandomi, A.H., Alavi, A.H., Ryan, C. (eds.) Handbook of Genetic Programming Applications, pp. 3–36. Springer, Cham (2015). https://doi.org/10.1007/978-3-319-20883-1_1

10. Mann, H.B., Whitney, D.R.: On a test of whether one of two random variables is stochastically larger than the other. Ann. Math. Statist. 18(1), 50–60 (1947)

11. Mano, M.M.: Digital Design. EBSCO Publishing Inc., Ipswich (2002)

12. Miller, J.F.: Cartesian Genetic Programming. Springer, Heidelberg (2011). https://doi.org/10.1007/978-3-642-17310-3

13. Miller, J.F., Smith, S.L.: Redundancy and computational efficiency in Cartesian Genetic Programming. IEEE Trans. Evol. Comput. 10(2), 167–174 (2006)

14. Miller, J.F., Thomson, P.: Cartesian genetic programming. In: Poli, R., Banzhaf, W., Langdon, W.B., Miller, J., Nordin, P., Fogarty, T.C. (eds.) EuroGP 2000. LNCS, vol. 1802, pp. 121–132. Springer, Heidelberg (2000). https://doi.org/10.1007/978-3-540-46239-2_9

15. O'Neill, M., Ryan, C.: Grammatical evolution. IEEE Trans. Evol. Comput. 5(4), 349–358 (2001)

16. Pereira, F.B., Machado, P., Costa, E., Cardoso, A.: Graph based crossover - a case study with the busy beaver problem. In: Proceedings of the Annual Conference on Genetic and Evolutionary Computation (GECCO), pp. 1149–1155. Morgan Kaufmann (1999)

17. Plump, D.: The design of GP 2. In: Proceedings of the Workshop on Reduction Strategies in Rewriting and Programming (WRS 2011), EPTCS, vol. 82, pp. 1–16 (2012). https://doi.org/10.4204/EPTCS.82.1

18. Plump, D.: From imperative to rule-based graph programs. J. Logical Algebraic Methods Program. 88, 154–173 (2017). https://doi.org/10.1016/j.jlamp.2016.12.001

19. Poli, R.: Evolution of graph-like programs with parallel distributed genetic programming. In: Bäck, T. (ed.) Proceedings of the International Conference on Genetic Algorithms, pp. 346–353. Morgan Kaufmann (1997)

20. Poli, R.: Parallel distributed genetic programming. In: Corne, D., Dorigo, M., Glover, F. (eds.) New Ideas in Optimization, pp. 403–431. McGraw-Hill (1999)

21. Ryan, C., Collins, J.J., Neill, M.O.: Grammatical evolution: evolving programs for an arbitrary language. In: Banzhaf, W., Poli, R., Schoenauer, M., Fogarty, T.C. (eds.) EuroGP 1998. LNCS, vol. 1391, pp. 83–96. Springer, Heidelberg (1998). https://doi.org/10.1007/BFb0055930

22. Skiena, S.S.: The Algorithm Design Manual, 2nd edn. Springer, London (2008). https://doi.org/10.1007/978-1-84800-070-4

23. Stanley, K.O., Miikkulainen, R.: Efficient reinforcement learning through evolving neural network topologies. In: Proceedings of the Annual Conference on Genetic and Evolutionary Computation (GECCO), pp. 569–577. Morgan Kaufmann Publishers Inc. (2002)

24. Stanley, K.O., Miikkulainen, R.: Evolving neural networks through augmenting topologies. Evol. Comput. **10**(2), 99–127 (2002)
25. Syvanen, M., Kado, C.I.: Horizontal Gene Transfer. Academic Press, London (2001)
26. Turner, A.J., Miller, J.F.: Introducing a cross platform open source Cartesian Genetic Programming library. Genet. Program Evolvable Mach. **16**(1), 83–91 (2015)
27. Vargha, A., Delaney, H.D.: A critique and improvement of the CL common language effect size statistics of McGraw and Wong. J. Educ. Behav. Stat. **25**(2), 101–132 (2000)
28. Walker, J.A., Miller, J.F.: Evolution and acquisition of modules in Cartesian Genetic Programming. In: Keijzer, M., O'Reilly, U.-M., Lucas, S., Costa, E., Soule, T. (eds.) EuroGP 2004. LNCS, vol. 3003, pp. 187–197. Springer, Heidelberg (2004). https://doi.org/10.1007/978-3-540-24650-3_17

Pruning Techniques for Mixed Ensembles of Genetic Programming Models

Mauro Castelli[1], Ivo Gonçalves[2,3],
Luca Manzoni[4(✉)], and Leonardo Vanneschi[1]

[1] NOVA IMS, Universidade Nova de Lisboa, 1070-312 Lisboa, Portugal
{mcastelli,lvanneschi}@novaims.unl.pt
[2] INESC Coimbra, DEEC, University of Coimbra,
Pólo 2, 3030-290 Coimbra, Portugal
icpg@dei.uc.pt
[3] CISUC, Department of Informatics Engineering, University of Coimbra,
3030-290 Coimbra, Portugal
[4] Dipartimento di Informatica, Sistemistica e Comunicazione,
Università degli Studi di Milano-Bicocca, 20126 Milano, Italy
luca.manzoni@disco.unimib.it

Abstract. The objective of this paper is to define an effective strategy for building an ensemble of Genetic Programming (GP) models. Ensemble methods are widely used in machine learning due to their features: they average out biases, they reduce the variance and they usually generalize better than single models. Despite these advantages, building ensemble of GP models is not a well-developed topic in the evolutionary computation community. To fill this gap, we propose a strategy that blends individuals produced by standard syntax-based GP and individuals produced by geometric semantic genetic programming, one of the newest semantics-based method developed in GP. In fact, recent literature showed that combining syntax and semantics could improve the generalization ability of a GP model. Additionally, to improve the diversity of the GP models used to build up the ensemble, we propose different pruning criteria that are based on correlation and entropy, a commonly used measure in information theory. Experimental results, obtained over different complex problems, suggest that the pruning criteria based on correlation and entropy could be effective in improving the generalization ability of the ensemble model and in reducing the computational burden required to build it.

1 Introduction

In the last few years, effort was dedicated to the definition and analysis of methods to exploit semantic awareness in GP [1], where the term semantics generally refers to the behavior of a program when executed on a set of training cases. Semantics-based methods provide a new conceptual view on GP and were successfully used to solve complex problems over different domains [2,3]. However,

© Springer International Publishing AG, part of Springer Nature 2018
M. Castelli et al. (Eds.): EuroGP 2018, LNCS 10781, pp. 52–67, 2018.
https://doi.org/10.1007/978-3-319-77553-1_4

standard syntax-based GP is also capable of obtaining competitive results in different fields [4,5]. In both cases, the application of a GP algorithm is generally performed with the objective of obtaining a final model able to fit the training data as best as possible. So far, a little research effort was dedicated to the construction of ensemble models based on GP [6] and this is somehow surprisingly considering the vast amount of literature where the advantages of ensemble methods are reported [7]. To formally define an ensemble model let us refer to a standard symbolic regression problem since this is the kind of application addressed in this study. In symbolic regression, the goal is to search for the symbolic expression $f(\boldsymbol{x})$ that best fits a particular training set $T = \{(\boldsymbol{x}_1, y_1), \ldots, (\boldsymbol{x}_n, y_n)\}$ of n input/output pairs with $\boldsymbol{x}_i \in \mathbb{R}^n$ and $y_i \in \mathbb{R}$. An ensemble of regression models is a set of symbolic expressions whose individual predictions are combined (typically by considering their median or a weighted sum) to predict the target y_i. Interestingly, ensembles are often much more accurate than the individual predictors that make them up [7]. The main reasons can be summarized as follows: (1) a learning algorithm performs a search in a space H of hypotheses to identify the best hypothesis in the space. When the number of training data is too small compared to the size of the hypothesis space, the learning algorithm can return different hypotheses in H and all of them present an error on the training instances. In such a situation, an ensemble can reduce the bias toward a particular hypothesis simply by considering different hypothesis and returning a prediction based on the predictions of *all* the hypothesis that made up the ensemble. (2) In the large part of the problems addressed by using machine learning techniques, the target function f cannot be represented by any of the hypotheses in H. In this case, an ensemble can expand the space of representable hypothesis by considering weighted sums of hypotheses drawn from H.

One of the main reasons for the poor attention dedicated to ensembles in the GP literature may be related to the fact that, since its inception, GP was considered a time-consuming process due to the computational complexity required by the evaluation of the fitness function. While this issue is not critical for semantics-based GP [8,9] and despite the availability of effective hardware that nowadays allows to perform fast parallel computations, the construction of ensemble models based on GP has not received the attention it deserves.

To answer this call, this study presents a method to effectively build ensembles of GP models. The method is designed in such a way to overcome the main limitations of the typical approach used for building an ensemble model, where different parallel GP populations are evolved and the final ensemble will be composed of the best models returned by each one of the these populations. In fact, while this approach is the predominant one across the literature [7,10], it is important to consider several issues that have a negative impact on the ensemble model developed with such an approach. One of the most important issues is the one related with generalization. Generalization refers to the ability of a model to perform well on unseen examples. This is a critical aspect of a model and the interest in studying generalization in GP has been recently increasing [9,11–18]. Its importance is related to the fact that, typically, the final user of a model wants to obtain satisfactory performance on new instances of the problem at hand, while the performance on the training cases is generally irrelevant. For this reason, this study takes into account different approaches that we expect to be beneficial in increasing the generalization ability of the ensemble model. The first idea comes from recent literature [19,20] where authors demonstrated that a blend of individuals created with standard syntax-based GP (STGP) and Geometric Semantic Genetic Programming (GSGP) results in a model with a better generalization ability with respect to the use of only one kind of solutions. With this in mind, to build the ensemble of GP models we run in parallel different GP populations, where some of them are evolved using STGP and others are evolved using GSGP (as done in [20]). One of the hypotheses of this study is that a blend of STGP and GSGP is beneficial (in terms of generalization) also in building an ensemble of GP models.

The second idea is related to the fact that running all the populations for a given number of generations may result in an unwanted behavior, where all the evolved individuals are semantically similar (i.e., they produce approximately the same outputs for all the different training cases). In such a situation, there would be a little advantage in using an ensemble with respect to the usage of a single model. In fact, as reported in [21] an ensemble should be composed of models that are accurate and *diverse*. In this context, two models are said to be diverse if they make different errors on new data. It is only in this way that it would be possible to achieve a better generalization ability with respect to the one achievable by considering a single model. To take into account this relevant aspect, the strategy presented in this work will evolve different parallel populations while some of them are removed by using different similarity-based pruning criteria. The criteria that are considered in this study are based on the level of entropy and correlation of the best models available in the populations. The use of these pruning criteria should guarantee some sort of diversity among the models used to build up the ensemble.

The paper is structured as follows: Sect. 2 briefly presents some previous studies related to the definition of ensemble models. Section 3 presents the strategy developed in this study to build an ensemble of GP models and the similarity-based criteria we defined. Section 4 contains the experimental study, including a presentation of the used test problems and of the experimental settings and a discussion of the obtained results. Finally, Sect. 5 concludes the paper and suggests possible avenues for future work.

2   Related Work

This section presents recent contributions related GSGP as well as existing studies involving ensemble models and GP.

The use of ensemble models and GP presents only a few contributions in the literature. One of the first studies dates back to 2000 when a study about the decomposition of regression error into bias and variance terms to provide insight into the generalization capability of modelling methods was proposed [6]. After an introduction to bias/variance decomposition of mean squared error, authors showed how ensemble methods such as bagging [22] and boosting [23] can reduce the generalization error in GP. Bagging and boosting were considered in the context of ensemble models for GP in [24]. In their work, authors presented an extension of GP by means of resampling techniques. By considering bagging and boosting, they manipulated the training data in order to improve the learning algorithm. In their work they extended GP by dividing a whole population into a set of sub-populations, each of which is evolved by using the bagging and boosting methods. Best individuals of each sub-population participate in voting to give a prediction on the unseen data. The performance of their approach was discussed and authors also showed the beneficial effect of the proposed technique in reducing bloat with respect to the standard GP algorithm. A study related to the suitability of EC techniques in building ensemble models for classification tasks was presented in [25], where authors presented the so-called Evolutionary Ensemble Learning (EEL) approach. The objective of the study was twofold: on one side they defined a new fitness function inspired by co-evolution to enforce the classifier diversity. Additionally, a new selection criterion based on the classification margin is proposed. The new selection criterion is used to extract the classifier ensemble from the final population or incrementally along the evolution. In the experimental phase, they showed the suitability of their approach when compared to a single-hypothesis evolutionary learning process.

Besides the aforementioned theoretical studies, ensemble models and GP were used to solve complex real-world problems, mainly related to classification tasks. In [26] authors demonstrated the suitability of GP as a base classifier algorithm in building ensembles for large-scale data classification. In particular, they showed that an ensemble of GP individuals is able to significantly outperform its counterparts built upon base classifiers that were trained with decision tree and logistic regression. Authors also claimed that the superiority of GP ensemble is partly attributed to the higher diversity, both in terms of the functional form as well as with respect to the variables defining the models, among the base classifiers upon which it was built on. In the same context of large-scale data classification, an extension of cellular genetic programming for data classification (CGPC) to induce an ensemble of predictors is presented in [27]. In their work authors developed two algorithms based on bagging and boosting and compared their performance with the one of CGPC. Results showed that the proposed approaches are able to deal with large datasets that do not fit in main memory, also producing better classification accuracy with respect to standard CGPC. The same authors proposed the use of GP ensemble for distributed intrusion detection systems [28]. The algorithm runs on a distributed hybrid multi-island model-based environment to monitor security-related activity within a network. Experiments showed the validity of the approach when compared to standard techniques for the task at hand. Other applications of ensemble methods to GP includes the use of querying-by-committee methods [29, 30] and of a divide-and-conquer strategy, in which ax solution need to work well only on a subset of the entire training set [31, 32].

With respect to ensembles of regression models, a quite recent contribution was proposed in [33]. The idea explored by the authors was to generate several regression models by concurrently executing multiple independent instances of a GP and, subsequently to analyze several strategies for fusing predictions from the multiple regression models. The study considered only small datasets due to memory constraints, but authors were able to draw interesting conclusions about the suitability of their approach in producing accurate predictions. Our study will differ from the one described in [33] in several ways: we do not put any constraint on the size of the datasets, we will consider models produced by different GP algorithms (blend of STGP and GSGP) and we define and use different similarity-based criteria that, by taking into account the information related to all the populations evolved, aim at improving the generalization ability of the final ensemble as well as reducing the computational effort. Hence, in the experiments described in this contribution and as explained in Sect. 3, the populations evolved are not independent of each other.

As reported in Sect. 1, one idea exploited in this study is to build an ensemble that consists of a blend of individuals produced by STGP and GSGP. GSGP is one of the newest methods to directly include semantic awareness in the search process [34]. The interested reader is referred to [1] for a description of the concept of semantics and its uses in GP, while [34, 35] present the semantic operators for GP used in this paper. Despite the plethora of studies investigating the role of semantics, this is still a hot topic in the field of GP. Particularly interesting with respect to our study is the work proposed in [19]. In their work, authors defined a simple yet effective algorithm for the initialization of a GP population inspired by the biological phenomenon of demes despeciation (i.e. the combination of demes of previously distinct species into a new population). In synthesis, the initial population of GP is created using the best individuals of a set of separate subpopulations, or demes, some of which run STGP and the others GSGP for few generations. Experimental results showed that this initialization technique outperforms GP with the traditional ramped half-and-half algorithm on six complex symbolic regression applications. Even more interesting, by using the proposed initialization technique, the GP process produces individuals with a better generalization ability than the ones obtained by initializing the population with the traditional ramped half-and-half algorithm. Hence, to construct the GP-based ensemble we build upon this idea and we expect to obtain a final ensemble with a better generalization ability with respect to its counterparts, where only STGP or GSGP are considered.

3   Method

This section describes the proposed system for building GP-based ensemble models. The main idea is to provide a pruning method to reduce the number of populations in an ensemble when they are exploring similar regions of the search space and, in some sense, they are possibly wasting computational effort to perform the same work two times. Therefore, we need a measure of similarity among solutions that allows the pruning procedure to take place. After each generation, all the best solutions for all the populations that are part of the ensemble are pairwise compared and, if two of them are deemed too similar, the worst performing one (in terms of fitness) is removed and the one remaining is now weighted more when calculating the semantics of the ensemble (that is a weighted sum of the semantics of the best solutions).

Formally, let $P_1, \ldots, P_n$ be n populations, and let $I_1, \ldots, I_n$ be their best individuals each one having semantics $s(I_i)$. Each population has associated a weight w_i (all the weights are equal to 1 after initialization) and the semantics of the entire ensemble is given by $\frac{1}{n} \sum_{i=1}^{n} w_i s(I_i)$. After each generation, a subroutine $D(I_i, I_j)$ that calculates the similarity between I_i and I_j is called for each of the best individuals of the populations. Since many similarity measures returns a real value measuring how similar the two individuals are, we obtain a Boolean answer by comparing the similarity measure with a threshold. After that, if True is returned (i.e., I_i and I_j are similar), the fitness $f(I_i)$ and $f(I_j)$ are compared and the population corresponding to the worst fitness is removed. For example, if $f(I_j)$ is the worst fitness, then P_j is removed from the ensemble and P_i will increase its weight from w_i to $w_i + w_j$. The main aspect that governs this process is the subroutine $D(I_i, I_j)$, and we are going to describe four different implementations of it based on two different notions of semantic similarity: entropy and correlation.

3.1 Correlation-Based Similarity

The idea of a correlation-based similarity is to consider two semantics as similar if the correlation among them is above a certain threshold. Let $s(I_i)$ and $s(I_j)$ be two semantics, i.e., two semantic vectors, and let $\rho_{i,j}$ be the Pearson correlation coefficient among them. Its value varies between -1 (negative correlation) to 1 (positive correlation). If the correlation is higher than 0.5 we consider the two individual similar enough. This threshold was selected after a preliminary tuning phase, where different values were tested across the benchmarks considered.

A variation of this method introduces a probability of being considered similar enough when the correlation coefficient goes above the threshold. The probability for $s(I_i)$ and $s(I_j)$ to be considered equal is set to $\rho_{i,j}$ (i.e., if $D(s(I_i), s(I_j))$ is greater than the threshold value, then true is returned with probability $\rho_{i,j}$). This will reduce the expected number of populations that will be removed from the ensemble while still assuring that very similar ones (i.e., with the correlation coefficient near one) will almost surely be removed.

3.2 Entropy-Based Similarity

The entropy-based similarity is based on the idea that, if two semantics are similar, it should be possible to infer the outputs of one based on the other, This is possible even if the relation is more complex (non-linear) than the one that can be captured by using the linear correlation coefficient. Given two semantics $s(I_i)$ and $s(I_j)$, we will denote their mutual entropy by $H(i, j)$ and their mutual information by $\mathcal{I}(i, j)$. Notice that to compute these values we need to provide discrete data, which is not the case for the semantics of GP individuals under regression problems. Therefore, we discretize them using $\sqrt{n}$ equally-sized bins, with n the length of the semantic vectors, and counting the number of elements that are present in each bin.

The similarity measure is based on the *variation of information*, that is, the metric $d(i,j) = H(i,j) - \mathcal{I}(i,j)$. To normalize it between 0 and 1 we use $D(i,j) = \frac{d(i,j)}{H(i,j)}$. This distance will be close to 1 if the two semantics are dissimilar and close to 0 otherwise. As a threshold the value 0.5 was chosen. Thus, two semantics with distance lower than 0.5 are considered similar enough. Also in this case, the threshold was selected after a preliminary tuning phase, where different values were tested.

As with the correlation-based similarity, it is possible to introduce a probabilistic variation, in which a population is selected for deletion with probability $1 - D(i,j)$ when $D(i,j) < 0.5$.

4   Experimental Settings and Results

Five datasets were considered for testing the performance of the GP-based ensemble. These datasets were already considered in previous GP studies. Hence, we summarize their main properties in Table 1 by reporting the number of independent variables, the name of the problem and a reference where readers may find detailed descriptions of the datasets. In these experimental phase, benchmarks with real-world data were chosen to establish the suitability of the proposed methods in a real-world setting.

Table 1. Description of the test problems. For each dataset, the number of features (independent variables) and the number of instances (observations) are reported.

Dataset	# Features	# Instances
Airfoil [36]	5	1502
Concrete [37]	8	1029
Protein Plasma Binding Level (PPB) [38]	626	131
Slump [39]	9	102
Yacht [40]	6	307

The objectives of the experiment are the following:

- to show that the pruning criteria can effectively remove some of the existing populations;
- that the removal of populations is performed in a way that does not deteriorate performance;
- that the pruning criteria perform better than simply randomly removing populations or selecting a smaller ensemble size.

Here, the notion of being better can be interpreted in two ways: either a performance improvement with respect to fitness or comparable fitness values obtained using a smaller number of populations. That is, the proposed pruning criteria either improve the results or reduce the computational burden.

The ensembles are composed, at the beginning, of 20 populations, half of them will evolve by using STGP and the other half by using GSGP. To generate the training set, for each run a global training set consisting of 70% of the problem instances was selected. The remaining 30% was used as the test set. From the training set, each population in the ensemble was provided with a local training set consisting of the same number of observations but obtained by randomly sampling with replacement from the global training set. This was performed for each of the 30 runs.

The following methods were compared:

- **Standard**. No deletion of populations.
- **Random**. Each population has a probability of 0.001 of being removed at each generation. At least one population will always remain.
- **Half**. At the start, half of the populations are selected (one half of them using STGP and the other half using GSGP) and no further deletion occurs.
- **Correlation**. The correlation-based similarity is used in the pruning.
- **Correlation-prob**. The probabilistic variation of the correlation-based similarity is used in the pruning.
- **Entropy**. The entropy-based similarity is used in the pruning.
- **Entropy-prob**. The probabilistic variation of the entropy-based similarity is used in the pruning.

The general parameters of the system are summarized in Table 2. The values of the parameters were selected taking into account the values already used in existing literature, where the datasets considered in this study were already used as benchmarks. For the sake of brevity, we report only the results obtained on the test set and avoid reporting the global training set results. To test the statistical significance of the results, we have used the single tailed Mann-Whitney U-test with the alternative hypothesis that the first series of fitness values is lower (i.e., better) than the second series. As a threshold for the p-value we selected $\alpha = 0.05$.

Table 3 show the results of the statistical tests, where the entry in row i and column j is the p-value of the Mann-Whitney U-test where the technique in the i-th row is compared against the one in the j-th column. A value less then 0.05 (highlighted in bold) indicates that we have accepted the alternative hypothesis and, thus, the i-th method produces lower (i.e., better) fitness values than the j-th one on the test data. Figures 2, 3, 4, 5 and 6 show, for the test problems considered, the fitness on the test set for the different methods. The average size of the ensemble across all generations and all runs is shown in Fig. 1.

The results on the airfoil dataset (Fig. 2 and Table 3) show that the standard ensemble method is the best performer. All the four proposed methods, however, perform better than simply halving or randomly removing populations from

Table 2. Parameters used in the experiments

Parameter	Value
Runs	30
Generations	1000
Population size	200
Training - Testing division	70% - 30%
Fitness	Root Mean Squared Error
Crossover probability	0.6
Mutation probability	0.3
Tree initialization	Ramped Half-and-Half, maximum depth 6
Function set	$+, -, *,$ and protected $/$
Terminal set	Input variables, no constants
Parent selection	Tournament of size 4
Elitism	Best individual always survives
Maximum tree depth	None
Ensemble size	20 (10 STGP and 10 STGP)

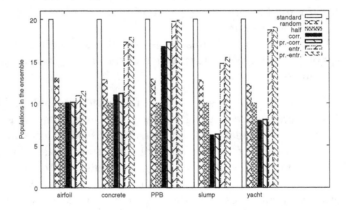

Fig. 1. The average number of populations in the ensemble for the considered methods

the ensemble. Figure 1 shows that those results were obtained while using, on average, about half of the populations than the standard method.

For the concrete dataset (Fig. 3 and Table 3), the standard method and all the four proposed methods perform in a similar way. Using only half of the populations or randomly removing them produces worse results. In this case, the two correlation-based methods also use about half of the populations employed by the standard method.

For the PPB dataset (Fig. 4 and Table 3) the correlation-based methods return the lowest (i.e., better) fitness but the difference is not statistically significant when compared to the standard and entropy-based methods. The other two

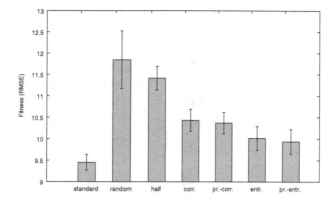

Fig. 2. The average fitness at the last generation in the airfoil dataset. The error bars are one standard deviation in length.

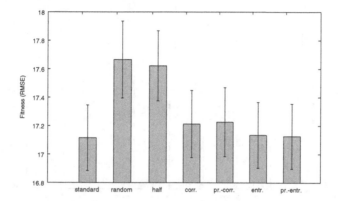

Fig. 3. The average fitness at the last generation in the concrete dataset. The error bars are one standard deviation in length.

methods are the worst performers. The best results are obtained by using quite a high number of populations for all the proposed methods. This might indicate that the PPB problem is well-suited to be solved with ensemble techniques and that additional populations help in producing better results.

The slump dataset (Fig. 5 and Table 3) has the standard and the entropy-based methods as the best performers. The worst performers are the correlation-based methods which, in this case, also employ a very low number of populations.

The yacht dataset (Fig. 6 and Table 3) is interesting since it provides an example in which two of the proposed methods, namely the correlation-based ones, perform the worse, while the entropy-based remains on-par with the standard method at the price of removing only a few populations.

To conclude this section, it is important to discuss the competitive advantage in considering a blend of STGP and GSGP. To evaluate this aspect, we compared the performance of the ensemble model built considering a blend of STGP and

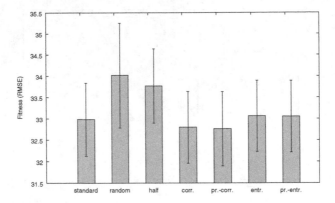

Fig. 4. The average fitness at the last generation in the PPB dataset. The error bars are one standard deviation in length.

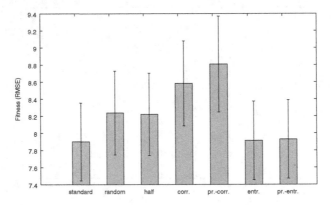

Fig. 5. The average fitness at the last generation in the slump dataset. The error bars are one standard deviation in length.

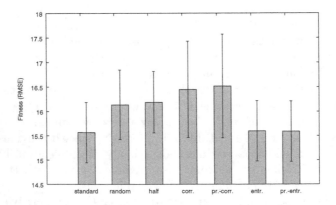

Fig. 6. The average fitness at the last generation in the yacht dataset. The error bars are one standard deviation in length.

Table 3. p-values for the statistical tests on the airfoil, concrete, PPB, slump, and yacht datasets (from top to bottom).

airfoil	standard	random	half	corr.	pr.-corr.	entr.	pr.-entr.
standard		**0.000**	**0.000**	**0.000**	**0.000**	**0.0000**	**0.001**
random	1.000		0.886	1.000	1.000	1.000	1.000
half	1.000	0.117		1.000	1.000	1.000	1.000
correlation	1.000	**0.000**	**0.000**		0.677	0.996	0.999
prob-correlation	1.000	**0.000**	**0.000**	0.329		0.990	0.998
entropy	1.000	**0.000**	**0.000**	**0.004**	0.010		0.735
prob-entropy	1.000	**0.000**	**0.000**	**0.001**	**0.002**	0.270	
concrete	standard	random	half	corr.	pr.-corr.	entr.	pr.-entr.
standard		**0.000**	**0.000**	0.190	0.156	0.432	0.476
random	1.000		0.720	0.999	0.999	1.000	1.000
half	1.000	0.285		0.999	0.999	1.000	1.000
correlation	0.814	**0.001**	**0.001**		0.456	0.749	0.777
prob-correlation	0.848	**0.001**	**0.002**	0.550		0.802	0.814
entropy	0.573	**0.000**	**0.000**	0.255	0.202		0.527
prob-entropy	0.529	**0.000**	**0.000**	0.228	0.190	0.479	
PPB	standard	random	half	corr.	pr.-corr.	entr.	pr.-entr.
standard		0.050	**0.034**	0.685	0.695	0.418	0.424
random	0.952		0.591	0.971	0.974	0.927	0.931
half	0.967	0.415		0.984	0.984	0.953	0.956
correlation	0.321	**0.030**	**0.016**		0.529	0.241	0.260
prob-correlation	0.310	**0.027**	**0.017**	0.476		0.232	0.241
entropy	0.588	0.075	**0.048**	0.763	0.772		0.509
prob-entropy	0.582	0.071	**0.045**	0.745	0.763	0.497	
slump	standard	random	half	corr.	pr.-corr.	entr.	pr.-entr.
standard		0.056	0.129	**0.013**	**0.002**	0.468	0.421
random	0.946		0.630	0.156	**0.053**	0.929	0.927
half	0.874	0.375		0.114	**0.028**	0.851	0.844
correlation	0.987	0.848	0.889		0.246	0.985	0.984
prob-correlation	0.998	0.949	0.973	0.759		0.997	0.997
entropy	0.538	0.073	0.152	**0.015**	**0.003**		0.441
prob-entropy	0.585	0.075	0.159	**0.017**	**0.003**	0.565	
yacht	standard	random	half	corr.	pr.-corr.	entr.	pr.-entr.
standard		0.050	**0.022**	**0.013**	**0.027**	0.468	0.479
random	0.952		0.398	0.370	0.381	0.946	0.949
half	0.979	0.608		0.421	0.515	0.972	0.972
correlation	0.987	0.636	0.585		0.650	0.983	0.985
prob-correlation	0.974	0.625	0.491	0.356		0.968	0.969
entropy	0.538	0.056	**0.029**	**0.018**	**0.033**		0.509
prob-entropy	0.527	0.053	**0.029**	**0.016**	**0.032**	0.497	

GSGP without pruning, against the ones obtained by considering only STGP and only GSGP. The results achieved on the five benchmarks (not reported here due to the page limit) show that no statistically significant difference exists with respect to the performance of the considered ensemble models on unseen instances. While this result seems to contradict recent studies (e.g., [19]), a deeper analysis is needed. In particular, it would be interesting to study how the number of models in the final ensemble affects the generalization ability of the three systems (only STGP, only GSGP, blend of STGP and GSGP) and to analyze the impact of the pruning techniques on the generalization error. Despite the fact that no statistically significant differences can be noticed, using a blend of STGP and GSGP might still be important and non detrimental: results show that the diversity of the model is generally greater than the one observable by considering a pool of individuals obtained with only GSGP (or STGP) and, additionally, the pruning criteria can determine what are the populations and models that should evolve. Hence, it might be advisable to use both GP systems to evolve the weak models and to let the pruning criteria to select the one needed to create a competitive ensemble model. As for the advantage in term of runtime and number of fitness evaluations, it is important to notice that a pruning method that, for example, reduces on average the number of distinct populations used by one third, also reduces the number of fitness evaluations by one third and, since the computation of the similarity is usually not the most computationally intensive part of the GP algorithm, a similar improvement is reflected in the reduction of the runtime.

5   Conclusions

Machine learning literature has deeply investigated ensemble models, reporting their advantages with respect to the use of a weak learner. In particular, ensemble models are characterized by different features that could improve the performance of a learning technique: they are able to reduce the variance, the average out biases, they can cover a larger area of the hypothesis space with respect to a single model, and they generally present a better generalization than a single model. Nonetheless, ensemble models were not deeply investigated in the field of GP. This paper answered this call by proposing a strategy to build ensemble models by using genetic programming. Two different GP versions are taken into account: standard syntax-based GP (STGP) and a GP system (GSGP) able to directly include semantic awareness in the search process by using geometric semantic operators.

The main objective of the study was to understand whether an ensemble model made of a blend of GP individuals evolved by both STGP and GSGP can be effectively pruned using different criteria based on correlation and entropy. These criteria are used to avoid the construction of an ensemble where weak learners are semantically too similar. In fact, as existing literature suggested, an ensemble model should be made of accurate and diverse models. The strategies developed were tested over different benchmark problems already considered in the GP literature. Results are interesting and, while they do not allow to draw a general conclusion about the superiority of a criterion with respect to the other ones, they show that considering a similarity criterion when constructing a GP ensemble can help in maintaining the generalization ability of the resulting model while reducing the computational effort.

This work represents a preliminary study and several future works are planned: first of all, a study aimed at determining the optimal number of weak learners for optimizing the performance of the ensemble over unseen data represent a priority. This would allow practitioners to use the system without the need to determine the number of weak learners, a parameter that has an impact on the performance of the ensemble. Furthermore, the design of pruning criteria that are parameter independent is an important future work to save the user the time required for tuning them. Additionally, it would be interesting to pursue the study of the interaction between syntax and semantics in GP, a very important topic in the field, that is still not well understood.

Acknowledgements. This work was also financed through the Regional Operational Programme CENTRO2020 within the scope of the project CENTRO-01-0145-FEDER-000006.

References

1. Vanneschi, L., Castelli, M., Silva, S.: A survey of semantic methods in genetic programming. Genet. Program. Evolvable Mach. **15**(2), 195–214 (2014)
2. Castelli, M., Vanneschi, L., Felice, M.D.: Forecasting short-term electricity consumption using a semantics-based genetic programming framework: the South Italy case. Energy Econ. **47**, 37–41 (2015)
3. Castelli, M., Castaldi, D., Giordani, I., Silva, S., Vanneschi, L., Archetti, F., Maccagnola, D.: An efficient implementation of geometric semantic genetic programming for anticoagulation level prediction in pharmacogenetics. In: Correia, L., Reis, L.P., Cascalho, J. (eds.) EPIA 2013. LNCS (LNAI), vol. 8154, pp. 78–89. Springer, Heidelberg (2013). https://doi.org/10.1007/978-3-642-40669-0_8
4. Yoo, S., Xie, X., Kuo, F.C., Chen, T.Y., Harman, M.: Human competitiveness of genetic programming in spectrum-based fault localisation: theoretical and empirical analysis. ACM Trans. Softw. Eng. Methodol. **26**(1), 4:1–4:30 (2017)
5. Picek, S., Mariot, L., Leporati, A., Jakobovic, D.: Evolving s-boxes based on cellular automata with genetic programming. In: Proceedings of the Genetic and Evolutionary Computation Conference Companion, GECCO 2017, pp. 251–252. ACM, New York (2017)
6. Keijzer, M., Babovic, V.: Genetic programming, ensemble methods and the bias/variance tradeoff – introductory investigations. In: Poli, R., Banzhaf, W., Langdon, W.B., Miller, J., Nordin, P., Fogarty, T.C. (eds.) EuroGP 2000. LNCS, vol. 1802, pp. 76–90. Springer, Heidelberg (2000). https://doi.org/10.1007/978-3-540-46239-2_6
7. Dietterich, T.G.: Ensemble methods in machine learning. In: Kittler, J., Roli, F. (eds.) MCS 2000. LNCS, vol. 1857, pp. 1–15. Springer, Heidelberg (2000). https://doi.org/10.1007/3-540-45014-9_1

8. Castelli, M., Silva, S., Vanneschi, L.: A C++ framework for geometric semantic genetic programming. Genet. Program. Evolvable Mach. **16**(1), 73–81 (2015)

9. Gonçalves, I., Silva, S., Fonseca, C.M., Castelli, M.: Unsure when to stop? Ask your semantic neighbors. In: Proceedings of the Genetic and Evolutionary Computation Conference, pp. 929–936. ACM (2017)

10. Polikar, R.: Ensemble learning. In: Zhang, C., Ma, Y. (eds.) Ensemble Machine Learning, pp. 1–34. Springer, Boston (2012). https://doi.org/10.1007/978-1-4419-9326-7_1

11. Gonçalves, I.: An exploration of generalization and overfitting in genetic programming: standard and geometric semantic approaches. Ph.D. thesis, Department of Informatics Engineering, University of Coimbra, Portugal (2017)

12. Chen, Q., Xue, B., Shang, L., Zhang, M.: Improving generalisation of genetic programming for symbolic regression with structural risk minimisation. In: Proceedings of the 2016 on Genetic and Evolutionary Computation Conference, pp. 709–716. ACM (2016)

13. Gonçalves, I., Silva, S., Fonseca, C.M.: On the generalization ability of geometric semantic genetic programming. In: Machado, P., Heywood, M.I., McDermott, J., Castelli, M., García-Sánchez, P., Burelli, P., Risi, S., Sim, K. (eds.) EuroGP 2015. LNCS, vol. 9025, pp. 41–52. Springer, Cham (2015). https://doi.org/10.1007/978-3-319-16501-1_4

14. Kommenda, M., Affenzeller, M., Burlacu, B., Kronberger, G., Winkler, S.M.: Genetic programming with data migration for symbolic regression. In: Proceedings of the Companion Publication of the 2014 Annual Conference on Genetic and Evolutionary Computation, pp. 1361–1366. ACM (2014)

15. Gonçalves, I., Silva, S.: Balancing learning and overfitting in genetic programming with interleaved sampling of training data. In: Krawiec, K., Moraglio, A., Hu, T., Etaner-Uyar, A.Ş., Hu, B. (eds.) EuroGP 2013. LNCS, vol. 7831, pp. 73–84. Springer, Heidelberg (2013). https://doi.org/10.1007/978-3-642-37207-0_7

16. Gonçalves, I., Silva, S., Melo, J.B., Carreiras, J.M.B.: Random sampling technique for overfitting control in genetic programming. In: Moraglio, A., Silva, S., Krawiec, K., Machado, P., Cotta, C. (eds.) EuroGP 2012. LNCS, vol. 7244, pp. 218–229. Springer, Heidelberg (2012). https://doi.org/10.1007/978-3-642-29139-5_19

17. Gonçalves, I., Silva, S.: Experiments on controlling overfitting in genetic programming. In: Proceedings of the 15th Portuguese Conference on Artificial Intelligence: Progress in Artificial Intelligence, EPIA 2011 (2011)

18. Castelli, M., Manzoni, L., Silva, S., Vanneschi, L.: A quantitative study of learning and generalization in genetic programming. In: Silva, S., Foster, J.A., Nicolau, M., Machado, P., Giacobini, M. (eds.) EuroGP 2011. LNCS, vol. 6621, pp. 25–36. Springer, Heidelberg (2011). https://doi.org/10.1007/978-3-642-20407-4_3

19. Vanneschi, L., Bakurov, I., Castelli, M.: An initialization technique for geometric semantic GP based on demes evolution and despeciation. In: 2017 IEEE Congress on Evolutionary Computation (CEC), pp. 113–120. IEEE (2017)

20. Vanneschi, L., Galvão, B.: A parallel and distributed semantic genetic programming system. In: 2017 IEEE Congress on Evolutionary Computation (CEC), pp. 121–128. IEEE (2017)

21. Hansen, L.K., Salamon, P.: Neural network ensembles. IEEE Trans. Pattern Anal. Mach. Intell. **12**(10), 993–1001 (1990)

22. Breiman, L.: Bagging predictors. Mach. Learn. **24**(2), 123–140 (1996)

23. Freund, Y., Schapire, R.E., et al.: Experiments with a new boosting algorithm. In: Icml, vol. 96, pp. 148–156 (1996)

24. Iba, H.: Bagging, boosting, and bloating in genetic programming. In: Proceedings of the 1st Annual Conference on Genetic and Evolutionary Computation, vol. 2, pp. 1053–1060. Morgan Kaufmann Publishers Inc. (1999)
25. Gagné, C., Sebag, M., Schoenauer, M., Tomassini, M.: Ensemble learning for free with evolutionary algorithms? In: Proceedings of the 9th Annual Conference on Genetic and Evolutionary Computation, pp. 1782–1789. ACM (2007)
26. Zhang, Y., Bhattacharyya, S.: Genetic programming in classifying large-scale data: an ensemble method. Inf. Sci. **163**(1), 85–101 (2004)
27. Folino, G., Pizzuti, C., Spezzano, G.: GP ensembles for large-scale data classification. IEEE Trans. Evol. Comput. **10**(5), 604–616 (2006)
28. Folino, G., Pizzuti, C., Spezzano, G.: GP ensemble for distributed intrusion detection systems. In: Singh, S., Singh, M., Apte, C., Perner, P. (eds.) ICAPR 2005. LNCS, vol. 3686, pp. 54–62. Springer, Heidelberg (2005). https://doi.org/10.1007/11551188_6
29. Isele, R., Bizer, C.: Active learning of expressive linkage rules using genetic programming. Web Semant. Sci. Serv. Agents World Wide Web **23**, 2–15 (2013)
30. Bartoli, A., De Lorenzo, A., Medvet, E., Tarlao, F.: Active learning of regular expressions for entity extraction. IEEE Trans. Cybern. 1–14 (2017)
31. Pappa, G.L., Freitas, A.A.: Evolving rule induction algorithms with multi-objective grammar-based genetic programming. Knowl. Inf. Syst. **19**(3), 283–309 (2009)
32. Bartoli, A., De Lorenzo, A., Medvet, E., Tarlao, F.: Learning text patterns using separate-and-conquer genetic programming. In: Machado, P., Heywood, M.I., McDermott, J., Castelli, M., García-Sánchez, P., Burelli, P., Risi, S., Sim, K. (eds.) EuroGP 2015. LNCS, vol. 9025, pp. 16–27. Springer, Cham (2015). https://doi.org/10.1007/978-3-319-16501-1_2
33. Veeramachaneni, K., Derby, O., Sherry, D., O'Reilly, U.M.: Learning regression ensembles with genetic programming at scale. In: Proceedings of the 15th Annual Conference on Genetic and Evolutionary Computation, pp. 1117–1124. ACM (2013)
34. Moraglio, A., Krawiec, K., Johnson, C.G.: Geometric semantic genetic programming. In: Coello, C.A.C., Cutello, V., Deb, K., Forrest, S., Nicosia, G., Pavone, M. (eds.) PPSN 2012. LNCS, vol. 7491, pp. 21–31. Springer, Heidelberg (2012). https://doi.org/10.1007/978-3-642-32937-1_3
35. Vanneschi, L., Castelli, M., Manzoni, L., Silva, S.: A new implementation of geometric semantic GP and its application to problems in pharmacokinetics. In: Krawiec, K., Moraglio, A., Hu, T., Etaner-Uyar, A.Ş., Hu, B. (eds.) EuroGP 2013. LNCS, vol. 7831, pp. 205–216. Springer, Heidelberg (2013). https://doi.org/10.1007/978-3-642-37207-0_18
36. Brooks, T., Pope, D., Marcolini, A.: Airfoil self-noise and prediction. Technical report, NASA RP-1218 (1989)
37. Castelli, M., Vanneschi, L., Silva, S.: Prediction of high performance concrete strength using genetic programming with geometric semantic genetic operators. Expert Syst. Appl. **40**(17), 6856–6862 (2013)
38. Castelli, M., Vanneschi, L., Popovič, A.: Parameter evaluation of geometric semantic genetic programming in pharmacokinetics. Int. J. Bio-Inspired Comput. **8**(1), 42–50 (2016)
39. Yeh, I.-C.: Simulation of concrete slump using neural networks. Constr. Mater. **162**(1), 11–18 (2009)
40. Ortigosa, I., Lopez, R., Garcia, J.: A neural networks approach to residuary resistance of sailing yachts prediction. In: Proceedings of the International Conference on Marine Engineering MARINE, vol. 2007, p. 250 (2007)

Analyzing Feature Importance
for Metabolomics Using Genetic
Programming

Ting Hu[1]([✉])(iD), Karoliina Oksanen[1], Weidong Zhang[2,3], Edward Randell[2],
Andrew Furey[2], and Guangju Zhai[2]

[1] Department of Computer Science, Memorial University,
St. John's, NL A1B 3X5, Canada
ting.hu@mun.ca
[2] Faculty of Medicine, Memorial University, St. John's, NL A1B 3V6, Canada
[3] School of Pharmaceutical Sciences, Jilin University, Jilin, Changchun, China

Abstract. The emerging and fast-developing field of metabolomics examines the abundance of small-molecule metabolites in body fluids to study the cellular processes related to how the human body responds to genetic and environmental perturbations. Considering the complexity of metabolism, metabolites and their represented cellular processes can correlate and synergistically contribute to a phenotypic status. Genetic programming (GP) provides advanced analytical instruments for the investigation of multifactorial causes of metabolic diseases. In this article, we analyzed a population-based metabolomics dataset on osteoarthritis (OA) and developed a Linear GP (LGP) algorithm to search classification models that can best predict the disease outcome, as well as to identify the most important metabolic markers associated with the disease. The LGP algorithm was able to evolve prediction models with high accuracies especially with a more focused search using a reduced feature set that only includes potentially relevant metabolites. We also identified a set of key metabolic markers that may improve our understanding of the biochemistry and pathogenesis of the disease.

Keywords: Metabolomics · Osteoarthritis · Biomarker discovery
Genetic programming · Classification

1 Introduction

Systems biology is an emerging research field that takes a holistic approach to modeling complex biological systems rather than examining different levels of biological systems separately [1–3]. It requires collaborative efforts from disciplines including biomedicine, statistics, and computer science. Systems biology approaches embrace the complexity of biological systems and focus on modeling the interactions among multiple components including genome, transcriptome,

© Springer International Publishing AG, part of Springer Nature 2018
M. Castelli et al. (Eds.): EuroGP 2018, LNCS 10781, pp. 68–83, 2018.
https://doi.org/10.1007/978-3-319-77553-1_5

proteome, and metabolome [4–6]. By integrating a variety of "omics" data, systems biology for human disease studies aims at better understanding the pathogenesis of common diseases, discovering biomarkers that can help predict early disease onset, progression, and severity, and identifying new drug targets [7,8].

Integrative data analysis and mining for systems biology often include hundreds to thousands of variables such as genes, proteins, and metabolites [9], in order to find the most relevant biomarkers that can explain a specific phenotype or disease. Most conventional tools adopt a univariate analysis strategy and examine one variable at a time on its individual association with the disease. This may overlook the intertwined relationships among multiple variables that contribute to the disease. Thus, retooling for systems biology is needed such that a large set of variables can be analyzed simultaneously on their synergistic effects [10,11]. However, the high dimensionality has imposed both methodological and computational challenges since learning algorithms that can model the complex non-linear relationships of multiple variables are yet to be explored, and searching combinations of variables becomes prohibitive as the search space grows exponentially with the number of variables.

Machine learning and heuristic search algorithms, including principal component analysis [12], artificial neural networks [13], and random forest [14], have seen increasing and successful applications in omics data mining for biomarker discovery. However, despite a few attempts [15,16], genetic programming, as a powerful learning and modeling algorithm, has not caught up with other comparable algorithms in wide applications.

Genetic programming (GP) holds great potentials for systems biology research. First, it can construct highly non-linear models of multiple variables (features) that can best predict a phenotypic or disease outcome using arithmetic functions, Boolean functions, and conditional statements. Second, the selection of relevant features in a model classifier is achieved automatically in GP. This feature selection process is embedded in model construction such that the inclusion of a feature is decided based on the classification performance of the model. Such an automatic and embedded feature selection mechanism distinguishes GP from many approaches that select features and construct classification models in separate steps. Third, the stochastic population-based search property of evolutionary algorithms allows to generate multiple best classification models. This provides a diverse set of classification models for subsequent interpretation and feature importance analysis.

In this study, we use a GP algorithm, specifically a Linear GP representation, to train classification models and to identify key biomarkers for metabolomics, in order to demonstrate the power of GP in the coming era of systems biology and big biomedical data research.

Recent developments in the field of metabolomics provide an array of new tools for the study of human diseases. A large number of small-molecule metabolites from body fluids or tissues can be quantitatively detected simultaneously, which promises an immense potential for early diagnosis, therapy monitoring and understanding the pathogenesis of complex diseases [17]. Metabolites are intermediate and end products of various cellular processes and their levels of concentration serve as a good indicator of a sequence of biological systems in response to genetic and environmental influences. This can, in turn, help us better understand the diseases and develop new drug treatments.

We use population-based metabolomics data where two phenotypically distinguished individuals, i.e., diseased cases and healthy controls, are recruited and their blood samples are collected to measure the concentration levels of a variety of metabolites. Classification models are then evolved and trained using GP algorithm. We adopt a two-round design where GP uses the full set of metabolites in the initial round of model exploration and selects a subset of potentially more relevant metabolites for the second round of more focused search. The importance of metabolites in terms of their contribution to the disease is then assessed based on their occurrence frequencies in the final best classification models.

2 Methods

2.1 Metabolomics Data on Osteoarthritis

Osteoarthritis (OA) is a slowly progressive joint disease and is the most common form of arthritis. It occurs when the protective cartilage on the ends of bones breaks down often because of mechanical stress or biochemical alterations. It causes a substantial morbidity and disability in the elderly populations, and imposes a great economic burden on our society [18,19]. Despite high prevalence and societal impact, there is no medication that can cure it, or reverse or halt the disease progression, partly because its pathogenesis is still unclear and there is no reliable method that can be used for early OA diagnosis.

In this study, we used a OA metabolomics dataset from the Newfoundland Osteoarthritis Study (NFOAS) [20,21]. The goal of the NFOAS is to identify novel genetic, epigenetic, and biochemical markers for OA, in order to better understand the diseases and to develop new drug treatment. In the NFOAS, knee OA patients who underwent a total knee replacement surgery due to primary OA were recruited. Healthy controls were selected from volunteering participants.

Both cases and controls were from the same source population. Knee OA diagnosis was made based on the American College of Rheumatology clinical criteria for the classification of idiopathic OA of the knee [22] and the judgment of the attending orthopedic surgeons. Controls were individuals without self-reported family doctor diagnosed knee OA based on their medical information collected by a self-administered questionnaire. A total number of 153 OA cases and 236 healthy controls were collected.

Blood samples were collected after at least 8 hours of fasting and plasma was separated from blood using the standard protocol. Metabolic profiling was performed on plasma using the Waters XEVO TQ MS system (Waters Limited, Mississauga, Ontario, Canada) coupled with Biocrates AbsoluteIDQ p180 kit, which measures 186 metabolites including 90 glycerophospholipids, 40 acylcarnitines (1 free carnitine), 21 amino acids, 19 biogenic amines, 15 sphingolipids and 1 hexose (above 90 percent is glucose). The details of the 186 metabolites and the metabolic profiling method were described in a previous publication [23]. Over 90% of the metabolites (167/186) were successfully determined in each sample.

The study protocol was approved by the Health Research Ethics Authority (HREA) of Newfoundland and Labrador with reference number 11.311 and a written consent was obtained from all the participants.

We followed a two-stage design and divided the samples randomly into *discovery* and *replication* datasets, such that our genetic programming algorithm can be applied separately to the two datasets and only the key features (metabolites) successfully replicated were reported. Since samples were collected and their metabolite concentrations were measured in various batches, certain biases can exist when samples from different batched were compared. We performed batch corrections to remove such biases by multiplying each metabolite concentration value by the ratio of the overall mean and the batch mean for that metabolite. In addition, age and BMI are known factors correlated with OA. Therefore, the residual of a linear regression using attributes age and BMI was applied to remove any partial correlations as a result of those two factors, and to adjust the data for subsequent analysis. Finally, each metabolite concentration value was normalized to zero mean and unit variance across the population.

2.2   Linear Genetic Programming Algorithm

Linear genetic programming (LGP) encodes evolutionary individuals as imperative programs that are executed sequentially [24]. Although LGP follows a linear instructional structure, it is very powerful and capable of modeling complex nonlinear relationships among multiple attributes. Comparing to the more traditional representation of trees, such an instructional structure of LGP enables fast execution and thus speedy fitness evaluation. Therefore, LGP has gained increasing popularity being applied to a variety of modeling and classification problems [25–27].

In the current study, an instruction of an LGP program can be either an assignment statement or a conditional statement. An assignment statement manipulates values stored in calculation registers by applying arithmetic operations such as addition, subtraction, multiplication, division, and the exponential function. We use `if-then` statements to change the flow of program execution by skipping one subsequent instruction when the condition in the `if` statement is false.

Feature registers contain input values of corresponding variables from data samples, and calculation registers are used to enhance the computational capacity of LGP programs. A feature register can only serve as an operand on the right-hand side of an assignment statement, while a calculation register can be used as an operand or a return on the left-hand side of an assignment statement. The calculation register `r[0]` is designated as the output register, and its final stored value is the outcome of the entire program. Since we consider a classification problem in the current study, the Sigmoid function will be applied to `r[0]`.

If $S(r[0])$ is greater than or equal to 0.5, the sample is predicted as diseased (class one), otherwise, the sample is predicted as healthy (class zero).

Therefore, an LGP program represents a classification model that takes a data sample with a set of feature values (metabolite concentration levels) as input, and outputs the predicted class status (diseased or healthy) of this sample. An example LGP program with eight instructions is given below.

```
if r[1]> r[5]
    then r[0] = r[7] + 5
r[4] = r[2] / r[0]
if r[0] > 4
    then if r[3] < 10
        then r[6] = r[3] - r[5]
r[2] = r[5] * r[5]
r[0] = r[2] + r[7]
```

At the initial generation, a population of LGP programs is generated randomly. The fitness of each program is evaluated using mean classification error (MCE), computed as the average number of incorrectly classified training samples. A set of programs are chosen as parents based on their fitness, and variation operators, including mutation and recombination, are applied to them. A micro mutation alters an element of a randomly picked instruction, i.e., replacing a return or an operand register by a randomly generated one or replacing the operator. A macro mutation deletes a randomly chosen instruction or inserts a randomly generated instruction. Recombination swaps segments of instructions of two parent programs. Survival selection picks fitter programs to form the population for the next generation. Such an evolution process iterates for a certain number of generations, and the program with the lowest MCE at the end is output as the final best model of a run.

In our study, the LGP algorithm is implemented using the Julia programming language [28]. The main parameters used in the implementation are shown in Table 1. A five-fold cross-validation strategy was used to prevent overfitting. That is, the data samples are randomly divided into five partitions, and each partition serves as the testing set once while the remaining four partitions are input to the LGP algorithm as the training set. Therefore, for each implementation, the algorithm produces five best classification models based on the five testing sets.

2.3   Full vs. Focused Feature Analysis

The goal of our metabolomics study is to identify key metabolites that can best explain the phenotypic class, i.e., diseased or healthy. The importance of a metabolite (feature), can be assessed by computing its occurrence frequency in the best classification models found by the LGP algorithm. Such an occurrence frequency measures how often a feature appears in the final outcome model of an LGP run, and thus reflects its contribution to the correct classification of the disease status.

Table 1. LGP parameter configurations for classification on metabolomics data.

Fitness function	Mean classification error (MCE)
Program length	$[1, 500]$
Number of calculation registers	150
Operator set	$\{+, -, \times, \div, x^y, \text{if} <, \text{if} >\}$
Constant set	$\{1, 2, 3, 4, 5, 6, 7, 8, 9, 10\}$
Population size	500
Mutation rate	0.1
Mutation operators	Micro and macro to effective instructions
Crossover rate	0.9
Parent selection	Tournament with size 16
Survival selection	Truncation
Number of generations	500
Number of runs	200

For the first round of analysis, the LGP algorithm is run using the full feature set of 167 metabolites on both the discovery and replication datasets using 200 distinct seed values for the random number generator. Each run gives five different best classification models as a result of the five-fold cross-validation. Therefore, our implementation produces a total of 1000 best classification models.

We investigate the resulting classification models by calculating various statistics of the fitness (MCE) values, sensitivity, specificity and area under the curve (AUC) as computed on the testing fold for each run. In addition, we inspect the models by counting how often each of the 167 metabolites appears as a predictive variable in the set of 1000 best models.

Note that although a total of 167 metabolites are measured in the OA metabolomics data, not all of them are relevant to the disease. In machine learning, removing irrelevant features can speed up the training process and improve the prediction accuracy of the models [29]. Therefore, we perform the second round of analysis by only using a focused subset of metabolites. The focused subset of metabolites is defined as the metabolites that have occurrence frequencies higher than the average among all 167 metabolites. We re-run the LGP algorithm using such focused feature sets on both the discovery and replication datasets, and investigate if reducing the number of features can improve the prediction performance.

3 Results

3.1 Best Models Found Using Full Feature Set

First, we investigate the 1000 best models found by the LGP algorithm on the discovery dataset using the full set of 167 metabolites. The statistics of the

Table 2. Statistics of the classification performance of the 1000 best models (discovery, full feature set).

	MCE	Sensitivity	Specificity	AUC
Mean	0.367	0.684	0.584	0.663
Median	0.367	0.667	0.600	0.667
Min	0.067	0.200	0.200	0.320
Max	0.667	1.000	0.933	0.947
Std dev	0.095	0.146	0.142	0.110
5% confidence	0.181	0.398	0.305	0.447
95% confidence	0.553	0.970	0.862	0.879

Table 3. Statistics of the classification performance of the 1000 best models (replication, full feature set).

	MCE	Sensitivity	Specificity	AUC
Mean	0.357	0.685	0.601	0.664
Median	0.367	0.667	0.600	0.664
Min	0.100	0.267	0.067	0.309
Max	0.667	1.000	1.000	0.960
Std dev	0.103	0.140	0.169	0.118
5% confidence	0.156	0.411	0.271	0.432
95% confidence	0.558	0.958	0.932	0.895

classification performance of those 1000 best models are shown in Table 2. The best classifier can achieve a mean classification error (MCE) as low as 0.067, and the area under the curve (AUC) as 0.947. This demonstrates the effectiveness of using the LGP algorithm to train a classifier for metabolomics studies.

We look at the distributions of the fitness (MCE) and the number of effective features of those 1000 best models (Fig. 1). The majority of those 1000 best models have an MCE in the range of [0.3, 0.5]. A feature is effective if it takes a role modifying the value stored in the output register when the LGP program, i.e., classification model, is executed to make a prediction. Although any subsets of those 167 metabolites can be chosen by a classification model, the LGP algorithm selects the most relevant features as the result of the evolutionary learning process. The majority of those 1000 best models have between 25 and 40 effective features. Figure 2 shows that the fitness and the number of effective features are not correlated (Spearman's correlation test $\rho = 0.044$ with a significance level $p = 0.16$).

The same analysis is then repeated on the replication dataset, and the statistics of the classification performance of the 1000 best models found by LGP

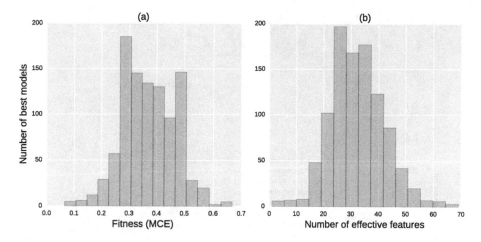

Fig. 1. Distributions of (a) the fitness and (b) the number of effective features for the 1000 best models (discovery, full feature set).

are shown in Table 3. We see that using the discovery and replication datasets achieve comparable classification performance.

3.2 Best Models Found Using Focused Feature Sets

For the second round of analysis, we reduce the feature set and only provide a more relevant subset of features to the LGP algorithm in order to perform a more focused classification model construction. In our study, the relevance, or importance, of a metabolite is assessed using its occurrence frequency in the 1000 best models, i.e., the number of times a metabolite appears in the 1000 best models as an effective feature. We follow the intuition that if a metabolite appears often in the evolved best models, it may play an important role explaining the disease.

Figure 3(a) shows the distribution of metabolite occurrence frequency in the 1000 best models using the discovery dataset. The majority of metabolites have occurrence frequencies between 170 and 220. The mean of the distribution is 193.562, and we use that as the threshold to select the focused feature set. That is, the focused feature set only includes 75 metabolites that have occurrence frequencies higher than or equal to the average value of 193.562. The distribution of metabolite occurrence frequency in the best models using the replication dataset is shown in Fig. 4(a). The mean of the distribution is 191.898, and similarly, we use it as the threshold to select the replication focused feature set with 60 metabolites for the second round of analysis.

The statistics of the classification performance using focused feature sets are shown in Tables 4 and 5 for the discovery and replication datasets respectively. Comparing to Tables 2 and 3, we can see that the classification performance is improved by examining all statistics. Specifically, the average MCE is reduced from 0.367 to 0.317 and the average AUC is improved from 0.663 to 0.714 for

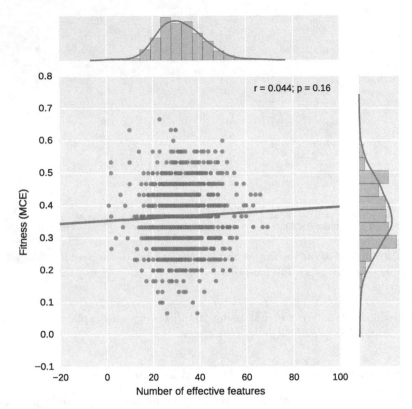

Fig. 2. Correlation of the fitness and the number of effective features in the best prediction models (discovery, full feature set). Each data point represents one of the 1000 best classification models found by LGP. The solid line provides a visual guide on the correlation between the fitness and the number of effective features.

discovery dataset, and from 0.357 to 0.286 and from 0.664 to 0.740 for replication dataset respectively. The improvement of the classification performance by reducing the feature set indicates that our LGP algorithm is able to identify important and relevant metabolites that can better explain the disease of OA.

Moreover, the best classifier among the 1000 evolved models can achieve an MCE as low as 0.067 and an AUC as high as 0.971 for the discovery dataset and 0.067 and 1 for the replication dataset respectively. Given the complexity of the disease, this suggests the effectiveness of using the LGP algorithm to infer the underlying highly non-linear interacting relationships of multiple metabolites that are associated with the disease.

3.3 Identification of Key Metabolic Markers

The goal of our informatics study is to provide a list of important metabolites for future biological validation, such that we can better understand the etiology of the disease and better design its drug treatments. To estimate the importance of each metabolite, we examine its occurrence frequency in both the discovery and

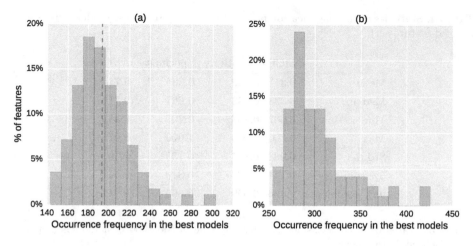

Fig. 3. Distributions of feature occurrence frequency in the 1000 best models on (a) the full set of 167 features and (b) the focused set of 75 features (discovery). In (a), the vertical dashed line represents the mean of the distribution.

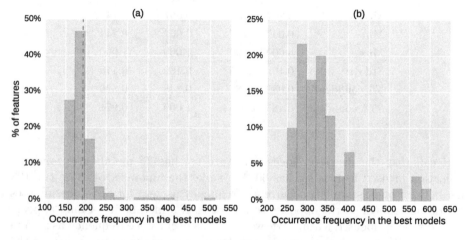

Fig. 4. Distributions of feature occurrence frequency in the 1000 best models on (a) the full set of 167 features and (b) the focused set of 60 features (replication). In (a), the vertical dashed line represents the mean of the distribution.

replication datasets. Figures 3(b) and 4(b) show the distributions of metabolite occurrence frequencies in both datasets in the second round of a more focused classification model construction using reduced feature sets. Comparing to using the full feature sets (Figs. 3(a) and 4(a)), there are more metabolites having much higher occurrence frequencies in the best models. The explanation could be that by removing irrelevant features, our LGP algorithm is able to pick up more important features through a more focused search.

Table 4. Statistics of the classification performance of the 1000 best models (discovery, focused feature set).

	MCE	Sensitivity	Specificity	AUC
Mean	0.317	0.732	0.635	0.714
Median	0.333	0.733	0.667	0.718
Min	0.067	0.267	0.200	0.353
Max	0.600	1.000	1.000	0.971
Std dev	0.088	0.137	0.135	0.103
5% confidence	0.144	0.464	0.370	0.512
95% confidence	0.490	0.999	0.899	0.917

Table 5. Statistics of the classification performance of the 1000 best models (replication, focused feature set).

	MCE	Sensitivity	Specificity	AUC
Mean	0.286	0.751	0.678	0.740
Median	0.267	0.733	0.667	0.744
Min	0.067	0.267	0.067	0.244
Max	0.600	1.000	1.000	1.000
Std dev	0.102	0.135	0.169	0.118
5% confidence	0.086	0.487	0.348	0.509
95% confidence	0.485	1.015	1.009	0.971

Recall that the discovery focused feature set has 75 metabolites, and the replication focused feature set has 60. We make the union set of those two (98 metabolites) and assign the occurrence frequency as zero for those metabolites that do not appear in the opposite set. That is, if a metabolite A only appears in the discovery focused feature set, we treat A's occurrence frequency as zero in replication. We then show the occurrence frequencies of those metabolites in the union set of discovery and replication (Fig. 5) in order to identify key metabolites whose importance can be both discovered and replicated.

By using a threshold of 0.3 on both axes, we identify 17 key metabolites at the right-upper corner of the scatter plot (Fig. 5). Those 17 key metabolites include the ones that have been reported previously with a strong association with the disease of OA, as well as the ones that haven't been linked to the disease in the literature yet but hold great potentials improving our understanding of the disease. Those new discoveries are particularly interesting since with further biological validation, they could help identify metabolic processes that are potentially related to the disease. The biology of those 17 key metabolites will be explained in more detail in the Discussion section.

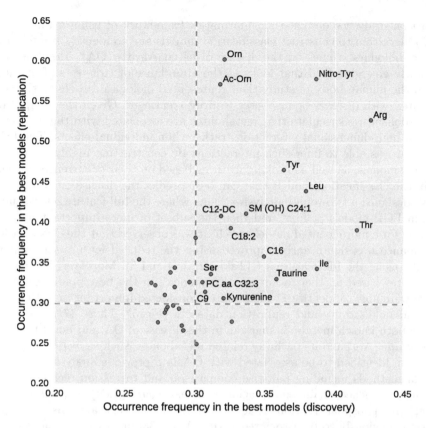

Fig. 5. Scatter plot of normalized metabolite occurrence frequencies in the best models using the focused feature sets. Each data point represents a metabolite. The x-axis is its occurrence frequency in the discovery dataset, and the y-axis is that in the replication dataset. Dashed lines define a set of 17 key metabolites that have higher occurrence frequencies in both datasets comparing to the rest of the features.

4   Discussion

The advancing of biomedical and computational technologies has brought about a new era for systems biology research, where abundant and various types of data become available for quantitative analysis for us to better understand the biology of living systems. The underlying causes of complex human diseases are often multifactorial such that intelligent learning algorithms are needed to identify the combinations of the most relevant biomarkers from hundreds to thousands of biological variables.

Machine learning techniques are often employed for modeling the complex non-linear relationships of combinations of biomarkers and the disease outcome, thanks to their robust heuristic search and learning abilities. However, genetic programming (GP), positioned at the intersection of machine learning and evolutionary computing, has not seen wide applications in systems biology.

In this study, we designed an informatics framework of using a Linear GP (LGP) algorithm to construct classification models and to identify key features for metabolomics studies on the disease of osteoarthritis (OA). Metabolomics is a newly emerging field that looks at the abundance of large sets of metabolites in the human body to study their represented biological processes that are associated with diseases or responses to drug treatment. Given the complexity of metabolism, we speculate that metabolites are associated with the disease in terms of high-dimensional interactions rather than individual effects. The LGP algorithm was able to infer such interactions by constructing highly non-linear symbolic models, as well as ranking features based on their occurrence frequencies in the classification models that can best predict the disease outcome.

We designed a two-round analysis scheme where the full feature set was used to train LGP models at first, and then the subset of more important features was used for a more focused model search. It was observed that the classification performance was significantly improved using the reduced feature set comparing with using the full feature set (Tables 2, 3, 4 and 5). Moreover, by ranking metabolites based on their occurrence frequencies in the best prediction models, we were able to identify 17 metabolites considered important in both of the independent discovery and replication datasets (Fig. 5). Those 17 metabolites include both known metabolic markers in the disease of OA and novel findings.

Arginine (Arg) and its pathway related metabolites, such as *ornithine* (Orn), have been identified to be associated with OA in a previous analysis using traditional methods including pairwise comparison and regression technique [30]. Similarly, branched chain amino acids such as *leucine* (Leu), several acylcarnitines and phosphatidylcholines identified in the current analysis were also reported previously to be associated with OA [31–33] or OA classification [34]. Importantly, the current analysis identified several novel metabolic markers that were otherwise missed by using traditional analytic methods. *Taurine* is the most abundant free amino acid in humans, and may play an important role in inflammation associated with oxidative stress [35], which has been implicated in the pathogenesis of OA [36]. *Taurine* has been reported to be associated with rheumatoid arthritis [37], suggesting taurine might be a novel marker to monitor disease progression of OA but not a diagnosis. *Nitrotyrosine* (Nitro-Tyr) is also associated with oxidative damage and has been found to be associated with aging and the development of OA in cartilage samples from both monkeys and humans [38]. *Kynurenine* pathway from tryptophan generates compounds which can act on glutamate receptors in peripheral tissues or modulate free radical activity and have been implicated in rheumatoid arthritis [39]. Together, these novel findings suggest the involvement of oxidative stress associated metabolic pathways in OA. Further investigations in independent cohorts are warranted to confirm these findings.

Our study demonstrates the power of a GP algorithm in complex classification model search and automatic feature selection for systems biology research. We have entered a golden era for bioinformatics research where large volumes of data that capture the different levels of biological systems are becoming available and are in need of intelligent and powerful learning algorithms that embrace the complexity of biological systems. We hope this small step can encourage more interdisciplinary communications between evolutionary computing and biomedicine and more explorations on the research front of evolutionary algorithm applications.

Acknowledgments. This research was supported by Newfoundland and Labrador Research and Development Corporation (RDC) Ignite Grant 5404.1942.101 and the Natural Science and Engineering Research Council (NSERC) of Canada Discovery Grant RGPIN-2016-04699 to TH. GZ acknowledges grants from Canadian Institute of Health Research (CIHR), Newfoundland and Labrador Research and Development Corporation (RDC) and Memorial University. We thank all the study participants who made this study possible and all the Operation Room staff at Eastern Health General Hospital and St. Clare's Hospital who helped for collecting samples.

References

1. Kitano, H.: Systems biology: a brief overview. Science **295**(5560), 1662–1664 (2002)
2. Kitano, H.: Computational systems biology. Nature **420**(6912), 206–210 (2002)
3. Ideker, T., Galitski, T., Hood, L.: A new approach to decoding life: systems biology. Annu. Rev. Genom. Hum. Genet. **2**(1), 343–372 (2001)
4. Cusick, M.E., Klitgord, N., Vidal, M., Hill, D.E.: Interactome: gateway into systems biology. Hum. Mol. Genet. **14**(suppl 2), R171–181 (2005)
5. Bruggeman, F.J., Westerhoff, H.V.: The nature of systems biology. Trends Microbiol. **15**(1), 45–50 (2007)
6. Shim, S.H.: Cell imaging: an intracellular dance visualized. Nature **546**, 39–40 (2017)
7. Wang, K., Lee, I., Carlson, G., Hood, L., Galas, D.: Systems biology and the discovery of diagnostic biomarkers. Dis. Markers **28**(4), 199–207 (2010)
8. Butcher, E.C., Berg, E.L., Kunkel, E.J.: Systems biology in drug discovery. Nat. Biotechnol. **22**(10), 1253–1259 (2004)
9. Li, Y., Chen, L.: Big biological data: challenges and opportunities. Genom. Proteomics Bioinf. **12**(5), 187–189 (2014)
10. Alfieri, R., Milanesi, L.: Multi-level data integration and data mining in systems biology. In: Handbook of Research on Systems Biology Applications in Medicine, pp. 476–496. IGI Global (2009)
11. Sugimoto, M., Kawakami, M., Robert, M., Soga, T., Tomita, M.: Bioinformatics tools for mass spectroscopy-based metabolomic data processing and analysis. Curr. Bioinf. **7**(1), 96–108 (2012)
12. Hotelling, H.: Analysis of a complex of statistical variables into principal components. J. Educ. Psychol. **24**(6), 417 (1933)
13. Bishop, C.M.: Neural Networks for Pattern Recognition. Oxford University Press, Oxford (1995)
14. Breiman, L.: Random forests. Mach. Learn. **45**(1), 5–32 (2001)
15. Worzel, W.P., Yu, J., Almal, A.A., Chinnaiyan, A.M.: Applications of genetic programming in cancer research. Int. J. Biochem. Cell Biol. **41**(2), 405–413 (2009)
16. Kandpal, M., Kalyan, C.M., Samavedham, L.: Genetic programming-based approach to elucidate biochemical interaction networks from data. IET Syst. Biol. **7**(1), 18–25 (2013)

17. Gowda, G.N., Zhang, S., Gu, H., Asiago, V., Shanaiah, N., Raftery, D.: Metabolomics-based methods for early disease diagnostics. Expert Rev. Mol. Diagn. **8**(5), 617–633 (2008)
18. WHO Scientic Group: the burden of musculoskeletal conditions at the start of the new millennium. WHO Technical Report Series 919, 218 (2003)
19. Reginster, J.Y.: The prevalence and burden of arthritis. Rheumatology **41**, 3–6 (2004)
20. Zhai, G., Aref-Eshghi, E., Rahman, P., Zhang, H., Martin, G., Furey, A., Green, R.C., Sun, G.: Attempt to replicate the published osteoarthritis-associated genetic variants in the newfoundland & labrador population. J. Orthop. Rheumatol. **1**(3), 5 (2014)
21. Hu, T., Zhang, W., Fan, Z., Sun, G., Likhodi, S., Randell, E., Zhai, G.: Metabolomics differential correlation network analysis of osteoarthritis. Pac. Symp. Biocomput. **21**, 120–131 (2016)
22. Altman, R., Alarcon, G., Appelrouth, D., Bloch, D., Borenstein, D., Brandt, K., Brown, C., Cooke, T.D., et al.: The american college of rheumatology criteria for the classification and reporting of osteoarthritis of the hip. Arthritis Rheum. **34**(5), 505–514 (1991)
23. Zhang, W., Likhodii, S., Aref-Eshghi, E., Zhang, Y., Harper, P.E., Randell, E., Green, R., Martin, G., Furey, A., Sun, G., Rahman, P., Zhai, G.: Relationship between blood plasma and synovial fluid metabolite concentrations in patients with osteoarthritis. J. Rheumatol. **42**(5), 859–865 (2015)
24. Brameier, M.F., Banzhaf, W.: Linear Genetic Programming. Springer, New York (2007)
25. Brameier, M.F., Banzhaf, W.: A comparison of linear genetic programming and neural networks in medical data mining. IEEE Trans. Evol. Comput. **5**(1), 17–26 (2001)
26. Guven, A.: Linear genetic programming for time-series modeling of daily flow rate. J. Earth Syst. Sci. **118**(2), 137–146 (2009)
27. Song, D., Heywood, M.I., Zincir-Heywood, A.N.: A linear genetic programming approach to intrusion detection. In: Cantú-Paz, E. (ed.) GECCO 2003. LNCS, vol. 2724, pp. 2325–2336. Springer, Heidelberg (2003). https://doi.org/10.1007/3-540-45110-2_125
28. Bezanson, J., Edelman, A., Karpinski, S., Shah, V.B.: Julia: a fresh approach to numerical computing. CoRR abs/1411.1607 (2014). http://arxiv.org/abs/1411.1607
29. Guyon, I., Elisseeff, A.: An introduction to variable and feature selection. J. Mach. Learn. Res. **3**, 1157–1182 (2003)
30. Zhang, W., Sun, G., Likhodii, S., Liu, M., Aref-Eshghi, E., Harper, P.E., Martin, G., Furey, A., Green, R., Randell, E., Rahman, P., Zhai, G.: Metabolomic analysis of human plasma reveals that arginine is depleted in knee osteoarthritis patients. Osteoarthr. Cartil. **24**, 827–834 (2016)
31. Zhai, G., Wang-Sattler, R., Hart, D.J., Arden, N.K., Hakim, A.J., Illig, T., Spector, T.D.: Serum branched-chain amino acid to histidine ratio: a novel metabolomic biomarker of knee osteoarthritis. Ann. Rheum. Dis. **69**(6), 1227–1231 (2010)
32. Zhang, W., Sun, G., Likhodii, S., Aref-Eshghi, E., Harper, P.E., Randell, E., Green, R., Martin, G., Furey, A., Rahman, P., Zhai, G.: Metabolomic analysis of human synovial fluid and plasma reveals that phosphatidylcholine metabolism is associated with both osteoarthritis and diabetes mellitus. Metabolomics **12**, 24 (2016)

33. Zhang, W., Sun, G., Aitken, D., Likhodii, S., Liu, M., Martin, G., Furey, A., Randell, E., Rahman, P., Jones, G., Zhai, G.: Lysophosphatidylcholines to phosphatidylcholines ratio predicts advanced knee osteoarthritis. Rheumatology **55**(9), 1566–1574 (2016)

34. Zhang, W., Likhodii, S., Zhang, Y., Aref-Eshghi, E., Harper, P.E., Randell, E., Green, R., Martin, G., Furey, A., Sun, G., Rahman, P., Zhai, G.: Classification of osteoarthritis phenotypes by metabolomics analysis. BMJ Open **4**, e006286 (2014)

35. Marcinkiewicz, J., Kontny, E.: Taurine and inflammatory diseases. Amino Acids **46**(1), 7–20 (2014)

36. Loeser, R.F.: Aging and osteoarthritis: the role of chondrocyte senescence and aging changes in the cartilage matrix. Osteoarthr. Cartil. **17**(8), 971–979 (2009)

37. Kontny, E., Wojtecka-ŁUkasik, E., Rell-Bakalarska, K., Dziewczopolski, W., Maśliński, W., Maślinski, S.: Impaired generation of taurine chloramine by synovial fluid neutrophils of rheumatoid arthritis patients. Amino Acids **23**(4), 415–418 (2002)

38. Loeser, R.F., Carlson, C.S., Carlo, M.D., Cole, A.: Detection of nitrotyrosine in aging and osteoarthritic cartilage: correlation of oxidative damage with the presence of interleukin-1β and with chondrocyte resistance to insulin-like growth factor 1. Arthritis Rheumatol. **46**(9), 2349–2357 (2002)

39. Forrest, C.M., Kennedy, A., Stone, T.W., Stoy, N., Darlington, L.G.: Kynurenine and neopterin levels in patients with rheumatoid arthritis and osteoporosis during drug treatment. In: Allegri, G., Costa, C.V.L., Ragazzi, E., Steinhart, H., Varesio, L. (eds.) Developments in Tryptophan and Serotonin Metabolism. AEMB, vol. 527, pp. 287–295. Springer, Boston (2003). https://doi.org/10.1007/978-1-4615-0135-0_32

Generating Redundant Features with Unsupervised Multi-tree Genetic Programming

Andrew Lensen(✉)📛, Bing Xue, and Mengjie Zhang

School of Engineering and Computer Science, Victoria University of Wellington,
PO Box 600, Wellington 6140, New Zealand
{Andrew.Lensen,Bing.Xue,Mengjie.Zhang}@ecs.vuw.ac.nz

Abstract. Recently, feature selection has become an increasingly important area of research due to the surge in high-dimensional datasets in all areas of modern life. A plethora of feature selection algorithms have been proposed, but it is difficult to truly analyse the quality of a given algorithm. Ideally, an algorithm would be evaluated by measuring how well it removes known bad features. Acquiring datasets with such features is inherently difficult, and so a common technique is to add synthetic bad features to an existing dataset. While adding noisy features is an easy task, it is very difficult to automatically add complex, redundant features. This work proposes one of the first approaches to generating redundant features, using a novel genetic programming approach. Initial experiments show that our proposed method can automatically create difficult, redundant features which have the potential to be used for creating high-quality feature selection benchmark datasets.

Keywords: Genetic programming · Feature creation
Feature construction · Feature selection · Mutual information
Evolutionary computation

1 Introduction

Feature Selection (FS) techniques aim to remove features from a dataset which are less useful than others. [1] Removing such features can improve the results of the data mining task being performed on the dataset, as well as making the results and/or model produced more interpretable and less complex. Features that should be removed are usually categorised as *irrelevant* or *redundant* features [2].

Irrelevant (or *noisy*) features are those which add little or no meaningful value to a dataset. In the worst case, an irrelevant feature may actually mislead the data mining process, when it contradicts the information given by other "correct" features. Removing such features reduces the search space of the data mining task, generally improving performance [1]. Redundant features ($r.fs$) share a high

© Springer International Publishing AG, part of Springer Nature 2018
M. Castelli et al. (Eds.): EuroGP 2018, LNCS 10781, pp. 84–100, 2018.
https://doi.org/10.1007/978-3-319-77553-1_6

amount of information overlap with other features. Removing r.fs can simplify the solutions found (as only one of a set of r.fs is needed), while again reducing the search space of the data mining task [2]. In certain cases, results may also be improved by reducing the bias towards a set of very similar features.

Many FS algorithms have been proposed, which are usually evaluated based on how well they can reduce the feature set size, while maintaining (or improving) the results of the data mining task. One technique used to compare FS algorithms is to purposefully add "bad" features to a dataset, so that a FS algorithm can be evaluated based on how well it removes those known bad features. Introducing irrelevant features to a dataset is quite straightforward—choose some stochastic noise generator, and generate a number of noisy features. Introducing r.fs, however, is much trickier, as discussed below.

Perhaps the most naive way of creating a r.f (Y) from a given *source* feature (X) is to multiply each feature value of X by some multiple α, such that the ith value of Y is computed as $Y_i = \alpha X_i$. By varying α, one can easily generate any given number of r.fs based on X. A particularly straightforward method is to simply duplicate features (i.e. let $\alpha = 1$)—but these are trivial to remove. To make the redundancy weaker, one can introduce some bias (β) such as adding a constant value to each Y_i, e.g. $Y_i = \alpha X_i + \beta$. However, such approaches have a number of serious limitations.

The above types of r.fs have very simple redundancies that do not represent realistic interactions between features in real data mining problems. For example, in a dataset of people, two potential features may be an individual's age and income. It is generally true that the older a person, the more they earn, and so we may expect these features to be linearly redundant. However, a child is likely to have no income regardless of their exact age, and a pensioner is likely to have a similar income to others aged over 65. While these two features are certainly partially redundant, the interaction is clearly more complex. In most datasets, the redundancy between two features tend to be even more complex still. Removing r.fs that have linear redundancies is also quite a trivial FS problem, and so is not an adequate challenge for non-trivial FS algorithms. For example, a greedy algorithm which uses Pearson's correlation can easily find groups of linearly-redundant features by measuring the correlation of each feature to those already selected.

There is hence an obvious need to have methods available to generate r.fs with (arbitrarily) complex interactions in order to benchmark FS methods more effectively. There has been very little work in the literature that has investigated how to automatically generate non-trivial r.fs. One common method that is used to automatically create functions to perform a particular task is Genetic Programming (GP), an Evolutionary Computation (EC) technique which evolves tree-like functions (*programs*) with a flexible structure. We believe that GP has the potential to evolve functions to produce r.fs, by taking a source feature as the program's input, and producing a r.f as the program's output.

1.1   Goals

In this paper, we propose the first approach to automatically generating r.fs, using Genetic Programming for Redundant Feature Creation (GPRFC). The proposed method uses GP to automatically generate functions to produce new r.fs from a given source feature, by using a multi-tree GP representation with a Mutual Information (MI)-based fitness function. This paper will:

- Introduce a novel multi-tree GP representation for automatically evolving multiple redundant features from a source feature.
- Formulate an appropriate fitness function for evolving high-quality redundant features, using mutual information as a proxy for measuring redundancy.
- Provide evidence that the redundant features created are non-trivial and highly redundant.
- Analyse a sample of the created redundant features to investigate how their design may introduce redundancy.

2   Background

This section will introduce some core concepts of feature manipulation and mutual information, and briefly discuss some related work.

2.1   Feature Manipulation

Feature manipulation is the act of purposefully altering the feature set of a dataset in order to improve the outcomes of a machine learning task. The two most common categories of feature manipulation are feature selection (FS) and construction (FC) [1]. FS attempts to select an optimal subset of features in order to improve performance and decrease complexity, whereas FC improves performance by creating new, more powerful high-level features which combine multiple features in some way.

EC algorithms have seen significant success recently in their application to FS and FC problems, due to their ability to search a large search space effectively [3,4]. In particular, Particle Swarm Optimisation (PSO) and Genetic Algorithms (GA) have been widely used for FS, whereas tree-based GP has seen significant use in FC due to its dynamic model structure and ability to apply a variety of functions to the feature set.

2.2   Mutual Information

Mutual Information (MI) [5] is an important concept in the field of Information Theory. MI is used as a way to measure the amount of information shared by two variables (or features). In this way, it is a measure of the mutual dependence of two variables, and is one way to measure how redundant one feature is with respect to another—the higher the MI, the more redundant the features are said to be. MI is formalised as follows:

$$MI(X,Y) = H(X) + H(Y) - H(X,Y) \tag{1}$$

where the entropy of a feature X, $H(X)$, is defined as:

$$H(X) = - \sum_{x \in X} p(x) \times \log_2 p(x) \tag{2}$$

and the joint entropy of two features, X, Y, is:

$$H(X,Y) = - \sum_{x \in X} \sum_{y \in Y} p(x,y) \times \log_2 p(x,y) \tag{3}$$

Equation 1 can be expanded as follows:

$$MI(X,Y) = - \sum_{x \in X, y \in Y} p(x,y) \times \log_2 \frac{p(x,y)}{p(x)(y)} \tag{4}$$

The above definition of MI assumes that the two features have discrete values; in the case of continuous features (such as in this work), the below definition applies:

$$MI(X,Y) = \int_X \int_Y p(x,y) \times \log_2 \frac{p(x,y)}{p(x)(y)} dx \, dy \tag{5}$$

Calculating the MI of two continuous features requires knowing the marginal and joint probability density functions (*pdf*) of the two features. In practice, this is infeasible, as the feature values for a given feature can be thought of as only a sample of the underlying *pdf* [6]. As such, a number of MI estimators have been proposed for estimating the MI of two continuous features. One venerable method uses a nearest-neighbour estimation approach, which compares the similarity of neighbours for each instance across the two dimensions X and Y to gauge the strength of the relationship between X and Y [6]. We use this approach, implemented in the Java Information Dynamics Toolkit (JIDT) [7], in this work.

2.3   Related Work

As this is the first work to propose the use of an EC algorithm to automatically evolve redundant features, there is no directly related work to discuss. Instead, we will briefly survey the use of GP for FC, since this is the most related area of research to the ideas proposed in this paper.

A variety of tree-based GP approaches to FC have been proposed, including for problems such as classification and clustering [8,9]. Most work uses a representation where a single GP tree produces a single constructed feature, as the output of the tree. The input to the tree is generally the set of features, and an optional random value input. This representation has been extended so that multiple features may be constructed in a single GP individual, commonly using a multi-tree representation [9,10]. Other representations have also been proposed [4], including using multiple sub-trees as a set of constructed features [8,11], using specially-tailored node designs [12], cooperative co-evolutionary GP [13], and even by performing multiple GP runs (each producing a single constructed feature) [14]. These works share similarity with this paper in that they perform a transformation of the original feature space, but they do so in order to improve the performance of a data mining task, rather than to perform feature creation.

3   The Proposed Method: GPRFC

This section details the proposed method for automatically generating redundant features, including the GP representation, fitness function, and other important considerations made when designing the method.

3.1   Genetic Programming Representation

In this work we use a multi-tree GP representation, where each GP individual contains n distinct trees rather than a single tree. Each tree in an individual represents a single mapping (function) from the source feature (X), to a new redundant feature (Y). Using a multi-tree representation allows us to generate multiple r.fs per source feature, while encouraging each r.f to be distinct (less redundant) from each other r.f. By generating a variety of r.fs, we increase the diversity of the types of redundancies between the source and redundant features. For example, a r.f Y_1 may have a polynomial relationship with X, whereas a second r.f Y_2 could have an exponential or trigonometric relationship—both Y_1 and Y_2 are highly redundant with X, but less redundant with each other. This behaviour is encouraged by the fitness function, which will be discussed in more detail in Sect. 3.3.

3.2   Function and Terminal Sets

We use only a single terminal in this work: the source feature, X. We purposefully do not use a random value input (unlike many GP works), as such a value is unlikely to meaningfully increase MI, and increases the search space unnecessarily.

In designing the function and terminal sets, it is important to have a wide range of operators with distinct behaviours, so that a variety of redundancy relationships can be constructed in different trees. Based on this, we use a range of different arithmetic, trigonometric, and conditional operators as follows:

– Unary operators (taking one input): $\sin(a)$, $\tan(a)$, $\tanh(a)$, $\log(a)$, e^a, $\sqrt{a}$, a^2, a^3, $-a$. We purposefully exclude $cos(a)$ due to its similarity to $sin(a)$. While a^2 and a^3 can be easily constructed in a GP tree, we include them as useful "building blocks".

- Binary operators (with two inputs): $a + b$, $a \times b$, $\max(a, b)$, $\min(a, b)$, a^b. We exclude $a - b$ and $a \div b$ as they are the complements of addition and multiplication, and as they were found to negatively affect the learning process by easily producing constant values (i.e. $X - X = 0$, $X \div X = 1$).
- A single ternary operator, if, which outputs the second input if the first input is non-negative and the third input otherwise. This operator, in addition to max and min, allows complex conditional behaviour and non-continuous functions to be generated.

3.3   Fitness Function

Our proposed fitness function is based on the concept of Mutual Information, a measure of the dependency between two features. We use MI as a proxy to measure the redundancy of a generated feature: if the MI between the source and generated feature is high, the generated feature is said to be highly redundant. Hence, the MI between the source and each generated feature/tree should be **maximised**. In addition, we choose to **minimise** the MI between each pair of generated features. In doing so, we implicitly encourage a set of r.fs that are redundant in *different ways* to be generated—for example, if two r.fs both had linear redundancies with the source feature, they would also have a high MI between them. This decision automatically increases the complexity of the generated r.fs, which should also make them harder for FS algorithms to remove. We describe the formulation of the fitness function in detail below.

Let X be the source feature, I be the GP individual whose fitness is being measured, which contains a set of trees (T), where n is the number of trees. Let the "baseline" MI, Ψ (used as a normalisation factor), be defined as the output of the MI estimation algorithm for $\Psi = MI(X, X)$. In measuring the quality of I, we consider the **minimum** MI between any X and any r.f (called minSourceMI), as well as the **maximum** mean MI between any r.f and all other r.fs (called maxSharedMI). The quality of I is measured by how much more redundant the r.fs are with X than with each other, defined as follows:

$$\text{minSourceMI} = \min_{t \in T} \frac{MI(X, t)}{\Psi} \tag{6}$$

$$\text{maxSharedMI} = \max_{t \in T} \frac{\sum_{y \in T, y \neq t} \frac{MI(t, y)}{\Psi}}{n - 1} \tag{7}$$

$$\text{Quality}_I = \text{minSourceMI} - \text{maxSharedMI} \tag{8}$$

While this quality measure is expected to be suitable as a fitness function, it does not consider that having a minSourceMI below a certain threshold means that the r.fs produced are in fact not very redundant at all. In addition, generally a lower minSourceMI leads to a higher potential fitness, making the fitness function biased towards creating a set of r.fs which are very unrelated to each other, and only weakly related to the source feature. To remedy this, we introduce an additional component to the fitness function for when the minSourceMI is below some threshold, Θ, where Θ is the minimum "acceptable" redundancy between a r.f and X. In other words, individuals not meeting this criteria can be thought of as *infeasible solutions*. For these infeasible solutions, we do not consider the shared MI between r.fs to be important, as at least one of the r.fs is not acceptable. To encourage increasing the redundancy of each r.f in this scenario (i.e. encouraging the solution towards becoming feasible), we penalise individuals based on the mean MI between the source and each r.f:

$$\text{Penalty}_I = \frac{-1}{\text{meanSourceMI}} \tag{9}$$

$$\text{meanSourceMI} = \frac{\sum_{t \in T} \frac{MI(X,t)}{\Psi}}{n} \tag{10}$$

This penalty function is designed as such so that the higher the mean-SourceMI, the lower the penalty applied. Our fitness function is then the combination of these two functions:

$$\text{Fitness}_I = \begin{cases} \text{Quality}_I, & \text{if minSourceMI} \geq \Theta \\ \text{Penalty}_I, & \text{otherwise} \end{cases} \tag{11}$$

As the Penalty term of the fitness function is constrained to be less than 0, an individual with minSourceMI $\geq \Theta$ will nearly always be better than one that does not meet the Θ threshold. As our measurements of MI are normalised by Ψ, the threshold Θ can be chosen (roughly) from the range $[0, 1]$, where a value of $\Theta = 0$ corresponds to all r.fs being independent to X, and a value of $\Theta = 1$ corresponding to all r.fs being perfectly redundant with X. In practice, we found a Θ in the range $[0.6, 0.7]$ was a good choice for $n = 5$.

3.4   Further Considerations

A number of other factors had to be addressed in order to achieve good results with the proposed method. These are discussed in turn below.

To improve the consistency of the GP method, the source feature was scaled so that all values fall in the range $[0, 1]$. However, this meant that at least one source feature value would be exactly 0. An input of 0 to the GP tree was found to significantly affect training as it would often result in multiplication or division by 0 within the tree. The common occurrence of dividing by 0 was particularly troublesome, as it meant the tree would not produce a valid output, making the whole individual invalid. To remedy this, we added a small weighting to each feature value, of size ϵ, such that all feature values lie in $[0 + \epsilon, 1 + \epsilon]$. In this work, we setting found $\epsilon = 1 \times 10^{-3}$ to be suitable.

While the above scaling approach is expected to work well on artificially-generated datasets, it does not address an issue with many real-world classification datasets: duplicate feature values. Consider the example of a (real-world) dataset where a feature takes values in $\{1, 2, 3, 4\}$. Given there are only 4 unique inputs to a GP tree, the tree may only produce (at most) 4 unique outputs.

This greatly limits the ability of GP to learn to create multiple distinct r.fs as only very "coarse" r.fs can be generated (with low complexity). To address this performance limitation, we add a small amount of stochastic noise (using a constant seed) to each source feature value, so that each feature value is likely to be distinct. This is essentially equivalent to changing the input of the GP tree to be $X + \delta$, where δ is a small value which is consistent for a given value of X. As before, we ensure δ is strictly positive. The feature values are hence in the range $[0 + \delta, 1 + \delta]$, where we defined δ to be a random number between 0.001ϵ and ϵ. In both the above approaches, we still evaluate the MI between a r.f and X (i.e. when computing the fitness function) using the **original** (i.e. unscaled) feature values, to ensure we measure the true redundancy.

In addition to scaling the source feature, we also scale the constructed redundant features to lie in $[0, 1]$. This serves two purposes: it ensures the r.fs have "sensible" ranges, and so can be more easily visualised, and it also means they have the same range as the source feature, which is important for many algorithms such as k-nearest neighbour, k-means clustering etc. Finally, the redundant features are rounded to 5 decimal places, to prevent GP from evolving very sensitive features whose precision may be lost when saved to file or used in another algorithm.

Other Parameter Settings: We use a relatively high max tree depth of 15 and mutation rate of 40% (with crossover of 60%). Using a high max tree depth was found to encourage more complex trees to be formed, which tended to produce more complex features. Evaluating the larger trees is not significantly more costly, as the computation of MI is the most expensive part of the fitness evaluation. 40% mutation was used to encourage the generation of more diverse trees—however, crossover is still important to ensure that useful function "building blocks" are passed between different GP individuals. The population size was set to 1,024, and top-10 elitism was used, as standard. In this work, we used $n = 5$ trees as it was found to produce a reasonable balance between making a large number of r.fs and making highly diverse r.fs. Decreasing n will produce r.fs which are less redundant to each other, whereas increasing n will give more, but less distinct r.fs. Θ was set to 0.7 in this work based on empirical results.

4   Experiment Design

We tested the proposed GPRFC approach on a number of popular datasets, as listed in Table 1. These datasets include three classification datasets from the UCI repository [15], two of which are quite simple and easy to classify well (Iris and Wine), whereas the third (Vehicle) is more challenging. We also use two synthetic clustering datasets (10d10cE and 10d40cE), which have 10 and 40 clusters respectively and are generated using an Ellipsoidal cluster generator [16]. The datasets chosen all have a reasonably small number of features to reduce the number of GP runs required. For each dataset, 5 r.fs are created per source feature, to give a result of $d + 5d = 6d$ features for d source features. As the

Table 1. Datasets used in the experiments.

Name	No. Features	No. Instances	No. Classes/Clusters
Iris	4	150	3
Wine	13	178	3
Vehicle	18	846	4
10d10cE	10	2903	10
10d40cE	10	2023	40

feature creation approach uses GP, it is stochastic, and so at least 30 runs were performed on each dataset.

To evaluate the created r.fs, we used the classifiers, clusterers, and feature selection algorithms provided by the WEKA [17] package. We selected four varied and popular classifiers: the J48 Decision Tree (DT) algorithm, k-nearest neighbour (KNN, with $k = 3$), Naive Bayes (NB), and the Sequential Minimal Optimisation implementation of the Support Vector Machine (SVM). For clustering, we use 3 different varieties of clustering algorithms: k-means++, agglomerative clustering (the average-link variant), and the Expectation Maximisation (EM) algorithm.

5   Results and Discussion

As there are no known redundant feature creation methods which use a guided search to automatically find good r.fs, we are unable to directly compare GPRFC to a known baseline. Instead, we directly evaluate the quality of the r.fs created across the datasets in terms of the fitness achieved. We also investigate how the addition of the r.fs affects the performance of some common classification and clustering algorithms, and how well some simple feature selection algorithms are able to identify (and remove) the added r.fs, in order to evaluate the suitability of the proposed method for creating benchmark datasets.

5.1   Fitness

Table 2 shows the performance of GPRFC in terms of the average fitness achieved across the tested datasets. GPRFC achieves a high fitness on two of the three classification datasets: Iris and Vehicle. A mean fitness of 0.351 on Vehicle indicates that the typical created r.f is 35.1% more redundant with the source feature than the other created r.fs, for example, 75.1% MI with the source feature vs only 40% MI with the other created r.fs. The performance on the two synthetic clustering datasets is not as strong, but the created r.fs are still clearly more redundant with the source feature than each other.

In general, it appears that datasets containing fewer instances tend to have a higher standard deviation—perhaps as the fitness is more sensitive to any one

Table 2. Fitness achieved by GPRFC across all features on each dataset. Standard deviation is taken across the means for each feature. At least 30 runs were performed per feature per dataset.

Dataset	Mean	Std. Dev.
Iris	0.333	0.082
Wine	0.203	0.055
Vehicle	0.351	0.041
10d10c	0.106	0.010
10d40c	0.141	0.006

single feature value being altered during the evolutionary process. The fitness across the Iris dataset, which has the highest standard deviation, is shown in Table 3 for each feature. This table clearly shows that F2 has a much lower mean fitness than the other features, and so gives a high standard deviation on the Iris dataset. It is not obvious as to why GPRFC can learn more effectively on certain features. One explanation may be that as GPRFC produces functions that transform the feature space, features that have very dense feature value distributions are harder to transform with a high level of granularity, and so harder to optimise. However, further investigation is needed.

Table 3. Fitness achieved by GPRFC across 30 runs on the Iris dataset.

Feature	Mean	Std. Dev.
F0	0.362	0.050
F1	0.398	0.036
F2	0.213	0.029
F3	0.359	0.053

5.2   Classification Performance

The performance of a number of classifiers on the original datasets compared to the datasets with added r.fs ("augmented datasets") are shown in Table 4. In general, performance is very consistent between the original and augmented datasets—in most cases, dropping by 2–3%, or holding steady. Given that redundant features aren't inherently misleading to a classifier, it makes sense that performance may not drop much – though the classification model produced will certainly be more complex. Two major exceptions to this are on the KNN classifier, which had a decrease of around 5% and 11% accuracy on Iris and Vehicle respectively. This is likely due to the created r.fs not having the same distances between instances' feature values as the source features had. As KNN is a distance-based classifier, any addition of features which transform the feature space non-linearly will directly alter the distances between instances. Testing on

Table 4. Test classification accuracy on each of the datasets before ("Original") and after ("Augmented") the created r.fs were added. Each of the 30 runs of GPRFC produced one augmented dataset—hence, the mean and standard deviation accuracy on these 30 augmented datasets are reported. A split of 70% training to 30% test was used.

Method	Iris		Wine		Vehicle	
	Original	Augmented	Original	Augmented	Original	Augmented
DT	0.978	0.956 ± 0.004	0.981	0.977 ± 0.017	0.709	0.692 ± 0.025
KNN	1.000	0.947 ± 0.035	0.962	0.961 ± 0.028	0.720	0.613 ± 0.029
NB	0.978	0.964 ± 0.018	1.000	0.979 ± 0.018	0.465	0.490 ± 0.025
SVM	0.978	0.968 ± 0.020	0.981	0.974 ± 0.016	0.740	0.715 ± 0.018

more difficult or datasets with many more features may show a bigger decrease in performance, as the search space may become complex/large enough to better challenge classification algorithms. The small increase in performance on the Vehicle dataset with NB is not statistically significant.

5.3   Clustering Performance

The performance of three clustering algorithms on the original and augmented datasets was investigated, with the results shown in Table 5. As with the classification datasets, there is generally little change in performance—in fact, performance appears to slightly increase when adding the created r.fs. However, the clusters produced are more complex and less interpretable—with $6d$ features per instance compared to only d in the original datasets.

5.4   Feature Selection Results

Feature Ranking: A common technique used in supervised feature selection is to measure how well a given feature can be used to predict the class label

Table 5. Adjusted Rand Index of the clusters produced on each of the datasets before ("Original") and after ("Augmented") the created r.fs were added. Each of the 30 runs of GPRFC produced one augmented dataset—hence, the mean and standard deviation accuracy on these 30 augmented datasets are reported. k-means++ and EM are stochastic algorithms and so the mean of 30 runs per augmented dataset was used.

Method	10d10cE		10d40cE	
	Original	Augmented	Original	Augmented
k-means++	0.548	0.558 ± 0.023	0.445	0.491 ± 0.019
Agglomerative	0.495	0.528 ± 0.064	0.276	0.309 ± 0.046
EM	0.588	0.606 ± 0.014	0.433	0.520 ± 0.011

for a set of instances. Information Gain (IG) [5] is often used as a metric to measure this, using similar principles to MI. To see how "confusing" our created r.fs may to be a FS algorithm, we ranked the features of the median and best result of applying GPRFC to the Iris dataset, using IG as shown in Table 6. We use the Iris dataset as our example as it has the fewest features, and so can be analysed most easily. The majority of created r.fs have similar rankings to their source features, with the top half of the ranks taken by F2 and F3, and the bottom half by F0 and F1. This is unsurprising—given that the created r.fs share a high amount of information with the source features, they are likely to also have a similar ability to predict the class label. However, the r.fs do have small variances in their IG value compared to their source features: for example, on the median result, F2 has an IG of 1.418, and its r.fs have IG values between 0.864 and 1.367. On the best result, F2c and F2e actually have **better** IG than the source feature; F2's r.fs range in IG value from 0.827 to 1.456. These results indicate that while the created r.fs clearly share information with their source

Table 6. Features ranked by Information Gain (with respect to the class label) on the augmented datasets created by the median (a) and best (b) runs of GPRFC.

(a) Median		(b) Best	
Info Gain	Feature	Info Gain	Feature
1.418	F2	1.456	F2c
1.385	F3a	1.421	F2e
1.378	F3	1.418	F2
1.367	F2b	1.378	F3
1.274	F3c	1.313	F3a
1.216	F2a	1.295	F2b
1.163	F3e	1.214	F3d
0.976	F3d	1.098	F3c
0.918	F2d	1.077	F3e
0.908	F3b	1.057	F3b
0.864	F2e	0.918	F2d
0.705	F0a	0.827	F2a
0.698	F0	0.722	F0a
0.554	F2c	0.698	F0
0.376	F1e	0.597	F0e
0.376	F1b	0.597	F0d
0.376	F1	0.419	F0c
0.364	F0b	0.376	F1c
0.325	F1d	0.376	F1
0.177	F0c	0.198	F1d
0.118	F0d	0.158	F1b
0.098	F0e	0.089	F1a
0.000	F1a	0.000	F0b
0.000	F1c	0.000	F1e

features, they are still different enough that their redundancy is non-trivial to identify and they are likely to have an effect on the classification task.

Using a FS Algorithm: To further investigate how suitable the created r.fs are for benchmarking FS algorithms, we applied a basic FS algorithm to the same two augmented Iris datasets. We used the canonical Sequential Floating Forward Search (SFFS) [18], which is an extension to the Sequential Forward Search (SFS) algorithm. SFS starts with no features selected, and iteratively adds the best of the remaining unselected features, until performance is not improved by adding the next feature. SFFS follows the same procedure, but also performs a backwards search after each addition of a new feature. That is, it repetitively removes the worst feature in the selected subset, until performance is not improved by removing an additional feature. This floating search helps to avoid the FS algorithm from getting stuck in local optima, and makes SFFS one of the most commonly used deterministic FS algorithms.

In this work, we used a wrapper method where the SVM algorithm is used to classify the dataset for a given feature subset, and the accuracy of the results is used as the performance of the selected features. We use the SVM classifier as it had the highest performance in Table 4. Our SFFS implementation used a training set to train the SVM, and a validation set to test the performance of the SVM on unseen data. The performance of the validation set is the performance of the selected features during training. Finally, we use a separate unseen test set to measure the quality of the final selected features on unseen data. The training, validation, and test sets are 60%, 20% and 20% of the shuffled dataset respectively.

On the median dataset (for Iris), this FS method selects features [F2b,F3] with a test accuracy rate of 0.933. On the best dataset, [F0,F1a,F3] are selected with an accuracy of 0.967. On the original dataset, the FS method selects only F3, with an accuracy of 0.967. While the obtained classification accuracy on the augmented datasets is similar, the FS method clearly selects extraneous features, which gives a more complex model than that of when only a single feature is selected on the original dataset.

Given that obtaining good performance on Iris is easy, and so FS is also relatively easy, we performed a similar experiment on the Vehicle dataset to see if different behaviour occurs on a harder, higher-dimensional problem. On the original 18-dimensional vehicle dataset, the SFFS method (as described above) selects 12 features: [F0,F2,F5,F7,F8,F9,F10,F12,F13,F16,F17], with a test accuracy of 0.710. On the median augmented dataset however, it selects 7 features: [F3,F9b,F12d,F12e,F13,F17,F17e] with a test accuracy of only 0.473. The FS method has clearly struggled to find a good set of features, as it selects multiple redundant features while also failing to select many features that were selected in the original dataset. Furthermore, the training accuracy was reasonable similar for both datasets (0.769 and 0.686 for the original and augmented respectively), indicating that the created r.fs were able to mislead the FS algorithm well enough to prevent a well-generalised classifier from being produced. Further investigation

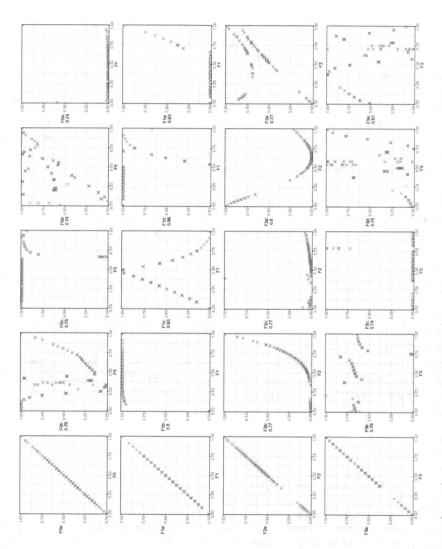

Fig. 1. Source Feature (x-axis) plotted against the five r.fs (y-axis) for each of F0 (1st row) to F3 (4th row) on Iris. The y-axis title is the name of each r.f and its source MI. Points are coloured red, green, or blue to indicate they belong to the setosa, versicolor, or virginica classes respectively. A small amount of jitter is added to each point to distinguish overlapping points. (Color figure online)

is needed to provide more quantitative evidence that GPRFC produces r.fs that make difficult benchmark datasets, but the preliminary results are a promising sign that the proposed method has potential.

6   Further Analysis

While we have shown that GPRFC is able to automatically produce a set of redundant features that have high MI with an original feature, it is not yet obvious **how** it is able to do so. To investigate this aspect, we plotted the created features against the original features for each of the 4 features on the Iris dataset, using the median result of the 30 runs of GPRFC. We choose to analyse Iris as it is the dataset with the smallest feature set. These plots are shown in Fig. 1.

The most striking observation of these plots is that the functions produced by GPRFC are incredibly varied—in fact, nearly every plot has a distinct appearance. The functions are also clearly complex, with no linear relationships apparent. A few functions are somewhat recognisable: for example, F2a (similar to a sine wave), F2b (a power curve), and F2d (a polynomial). The function evolved for F0a is similar in appearance to a sigmoid function, despite the sigmoid not being directly in the function set. Many of the remaining functions are more difficult to classify, as they either appear to have a number of different components (e.g. F0b, F3b), or have a majority of instances at a similar scale (e.g. F0e, F1b, F3a, F3c).

While generally each set of r.fs for a given source feature appear to be quite distinct, F3a and F3c appear to be very similar. This behaviour is counter-intuitive, as the fitness function directly penalises a r.f being similar to other r.fs for the same source feature. Indeed, F3a and F3c have a MI of 0.83—however, they each have very low MI (a maximum of 0.14) with the other r.fs (F3b/d/e), which means their average shared MI is still very low, at 0.34. The two trees corresponding to these two features are very similar (see Fig. 2). This issue may be alleviated by adapting the fitness function to consider the **worst-case**: that is, what is the highest value a given r.f shares with another r.f?

```
F3a = (pow (+ (exp (sqrt
(cube X)))
(neg (log X))) (tan (exp (+ (log X) (mul X X)))))

F3c = (pow (+ (exp (sqrt
(sqrt (tan (square (square (max (sin (exp X)) (sin X)))))))))
(neg (log X))) (tan (exp (+ (log X) (mul X X)))))
```

Fig. 2. The example trees produced by GPRFC for F3a and F3c. The entire first and third lines are shared by both trees.

7   Conclusion

This paper proposed the first approach to automatically evolving redundant features, using a Genetic Programming approach with a multi-tree representation, and a novel mutual information-based fitness function. The proposed GPRFC method was shown to generate high-quality and complex redundant features which are suitable for augmenting existing datasets for use in testing feature selection algorithms. We showed that good and interesting results could be achieved on both supervised and unsupervised problems. This paper represents the first piece of work in this area, but it already demonstrates the considerable potential of GP for this task. We hope that others in the GP community share our optimism, and we expect GP to ultimately be able to generate good benchmark data sets that can be used to test FS methods in data mining tasks such as classification, clustering and regression.

As GP has not been used for this sort of task previously, there is a number of different extensions that could be researched in the future. There is certainly scope for refining the fitness function further, in order to produce even more complex and distinct sets of r.fs. This work considered only one-to-one feature redundancies—the source feature to each of the r.fs in turn. More difficult/complex feature redundancy relationships could be formed by using a multivariate mutual information approach, where a set of multiple source features are used to create a set of r.fs, i.e. many-to-many redundancies. The GP representation could also be refined further, by investigating more rigorously which function set is most suitable to produce good r.fs, and evaluating how the number of trees used is best determined.

Acknowledgement. The authors would like to thank Tony Butler-Yeoman for his help in developing the initial ideas, and suggestions throughout the development of this work.

References

1. Liu, H., Motoda, H.: Feature Selection for Knowledge Discovery and Data Mining, vol. 454. Springer, Boston (2012). https://doi.org/10.1007/978-1-4615-5689-3
2. Tang, J., Alelyani, S., Liu, H.: Feature selection for classification: a review. In: Data Classification: Algorithms and Applications, pp. 37–64 (2014)
3. Xue, B., Zhang, M., Browne, W.N., Yao, X.: A survey on evolutionary computation approaches to feature selection. IEEE Trans. Evol. Comput. **20**(4), 606–626 (2016)
4. Espejo, P.G., Ventura, S., Herrera, F.: A survey on the application of genetic programming to classification. IEEE Trans. Syst. Man Cybern. Part C **40**(2), 121–144 (2010)
5. Jaynes, E.T.: Information theory and statistical mechanics. Phys. Rev. **106**(4), 620 (1957)
6. Kraskov, A., Stögbauer, H., Grassberger, P.: Estimating mutual information. Phys. Rev. E **69**(6), 066138 (2004)
7. Lizier, J.T.: JIDT: an information-theoretic toolkit for studying the dynamics of complex systems. Front. Rob. AI **1**, 11 (2014)

8. Tran, B., Xue, B., Zhang, M.: Genetic programming for feature construction and selection in classification on high-dimensional data. Memetic Comput. **8**(1), 3–15 (2016)
9. Lensen, A., Xue, B., Zhang, M.: GPGC: genetic programming for automatic clustering using a flexible non-hyper-spherical graph-based approach. In: Proceedings of the Genetic and Evolutionary Computation Conference, GECCO, pp. 449–456. ACM (2017)
10. Muni, D.P., Pal, N.R., Das, J.: Genetic programming for simultaneous feature selection and classifier design. IEEE Trans. Syst. Man. Cybern. Part B **36**(1), 106–117 (2006)
11. Ahmed, S., Zhang, M., Peng, L., Xue, B.: Multiple feature construction for effective biomarker identification and classification using genetic programming. In: Proceedings of the Genetic and Evolutionary Computation Conference, GECCO 2014, pp. 249–256. ACM, Vancouver (2014)
12. Zhang, Y., Zhang, M.: A multiple-output program tree structure in genetic programming. Technical report, Victoria University of Wellington, New Zealand (2004)
13. Lin, Y., Bhanu, B.: Evolutionary feature synthesis for object recognition. IEEE Trans. Syst. Man Cybern. Part C **35**(2), 156–171 (2005)
14. Neshatian, K., Zhang, M., Andreae, P.: A filter approach to multiple feature construction for symbolic learning classifiers using genetic programming. IEEE Trans. Evol. Comput. **16**(5), 645–661 (2012)
15. Lichman, M.: UCI machine learning repository (2013)
16. Handl, J., Knowles, J.D.: An evolutionary approach to multiobjective clustering. IEEE Trans. Evol. Comput. **11**(1), 56–76 (2007)
17. Hall, M.A., Frank, E., Holmes, G., Pfahringer, B., Reutemann, P., Witten, I.H.: The WEKA data mining software: an update. SIGKDD Explor. **11**(1), 10–18 (2009)
18. Pudil, P., Novovicová, J., Kittler, J.: Floating search methods in feature selection. Pattern Recogn. Lett. **15**(10), 1119–1125 (1994)

On the Automatic Design of a Representation for Grammar-Based Genetic Programming

Eric Medvet$^{(\boxtimes)}$ ⓘ and Alberto Bartoli ⓘ

Department of Engineering and Architecture, University of Trieste, Trieste, Italy
{emedvet,bartoli.alberto}@units.it

Abstract. A long-standing problem in Evolutionary Computation consists in how to choose an appropriate representation for the solutions. In this work we investigate the feasibility of synthesizing a representation *automatically*, for the large class of problems whose solution spaces can be defined by a context-free grammar. We propose a framework based on a form of meta-evolution in which individuals are candidate representations expressed with an ad hoc language that we have developed to this purpose. Individuals compete and evolve according to an evolutionary search aimed at optimizing such representation properties as redundancy, locality, uniformity of redundancy.

We assessed experimentally three variants of our framework on established benchmark problems and compared the resulting representations to human-designed representations commonly used (e.g., classical Grammatical Evolution). The results are promising in the sense that the evolved representations indeed exhibit better properties than the human-designed ones. Furthermore, while those improved properties do not result in a systematic improvement of search effectiveness, some of the evolved representations do improve search effectiveness over the human-designed baseline.

Keywords: Genotype-phenotype mapping · Grammatical evolution
Meta-evolution

1 Introduction

The choice of the representation of individuals in an Evolutionary Algorithm (EA) has been a central point in the field of Evolutionary Computation since its inception [22,29]. In many cases, that choice has been guided *a priori* by analogies with the biology, in which researchers looked for inspiration while designing their artificial evolutionary systems, on the assumption that Nature eventually succeeded as an effective search method [34]. On the other hand, the impact of the representation on the EA search effectiveness has also been widely studied *a posteriori*. In this respect, a common and well established practice consists in investigating any possible relationship between properties of the representation

© Springer International Publishing AG, part of Springer Nature 2018
M. Castelli et al. (Eds.): EuroGP 2018, LNCS 10781, pp. 101–117, 2018.
https://doi.org/10.1007/978-3-319-77553-1_7

such as, e.g., redundancy and locality [15, 23, 31], and higher level properties of the EA, e.g., neutrality [3] and evolvability [14].

Despite these efforts, it is fair to claim that both approaches (a priori and a posteriori) failed in clearly determining if and when a representation can guarantee the search effectiveness of an EA: copying from the Nature does not necessarily lead to a good design [7, 30] and there is not a clear view of which properties actually explain a good or a poor search effectiveness [1]. Indeed, the debate is still lively, with arguments ranging from (deemed) misuse of Nature analogies [35] to experimental-based (counter-)evidences [24] and outcomes including guidelines for the design of a representation [34] or directions for future research [29].

A case of particular interest is the one of indirect representations, i.e., those in which each individual is represented by means of a *genotype* and a *phenotype* and a mapping function exists for mapping the former to the latter. Practical motivations for choosing an indirect representation include the possibility of using standard genetic operators—whose behavior is well known—and, at the same time, tackling problems for which specific constraints act on the solutions (i.e., phenotypes). Moreover, indirect representations do have a counterpart in biology, where the form of living organisms depends on the result of a transcription process operating on encoded genetic material. Finally, indirect representation properties can be easily defined and studied both analytically and experimentally basing on the mapping function.

One of the most used EAs based on an indirect representation is Grammatical Evolution (GE) [25], a form of grammar-based Genetic Programming (GP) [10], which captures all the three aspects of indirect representations described above. First, GE allows tackling the large class of problems in which constraints on the solutions may be expressed by means of a context-free grammar (CFG). Second, according to its inventors, the overall GE framework was directly inspired by Nature [17]. Third, the properties of the GE genotype-phenotype mapping function have been widely studied [11, 31, 32]: indeed, those properties eventually served as main goals while designing new GE variants, essentially consisting in new mapping functions which were shown to be more effective than the original approach [9, 12].

In this work, we attempt to provide new insights on the long-standing, undercurrent topic of the choice of the representation. To this end, we consider the broad class of EAs corresponding to grammar-based GP and propose a novel approach for the automatic design of a representation driven by an evolutionary search aimed at optimizing the representation properties. Our proposal thus tries, in a sense, to merge the a priori and a posteriori approaches.

Our contribution consists of the following: (a) we define a class of representations in which the genotype is a variable-length bit string and the phenotype is a valid string w.r.t. a user-provided grammar; (b) we propose an evolutionary framework for searching the aforementioned space of representations; (c) we experimentally investigate the ability of the proposed framework to generate representations whose properties and search effectiveness are better than existing, established representations.

In detail, the class of representations is defined by a genotype-phenotype mapping function template whose variable parts are described with a language which we defined by means of a CFG. The mapping function template and the language are such that: (i) any representation in the resulting class is a valid genotype-phenotype mapping function—i.e., any input bit string is mapped to a valid phenotype in a finite number of steps; (ii) it is possible to express such existing and established representations as the original GE mapping [25] and the recently proposed HGE and WHGE [12]. Having defined the search space in terms of a CFG, we use a grammar-based evolutionary search method (CFG-GP [36]) which we augmented using a diversity promotion strategy in order to improve search effectiveness [13]. For driving the search, we use a fitness function measuring to which degree an individual (i.e., a genotype-phenotype mapping function) exhibits such mapping properties as redundancy, locality, uniformity of redundancy. We compute those measures on a large amount of mappings obtained from a sample grammar and a set of randomly generated genotypes.

We investigated 3 search variants differing in the fitness definition and optimization strategy (i.e., single-objective vs. multi-objective). We assessed each obtained representation experimentally not only in terms of the mapping properties, but also in terms of higher level EA properties (diversity) and of the search effectiveness achieved on a small set of benchmark problems previously used in the literature for assessing GE and its variants. The results are promising as some of the automatically generated representations are better than the existing ones. Although our findings do not imply that automatically-designed representations may fully surrogate carefully human-designed representations, they further corroborate the importance of representation properties and might ignite new research in the novel field of "self-evolving" evolutionary algorithms.

The remainder of the paper is organized as follows. In Sect. 2, we briefly survey the state-of-the-art. In Sect. 3, we introduce our genotype-phenotype mapping function template and the related CFG for describing its variable parts. In Sect. 4, we describe which are the properties we use to drive the evolution of the mapping function and how we compute them. In Sect. 5, we present and discuss the results of our experimental evaluation. Finally, in Sect. 6, we draw the conclusions.

2   Related Work

Broadly speaking, our proposal is a form of meta-evolution [6] (also known as hyper-heuristic [20] or self-adaptation [26]), where parts of an EA are chosen or tuned according to a second-level evolutionary search. In most cases, the literature focuses on specific EA parameters which can be optimized, rather than designed from scratch—e.g., mutation and crossover rate in Genetic Semantic Programming [2] or trial vector and control parameters in Differential Evolution [21]. The application of evolutionary computation to evolve (online or offline) components, rather than parameter values, of an EA is instead still believed to be in its infancy [29], in particular for representation and variation operators. For the former, the scarcity of research results may be explained by its hardness, as observed by De Jong [5]: "perhaps the most difficult and least understood area of EA design is that of adapting its internal representation."

Concerning the evolution of operators, the authors of [8] show how they evolved a general purpose mutation operator for Evolutionary Programming which outperforms existing operators on classes of functions (i.e., problems); they also experimentally show that a mutation operator evolved for a specific problem is better than a general purpose evolved operator. A similar goal is aimed at in [4], where a framework for the online evolution of the operators, together with the solutions, is proposed: as in the previously cited work, operators are represented as trees and evolved using GP. Similarly to the present work, [4] considers also other EA properties (diversity) other than search effectiveness as a criterion of analysis.

Concerning the automatic design or adaptation of representations, a proposal is presented in [28], where genotype-phenotype mapping for continuous optimization problems is considered. The authors show, using a proof-of-concept self-adaptation mechanism, that feed-forward neural networks can be used to represent and improve a genotype-phenotype mapping, also for problems of realistic complexity. Similarly to our work, the authors carefully consider redundancy and locality in their analysis.

Another view on automatic design of representation is given by [27], which again addresses the class of real-valued optimization problems: here, the representation is the way in which the real values are encoded using a bit string. With the premise that they focused only on (few) synthetic problems, due to the high computational costs implied by meta-evolution, the authors find that an evolved representation may improve the classical Gray encoding.

Also relevant w.r.t. our work are some proposals concerning grammar-based GP in which the grammar itself is evolved (or improved) online, during the evolution [18,38]. Despite the evolution of a new, general purpose representation was not among the goals of the cited papers (they rather attempt to discover more knowledge about the problem defined by the user-provided grammar by improving the grammar itself), they somehow demonstrate how a representation can change while still enforcing the problem-specific constraints on the solutions. In conclusion, to the best of our knowledge, our work is the first attempt of evolving a general purpose representation for a large class of problems, as the one addressable with grammar-based GP.

3   Representation Template

We consider a family of EAs with an indirect representation where the *genotype* $\hat{g}$ is a variable-length bit string and the *phenotype* $\hat{p}$ is a string of a language $\mathcal{L}(\mathcal{G})$ defined by a CFG $\mathcal{G} = (N, T, s_0, R)$, where: N is the set of non-terminal symbols, T is the set of terminal symbols (with $T \cap N = \emptyset$), $s_0 \in N$ is the starting symbol, and R is the set of production rules. We do not pose any constraint on components of the EA other than the representation (e.g., selection criteria for reproduction of removal of individuals, initialization). It is worth to note that many significant and widely used variants of GE (beyond its original version) belong to this family of EAs (e.g., πGE [16], HGE and WHGE [12]).

We define a *representation template*, i.e., a template of a mapping between a variable-length bit string (genotype) and a string in $\mathcal{L}(\mathcal{G})$ (phenotype), as follows. The mapping is based on the notion of *derivation tree* of a symbol s in $N \cup T$. Such a tree is rooted at s and the children of each non-terminal node $s' \in N$ are symbols (in the proper order) of one of the derivation options for s' in $\mathcal{G}$. The derivation tree is constructed with the algorithm specified below. The mapping occurs in two steps: the input genotype $\hat{g}$ is mapped to a derivation tree of the initial symbol s_0 of $\mathcal{G}$; the corresponding phenotype $\hat{p}$ is then obtained by concatenating, from the left to the right, the leaf nodes of the derivation tree.

Construction of a derivation tree is performed by a function $\text{MAP}(s, g, d)$, where s is a symbol of $T \cup N$, g is a bit string, and $d \in \mathbb{N}^+ \cup \{0\}$ is a positive number. This function essentially consists in three key steps: (i) choose one derivation option among the ones available for s, by invoking function $\text{CHOOSE}()$; (ii) obtain from g several bit strings, by invoking function $\text{DIVIDE}()$; (iii) recursively call itself for each symbol in the chosen derivation option, with the symbol, one of the bit strings previously obtained, a counter $d + 1$ of recursion depth as input parameters.

Functions $\text{CHOOSE}()$ and $\text{DIVIDE}()$ are *parameters* of $\text{MAP}()$ and their signature includes a bit string as input argument. Their domain consists of all the functions that can be defined by a language described in Sect. 3.1 that we developed. The search space for representations, thus, essentially consists in all the possible implementations for $\text{CHOOSE}()$ and $\text{DIVIDE}()$.

The mapping of $\hat{g}$ to a derivation tree of s_0 is done by invoking $\text{MAP}(s_0, \hat{g}, 0)$. The corresponding phenotype $\hat{p}$ is then obtained by concatenating the leaf nodes of the derivation tree.

In details, $\text{MAP}()$ is shown in Fig. 1 and works as follows. If s is a terminal node, the tree composed by a single node s is returned by $\text{MAP}(s, g, d)$, regardless of the values of g and d. Otherwise, the following steps are performed.

1. The derivation rule r_s for the input argument s is obtained.
2. A vector $e \in \mathbb{R}^{|r_s|}$ is built, where each element e_j is the product of the *expressiveness* of all the symbols in the jth option of r_s. The expressiveness of a symbol s' (denoted by $\text{EXPRESSIVENESS}(s')$ in Fig. 1) is a measure of the expressive power of s': we quantify expressiveness with the number of different derivation trees which can be obtained from s'. We limit the counting to derivation trees with a maximum d_{expr} depth (an implicit parameter of $\text{EXPRESSIVENESS}()$ and hence of the representation itself) in order to cope with non-finite languages, for which $\text{EXPRESSIVENESS}(s')$ may be infinite.
3. If the input argument d is greater than or equal to a predefined value d_{max} (a parameter of the representation), the index i of the chosen rule option is set to the value for which e_i is the lowest in e. Otherwise, i is set to the

Algorithm 1. The genotype-phenotype recursive mapping function, which is first invoked as MAP($s_0, \hat{g}, 0$).

---

function MAP(s, g, d)
   $t \leftarrow$ TREENODE(s)
   if $s \in N$ **then**   ▷ s is a non-terminal
      $r_s \leftarrow$ RULEFOR(s)
      for $j \in \{1, \ldots, |r_s|\}$ **do**
         $e_j \leftarrow \displaystyle\prod_{s' \in \text{SYMBOLS}(r_s, j)}$ EXPRESSIVENESS(s')
      end for
      $e \leftarrow (e_1, \ldots, e_{|r_s|})$
      if $d \geq d_{\max}$ **then**   ▷ maximum depth reached
         $i \leftarrow \arg\min_{j \in \{1, \ldots, |r_s|\}} e_j$
      else
         $i \leftarrow$ CHOOSE(g, e, d)
      end if
      $(s_1, \ldots, s_n) \leftarrow$ SYMBOLS(r_s, i)
      for $j \in \{1, \ldots, n\}$ **do**
         $e_j \leftarrow$ EXPRESSIVENESS(s_j)
      end for
      $e \leftarrow (e_1, \ldots, e_n)$
      $(g_1, \ldots, g_m) \leftarrow$ DIVIDE(g, e, d)
      for $j \in \{1, \ldots, n\}$ **do**   ▷ Append children
         APPENDCHILD(t, MAP($s_j, g_j, d + 1$))
      end for
   end if
   return t
end function

---

return value of a function CHOOSE() which takes as input g, e, d and returns a number that will be used at the next step for choosing one of the options of the derivation rule r_s.

4. The sequence of symbols $s_1, \ldots, s_n$ corresponding to the ith option of the r_s rule is obtained. We denote by SYMBOLS() the corresponding grammar look-up function in Fig. 1; SYMBOLS() is protected, i.e., it works for any i by using $\min(|r_s| - 1, \max(0, \lfloor i \rfloor))$ instead of the original argument i.

5. The vector e is reset to $(e_1, \ldots, e_n)$, where e_j is the expressiveness of s_j obtained at the previous step.

6. A sequence $(g_1, \ldots, g_m)$ of bit strings is set to the return value of a function DIVIDE() which takes as input g, e, d and returns a sequence of bit strings. Each of these bit strings will be used at the next step for constructing subtrees to be appended to the derivation tree being constructed.

7. For each symbol s_j in $s_1, \ldots, s_n$, the tree obtained by recursively invoking the MAP($s_j, g_j, d + 1$) is appended to the tree (initially) composed of the only node s, which is eventually returned. While performing this step, in case $j > m$ (i.e., if there are fewer bit strings than symbols to built the children of s), an empty bit string is passed to MAP() as g_j.

Regardless of the actual behavior of CHOOSE() and DIVIDE(), it can be easily seen that MAP() always returns a derivation tree (from which a valid phenotype is then obtained) in a finite number of steps. First, whenever the value of d (which is increased at each recursive invocation) reaches a threshold, the derivation option is chosen as the one with the lowest expressiveness, instead of by using the CHOOSE() function: since in any valid CFG, for any non-terminal symbol, there is at least one derivation option with a finite expressiveness, this guarantees that in a finite number of steps MAP() will be invoked with a terminal symbol $s \in T$. Second, regardless of the return value of CHOOSE(), a valid derivation is always chosen for s, since only options of r_s are considered.

3.1   Language for the Mapping Function

Functions CHOOSE() and DIVIDE() are parameters of the mapping function. The space of possible values for these parameters consists of all the functions that may be described by the CFG $\mathcal{G}_{\mathrm{MAP}}()$ specified in Fig. 1 and discussed below.

$\langle \text{mapper} \rangle ::= \langle n \rangle \, \langle \text{lg} \rangle$

$\langle n \rangle ::= \langle \text{const.n} \rangle \mid \langle \text{var.n} \rangle \mid \langle \text{fun.n.g} \rangle (\, \langle g \rangle \,) \mid \langle \text{fun.n.n.n} \rangle (\, \langle n \rangle \,, \langle n \rangle \,) \mid \langle \text{fun.n.ln} \rangle (\, \langle \text{ln} \rangle \,) \mid$
$\qquad \langle \text{fun.n.ln,n} \rangle (\, \langle \text{ln} \rangle \,, \langle n \rangle \,) \mid \langle \text{fun.n.lg} \rangle (\, \langle \text{lg} \rangle \,)$

$\langle \text{ln} \rangle ::= \langle \text{var.ln} \rangle \mid \langle \text{fun.ln.n} \rangle (\, \langle n \rangle \,) \mid \langle \text{fun.ln.n,n} \rangle (\, \langle n \rangle \,, \langle n \rangle \,) \mid \text{apply} (\, \langle \text{fun.n.g} \rangle \,, \langle \text{lg} \rangle \,)$

$\langle g \rangle ::= \langle \text{var.g} \rangle \mid \langle \text{fun.g.g,n} \rangle (\, \langle g \rangle \,, \langle n \rangle \,) \mid \langle \text{fun.g.lg,n} \rangle (\, \langle \text{lg} \rangle \,, \langle n \rangle \,)$

$\langle \text{lg} \rangle ::= \langle \text{fun.lg.g,n} \rangle (\, \langle g \rangle \,, \langle n \rangle \,) \mid \langle \text{fun.lg.lg,n} \rangle (\, \langle g \rangle \,, \langle \text{ln} \rangle \,) \mid \text{apply} (\, \langle \text{fun.g.g,n} \rangle \,, \langle \text{ln} \rangle \,, \langle g \rangle \,)$

$\langle \text{const.n} \rangle ::= 0 \mid 1 \mid 2 \mid 3 \mid 4 \mid 5 \mid 6 \mid 7 \mid 8 \mid 9$

$\langle \text{var.n} \rangle ::= \text{depth} \mid \text{g.count.r} \mid \text{g.count.rw}$

$\langle \text{var.g} \rangle ::= \text{g}$

$\langle \text{var.ln} \rangle ::= \text{ln}$

$\langle \text{fun.n.g} \rangle ::= \text{size} \mid \text{weight} \mid \text{weight.r} \mid \text{int}$

$\langle \text{fun.n.n,n} \rangle ::= + \mid - \mid {}^* \mid / \mid \%$

$\langle \text{fun.n.ln} \rangle ::= \text{length} \mid \text{max.index} \mid \text{min.index}$

$\langle \text{fun.n.ln,n} \rangle ::= \text{get}$

$\langle \text{fun.n.lg} \rangle ::= \text{length}$

$\langle \text{fun.ln.n} \rangle ::= \text{seq}$

$\langle \text{fun.ln.n,n} \rangle ::= \text{repeat}$

$\langle \text{fun.g.g,n} \rangle ::= \text{rotate.left} \mid \text{rotate.right} \mid \text{substring}$

$\langle \text{fun.g.lg,n} \rangle ::= \text{get}$

$\langle \text{fun.lg.g,n} \rangle ::= \text{split} \mid \text{repeat}$

$\langle \text{fun.lg.g,ln} \rangle ::= \text{split.w}$

Fig. 1. The CFG $\mathcal{G}_{\mathrm{MAP}}()$ defining the language for the CHOOSE() and DIVIDE() functions and hence for an instance of the genotype-phenotype mapping function template defined by MAP().

$\mathcal{G}_{\mathrm{MAP}}()$ includes terminal symbols representing numerical constants $(0, \ldots, 9)$, input arguments (g for g, ln for e, and depth for d), and functions (e.g., size returns the length of a bit string, weight returns the number of bits set to 1 in a bit string).

Names for the non-terminal symbols representing functions begin with fun and encode the signature of the function with a simple conventional rule. For example, $\langle \text{fun.lg.g,n} \rangle$ represents functions whose return value is of type $\langle \text{lg} \rangle$ and whose list of input arguments is of types $\langle g \rangle$ and $\langle n \rangle$. Non-terminal symbols other than functions represent data types: $\langle n \rangle$ represents numbers, $\langle \text{ln} \rangle$ represents sequences of numbers, $\langle g \rangle$ represents bit strings, $\langle \text{lg} \rangle$ represents sequences of bit strings. Thus, the above example represents functions whose return value is a number and that take a bit string followed by a number as input arguments.

Concerning terminal symbols that represent functions, as the size and weight described above, we omit a detailed description of the semantics, leaving it implicit in the name of the corresponding symbols. All the functions are type-protected, i.e., they guarantee that a correctly typed value is always returned—e.g., the number n in which a bit string g is split by the split function is internally adjusted as $\min(\ell(g), \max(1, \lfloor n \rfloor))$, where $\ell(g)$ is the length of the bit string g.

Symbols g.count.r and g.count.rw corresponds to accessing a global counter, the former reads the value of the counter while the latter reads and then increments its value. By "global" we mean that a single counter is maintained during the execution of both CHOOSE() and DIVIDE(); this counter is set to 0 when the enclosing MAP() is first called with parameters $s_0, \hat{g}, 0$. Including a global counter allows to express also genotype-phenotype mapping functions which are not inherently recursive, but can be expressed as recursive function thanks to the counter: the original GE mapping fits this case (see Fig. 2).

Non-terminal symbol ⟨mapper⟩ is the crucial component for expressing an instance of the genotype-phenotype mapping function, i.e., of functions CHOOSE() and DIVIDE(). This symbol can be derived only as a pair ⟨n⟩, ⟨lg⟩: the concatenation of the leaves of the derivation tree rooted at the left child of ⟨mapper⟩ is the function CHOOSE(); similarly, the right child represents the function DIVIDE().

As stated in the introduction, a key feature of our proposal is that it allows expressing such existing and established genotype-phenotype mapping functions as those used in GE, HGE, and WHGE. Indeed, Fig. 2 shows the CHOOSE() and DIVIDE() functions corresponding to (a slightly improved version of) GE and WHGE. Differently than the original GE mapping, this version does not require a mechanism for aborting the mapping when it looks endless (in [25] there is a maximum number of genotype reuses, i.e., wrappings), since that case is addressed by comparing d against $d_{\max}$ in MAP().

$$\text{GE} \quad \begin{array}{ll} \text{CHOOSE()} & \text{int(substring(rotate.left(g, *(gl.count.rw, 8)), 8))} \\ \text{DIVIDE()} & \text{repeat(g, length(ln))} \end{array}$$

$$\text{WHGE} \quad \begin{array}{ll} \text{CHOOSE()} & \text{max.index(apply(weight.r, split(g, length(ln))))} \\ \text{DIVIDE()} & \text{split.w(g, ln)} \end{array}$$

Fig. 2. CHOOSE() and DIVIDE() for the original GE mapping and for WHGE.

4   Properties-Driven Evolution

Since we defined the search space of the problem of the automatic design of a representation by means of the CFG $\mathcal{G}_{\text{MAP}}()$, we can tackle that problem using any grammar-based GP approach (e.g., GE, πGE, SGE, HGE, WHGE, CFG-GP), provided that we define a fitness function suitable for driving the search. In this work, we want a fitness function able to capture the degree to which a candidate representation $m \in \mathcal{L}(\mathcal{G}_{\text{MAP}}())$ exhibits the desired mapping properties.

Among the several properties of indirect representations which have been studied in the literature (see [22] for a comprehensive analysis), we considered redundancy, locality, and uniformity of redundancy—we actually considered non-locality and non-uniformity in order to conform to the semantics of "the lower, the better".

We measure the properties of a representation m basing on how m maps a predefined set G of genotypes to a corresponding set P of phenotypes using a predefined CFG $\mathcal{G}_{\text{learn}}$. That is, for each $\hat{g} \in G$ we construct $\hat{p} = m(\hat{g})$ by concatenating, from the left to the right, the leaf nodes of the derivation tree returned by $\text{MAP}(\hat{g}, s_0, 0)$, where $\text{MAP}()$ is the instance of the map function template corresponding to m and s_0 is the starting symbol of $\mathcal{G}_{\text{learn}}$. Having constructed P from G according to m, we quantify the properties of interest as follows.

The *redundancy* of m is measured as $1 - \frac{|G|}{|P|}$, i.e., one minus the ratio between the number $|G|$ of unique genotypes and the number $|P|$ of unique phenotypes.

The *locality* of m is measured as the Pearson correlation between the distances among genotypes and distances among phenotypes. More formally, let D^G be the sequence of $\frac{|G|(|G|-1)}{2}$ genotype distances (i.e., $d_{i,j}^G = d^G(\hat{g}_i, \hat{g}_j)$ is the distance between the ith and the jth elements of G, with $j < i$) and let D^P be the corresponding sequence of phenotype distances (i.e., $d_{i,j}^P = d^P(m(\hat{g}_i), m(\hat{g}_j))$). The locality is the Pearson correlation $\text{cor}(D^G, D^P)$ between D^G and D^P. As distances, we used the edit distance for both bit strings and strings of $\mathcal{L}(\mathcal{G}_{\text{learn}})$. The non-locality is measured as $1 - \frac{1 + \text{cor}(D^G, D^P)}{2}$, such that it is 0 when genotype and phenotype distances are perfectly correlated ($\text{cor}(D^G, D^P) = 1$), and 1 when they are inversely correlated ($\text{cor}(D^G, D^P) = -1$).

Finally, the *uniformity* of m is measured by means of the coefficient of variation of the size of the partitions of G for which every genotype in the partition corresponds to the same phenotype. More formally, let $G_1, \ldots, G_{|P|}$ the partitions of G such that, for each k, $\forall \hat{g}_i, \hat{g}_j \in G_k : m(\hat{g}_i) = m(\hat{g}_j)$, and let $S = |G_1|, \ldots, |G_{|P|}|$ contains the sizes of the partitions. The non-uniformity is the coefficient of variation $\frac{\sigma_S}{\mu_S}$ of S.

In order to define a criterion for driving the evolutionary search in the space of representations, we considered that, according to many studies, redundancy, locality, and uniformity appears to affect the effectiveness of the search in the respective order [14, 17, 23]. We hence explored three variants for driving the search for a representation: by minimizing redundancy only (single-objective), by minimizing redundancy and non-locality (multi-objective), and by minimizing redundancy, non-locality, and non-uniformity (multi-objective). We denote the respective *search variants* by R, R/NL, and R/NL/NU.

In all of our experiments, we used CFG-GP [36] as the evolutionary search algorithm, in a version augmented with the diversity promotion mechanism presented in [13] (with $n_{\text{partition}} = 10$ as partition size, phenotype equivalence as partitioning criterion, and youngest individual as parent representative selection criterion) and with the selection criteria for reproduction and removal of individuals based on the comparison between individuals according to the Pareto dominance.

5 Experiments and Discussion

We performed an experimental evaluation aimed at answering the two following research questions: RQ1: Can we evolve a representation which is better than the existing ones in terms of redundancy, locality, and uniformity? RQ1: Are the evolved representations also effective when used inside an actual EA?

In order to answer RQ1, we proceeded as follows. First, we executed a number $n_{\text{run}}^{\text{learning}} = 10$ of *learning runs* for each of our proposed variants R, R/NL, R/NL/NU. From each learning run we obtained a set of non-dominated representations (R/NL and R/NL/NU variants, multi-objective) and a set of representations with the same, minimal redundancy value (R variant, single-objective).

Second, for each learning run, we selected a subset of $n_{\text{repr}}^{\text{validation}} = 5$ representations for further analysis, as follows. We selected one representation randomly and then we selected iteratively, one at once, the $n_{\text{repr}}^{\text{validation}} - 1$ representations which are farthest from those already selected in terms of Euclidean distance on the fitness space (in case of ties we chose one representation at random).

Third, for each selected representation m, we performed a number $n_{\text{run}}^{\text{validation}} = 5$ of *validation runs* on each of the three validation problems specified below. That is, we solved each of those problems with representation m and the evolutionary search algorithm resulting from Table 1 (right).

In summary, we performed $3 \times 10 = 30$ learning runs and $3 \times 10 \times 5 \times 3 \times 5 = 2250$ validation runs. The software we developed for this experimentation is publicly available[1].

Table 1. Parameters for the evolutionary runs.

	Learning	Validation
Population size	500	500
Pop. initialization	Ramped half-and-half	Random
Generations	50	30
Max depth $d_{\max}$	14	9
Expressiveness depth d_{expr}	N. A.	2
Genotype size	N. A.	1024
Crossover rate	0.8	0.8
Crossover operator	CFG-GP crossover	Two-points same length
Mutation rate	0.2	0.2
Mutation operator	CFG-GP mutation	Bit flip w. $p_{\text{mut}} = 0.01$
Selection for reproduction	Tournament with size 3	Tournament with size 3
Selection for removal	Worst individual	Worst individual
Replacement	$m + m$ w. overlapping	$m + m$ w. overlapping

---

[1] https://github.com/ericmedvet/evolved-ge.

We structured learning runs as follows. We composed the set of genotypes G with the following steps: (i) we randomly generated a seed set of 10 bit strings, each of length equal to 256 bit; and, (ii) for each genotype in the seed set, we obtained other 9 genotypes by iteratively applying the bit-flip mutation operator (with $p_{mut} = 0.01$). The rationale was to obtain a uniform distribution of distances among the genotypes, useful in particular for measuring of the locality property. We used the CFG of the Pagie1 problem as grammar $\mathcal{G}_{learn}$ for mapping G to the corresponding set P of phenotypes. We set the parameters of the evolutionary search with CFG-GP in the space of representations as in Table 1 (left).

We used the following three benchmarks as validation problems: the K-Landscape synthetic problem [33] (with $k = 5$), the Pagie1 symbolic regression problem [19], and the Text generation synthetic problem [11]. Two of these benchmarks have been recommended as standard benchmarks for GP performance evaluation [37], whereas the last one (Text) has been designed specifically for assessing GE and presents a grammar of larger complexity.

Table 2 shows the property values for the evolved representations, averaged across the 5 selected representations for each of the 10 learning runs. The first three rows correspond to property values computed in the learning runs only (hence using the Pagie1 grammar only); the second three rows correspond to property values computed using the grammars of the 3 validation problems; the last three rows correspond to property values for the GE, HGE, and WHGE representations computed using the grammars of the 3 validation problems and can be used as *baseline*. We emphasize that all the baseline are *human-designed*, i.e., they are the result of dedicated research efforts.

Table 2. Representation properties.

	Search variant	Redundancy	Non-locality	Non-uniformity
Learn.	R	0		
	R/NL	0.095	0.032	
	R/NL/NU	0.797	0.319	0.077
Val.	R	0.266	0.291	0.284
	R/NL	0.247	0.28	0.292
	R/NL/NU	0.261	0.29	0.288
	GE	0.990	1.000	0.000
	HGE	0.620	0.403	2.211
	WHGE	0.410	0.412	2.689

It can be seen that, in general, property values for the evolved representation are much better than for the baselines. Furthermore, property values on the validation problems appear to be independent from the search variant (R vs. R/NL vs. R/NL/NU). We interpret this result as a combination of: (i) these values are computed on 3 grammars w.r.t. the one used for learning; and, (ii) for multi-objective fitness variants, the shown values tend to "average" different representations, i.e., points which are far away from each other in the fitness space.

In order to answer RQ2 we then examined the search effectiveness of the evolved representations. That is, we examined the fitness values for each of the validation problems when solved with the evolved representations and when solved with the baseline representations. Table 3 shows, for each validation problem, the final best fitness BF and the difference ΔBF between the final and initial best fitness—both BF and ΔBF are averaged, for each evolved representation and baseline, across the $n_{\text{run}}^{\text{validation}} = 5$ validations runs. Index ΔBF is relevant as it should capture the ability of the representation to actually improve the solution during the evolution. For each of the three search variants, Table 3 shows BF and ΔBF obtained with the best, mean, and worst representations among the $n_{\text{run}}^{\text{learning}} = 10$ learning runs with that search variant.

Table 3. Final best fitness BF and difference ΔBF between final and initial best fitness.

	Search variant	BF			ΔBF		
		Best	Mean	Worst	Best	Mean	Worst
KLand.-5	R	0.11	0.6	0.81	0.48	0.11	0
	R/NL	0.58	0.66	1	0.27	0.11	0
	R/NL/NU	0.55	0.7	1	0.33	0.06	0
	GE	1			0		
	HGE	0.58			0.06		
	WHGE	0.6			0.25		
Pagie1	R	3.42	338.66	4488.27	2440.7	400.88	2.16
	R/NL	3.32	114.39	1142.28	7975.03	579.07	0
	R/NL/NU	7.42	45.61	169.16	172.18	33.62	0
	GE	20.99			0		
	HGE	4.32			6.33		
	WHGE	2.75			6.86		
Text	R	6.5	65.12	176	10.5	3.93	0
	R/NL	7	88.06	176	154	25.23	0
	R/NL/NU	8.33	75.95	176	57	3.89	0
	GE	9.2			1.8		
	HGE	5.4			2.6		
	WHGE	5.4			3.2		

It can be seen that for each validation problem there is at least one evolved representation which is more effective than the GE baseline. On the other hand, all the human-designed baselines tend to perform better than the average evolved representation. It can also be seen that the R search strategy tends to be more effective than either R/NL or R/NL/NU: driving the evolution of the representation by redundancy only, thus, appears to be the more effective choice. It is interesting to note that the evolved representations tend to exhibit a much greater value for ΔBF than the baseline representations, that is, the evolved representations appear to be able to improve fitness during a search significantly.

The finding that the R strategy is more effective than either R/NL or R/NL/NU is confirmed also by Table 4. The table shows, for each problem and each of the three best representations obtained with each search variant, the average percentile rank of the final best fitness among all the validation runs (i.e., including other evolved representations) on that problem.

Table 4. Percentile ranks of three most effective representations for each search variant.

Search variant	n	KLand.-5	Pagie1	Text
R	1	0.003	0.086	0.003
	2	0.009	0.09	0.003
	3	0.316	0.021	0.003
R/NL	1	0.182	0.016	0.003
	2	0.059	0.252	0.035
	3	0.549	0.303	0.051
R/NL/NU	1	0.1	0.303	0.068
	2	0.311	0.303	0.103
	3	0.311	0.303	0.103

Finally, in order to gain further insights into the evolved representations, we analyzed the populations of the validation runs in terms of diversity at the level of phenotype and of fitness. We measured diversity as the rate of unique individuals in the initial and in the final population. Table 5 shows the results, for each search variant and for each baseline: for the evolved representations, diversity were computed averaging across runs and across representations with the same search variant—e.g., the 0.66 initial phenotype diversity for R on the KLandscapes-5 problem is obtained by averaging the phenotype diversities measured at the first generation of the $5 \times 5 = 25$ validation runs performed with R search variant on that problem.

It can be seen that the populations evolved with the evolved representations are, in general, more diverse, than those evolved with the baselines, both from the point of view of the phenotype and of the fitness. However, we believe that this effect might be a result of the generally better search effectiveness of the baselines, which could lead to faster convergence of the population towards one or few (possibly locally) optimal solutions. On the other hand, it is interesting to

Table 5. Initial and final diversities.

	Search variant	KLand.-5		Pagie1		Text	
		In.	Fin.	In.	Fin.	In.	Fin.
Phenotype	R	0.66	0.5	0.99	0.49	0.93	0.21
	R/NL	0.95	0.68	1	0.27	0.68	0.19
	R/NL/NU	0.41	0.29	0.37	0.06	0.28	0.06
	GE	0.01	0	0.01	0	0.05	0
	HGE	0.49	0.31	0.58	0.01	0.95	0.03
	WHGE	0.45	0.46	0.63	0.06	0.92	0.02
Fitness	R	0.44	0.19	0.97	0.47	0.13	0.01
	R/NL	0.83	0.04	0.98	0.16	0.04	0
	R/NL/NU	0.35	0.01	0.37	0.05	0.03	0
	GE	0.01	0	0.01	0	0	0
	HGE	0.42	0	0.42	0	0.13	0
	WHGE	0.33	0.02	0.54	0.05	0.12	0

note that the representations evolved with the R strategy appear more capable of preserving the population diversity: this finding confirms the interplay existing between redundancy and diversity, which has already been highlighted in previous works [11].

6   Concluding Remarks and Future Work

In the attempt of providing new insights into the long-standing problem of choosing the most appropriate representation for an EA, we have presented a method for the automatic synthesis of a representation for the large class of problems whose solutions spaces can be defined by a CFG. We have defined a representation template for genotype-phenotype mapping, in the form of a recursive function with two parameter functions that can be described using an ad hoc language that we have developed for this purpose. Our representation template is expressive enough to describe the classic GE mapping and more recent proposals such as HGE and WHGE; at the same time, our template is much more general and ensures that any instance representation is valid, i.e., it maps any input variable-length bit string to a string of the user-provided language in a finite number of steps. We used CFG-GP to evolve the representations expressed by our template with a multi-objective optimization of 3 crucial representation properties: redundancy, non-locality, and non-uniformity.

We executed a number of experiments and carefully assessed the evolved representations using human-designed representations proposed earlier in the literature, i.e., GE, HGE, and WHGE. The results show that our proposal indeed allows automatically designing a representation which exhibits better properties than the human-designed ones. However, only in few cases the evolved representations are also able to provide better search effectiveness. We hope that our work might open new research perspectives in the young field of automatic design of representations.

References

1. Altenberg, L.: Probing the axioms of evolutionary algorithm design: commentary on "On the mapping of genotype to phenotype in evolutionary algorithms" by Peter A. Whigham, Grant Dick, and James Maclaurin. Genet. Program. Evolvable Mach. **18**(3), 363–367 (2017)
2. Castelli, M., Manzoni, L., Vanneschi, L., Silva, S., Popovič, A.: Self-tuning geometric semantic genetic programming. Genet. Program. Evolvable Mach. **17**(1), 55–74 (2016)
3. Correia, M.B.: A study of redundancy and neutrality in evolutionary optimization. Evol. Comput. **21**(3), 413–443 (2013)
4. Cruz-Salinas, A.F., Perdomo, J.G.: Self-adaptation of genetic operators through genetic programming techniques. In: Proceedings of the Genetic and Evolutionary Computation Conference, GECCO 2017, pp. 913–920. ACM, New York (2017)
5. De Jong, K.: Parameter setting in EAs: a 30 year perspective. In: Lobo, F.G., Lima, C.F., Michalewicz, Z. (eds.) Parameter Setting in Evolutionary Algorithms, pp. 1–18. Springer, Heidelberg (2007). https://doi.org/10.1007/978-3-540-69432-8_1
6. Fogel, D.B., Fogel, L.J., Atmar, J.W.: Meta-evolutionary programming. In: 1991 Conference Record of the Twenty-Fifth Asilomar Conference on Signals, Systems and Computers, pp. 540–545. IEEE (1991)
7. Foster, J.A.: Taking "biology" just seriously enough: commentary on "On the mapping of genotype to phenotype in evolutionary algorithms" by Peter A. Whigham, Grant Dick, and James Maclaurin. Genet. Program. Evolvable Mach. **18**(3), 395–398 (2017)
8. Hong, L., Drake, J.H., Woodward, J.R., Özcan, E.: A hyper-heuristic approach to automated generation of mutation operators for evolutionary programming. Appl. Soft Comput. **62**, 162–175 (2018)
9. Lourenço, N., Pereira, F.B., Costa, E.: SGE: a structured representation for grammatical evolution. In: Bonnevay, S., Legrand, P., Monmarché, N., Lutton, E., Schoenauer, M. (eds.) EA 2015. LNCS, vol. 9554, pp. 136–148. Springer, Cham (2016). https://doi.org/10.1007/978-3-319-31471-6_11
10. Mckay, R.I., Hoai, N.X., Whigham, P.A., Shan, Y., O'Neill, M.: Grammar-based genetic programming: a survey. Genet. Program. Evolvable Mach. **11**(3–4), 365–396 (2010)
11. Medvet, E.: A comparative analysis of dynamic locality and redundancy in grammatical evolution. In: McDermott, J., Castelli, M., Sekanina, L., Haasdijk, E., García-Sánchez, P. (eds.) EuroGP 2017. LNCS, vol. 10196, pp. 326–342. Springer, Cham (2017). https://doi.org/10.1007/978-3-319-55696-3_21
12. Medvet, E.: Hierarchical grammatical evolution. In: Proceedings of the Genetic and Evolutionary Computation Conference, GECCO (2017)
13. Medvet, E., Bartoli, A., Squillero, G.: An effective diversity promotion mechanism in grammatical evolution. In: Proceedings of the Genetic and Evolutionary Computation Conference Companion, GECCO 2017, pp. 247–248. ACM, New York (2017)

14. Medvet, E., Daolio, F., Tagliapietra, D.: Evolvability in grammatical evolution. In: Proceedings of the Genetic and Evolutionary Computation Conference, GECCO 2017, pp. 977–984. ACM, New York (2017)
15. Miller, J.F., Smith, S.L.: Redundancy and computational efficiency in Cartesian genetic programming. IEEE Trans. Evol. Comput. **10**(2), 167–174 (2006)
16. O'Neill, M., Brabazon, A., Nicolau, M., Garraghy, S.M., Keenan, P.: πGrammatical evolution. In: Deb, K. (ed.) GECCO 2004. LNCS, vol. 3103, pp. 617–629. Springer, Heidelberg (2004). https://doi.org/10.1007/978-3-540-24855-2_70
17. O'Neill, M., Ryan, C.: Genetic code degeneracy: implications for grammatical evolution and beyond. In: Floreano, D., Nicoud, J.-D., Mondada, F. (eds.) ECAL 1999. LNCS (LNAI), vol. 1674, pp. 149–153. Springer, Heidelberg (1999). https://doi.org/10.1007/3-540-48304-7_21
18. O'Neill, M., Ryan, C.: Grammatical evolution by grammatical evolution: the evolution of grammar and genetic code. In: Keijzer, M., O'Reilly, U.-M., Lucas, S., Costa, E., Soule, T. (eds.) EuroGP 2004. LNCS, vol. 3003, pp. 138–149. Springer, Heidelberg (2004). https://doi.org/10.1007/978-3-540-24650-3_13
19. Pagie, L., Hogeweg, P.: Evolutionary consequences of coevolving targets. Evol. Comput. **5**(4), 401–418 (1997)
20. Pappa, G.L., Ochoa, G., Hyde, M.R., Freitas, A.A., Woodward, J., Swan, J.: Contrasting meta-learning and hyper-heuristic research: the role of evolutionary algorithms. Genet. Program. Evolvable Mach. **15**(1), 3–35 (2014)
21. Qin, A.K., Huang, V.L., Suganthan, P.N.: Differential evolution algorithm with strategy adaptation for global numerical optimization. IEEE Trans. Evol. Comput. **13**(2), 398–417 (2009)
22. Rothlauf, F.: Representations for genetic and evolutionary algorithms. In: Rothlauf, F. (ed.) Representations for Genetic and Evolutionary Algorithms, pp. 9–32. Springer, Heidelberg (2006). https://doi.org/10.1007/3-540-32444-5_2
23. Rothlauf, F., Goldberg, D.E.: Redundant representations in evolutionary computation. Evol. Comput. **11**(4), 381–415 (2003)
24. Ryan, C.: A rebuttal to Whigham, Dick, and Maclaurin by one of the inventors of grammatical evolution: commentary on "On the mapping of genotype to phenotype in evolutionary algorithms" by Peter A. Whigham, Grant Dick, and James Maclaurin. Genet. Program. Evolvable Mach. **18**, 385–389 (2017)
25. Ryan, C., Collins, J.J., Neill, M.O.: Grammatical evolution: evolving programs for an arbitrary language. In: Banzhaf, W., Poli, R., Schoenauer, M., Fogarty, T.C. (eds.) EuroGP 1998. LNCS, vol. 1391, pp. 83–96. Springer, Heidelberg (1998). https://doi.org/10.1007/BFb0055930
26. Saravanan, N., Fogel, D.B., Nelson, K.M.: A comparison of methods for self-adaptation in evolutionary algorithms. BioSystems **36**(2), 157–166 (1995)
27. Scott, E.O., Bassett, J.K.: Learning genetic representations for classes of real-valued optimization problems. In: Proceedings of the Companion Publication of the 2015 Annual Conference on Genetic and Evolutionary Computation, pp. 1075–1082. ACM (2015)
28. Simões, L.F., Izzo, D., Haasdijk, E., Eiben, A.E.: Self-adaptive genotype-phenotype maps: neural networks as a meta-representation. In: Bartz-Beielstein, T., Branke, J., Filipič, B., Smith, J. (eds.) PPSN 2014. LNCS, vol. 8672, pp. 110–119. Springer, Cham (2014). https://doi.org/10.1007/978-3-319-10762-2_11
29. Spector, L.: Introduction to the peer commentary special section on "On the mapping of genotype to phenotype in evolutionary algorithms" by Peter A. Whigham, Grant Dick, and James Maclaurin. Genet. Program. Evolvable Mach. **18**(3), 351–352 (2017)

30. Squillero, G., Tonda, A.: (over-)realism in evolutionary computation: commentary on "On the mapping of genotype to phenotype in evolutionary algorithms" by Peter A. Whigham, Grant Dick, and James Maclaurin. Genet. Program. Evolvable Mach. **18**, 391–393 (2017)

31. Thorhauer, A.: On the non-uniform redundancy in grammatical evolution. In: Handl, J., Hart, E., Lewis, P.R., López-Ibáñez, M., Ochoa, G., Paechter, B. (eds.) PPSN 2016. LNCS, vol. 9921, pp. 292–302. Springer, Cham (2016). https://doi.org/10.1007/978-3-319-45823-6_27

32. Thorhauer, A., Rothlauf, F.: On the locality of standard search operators in grammatical evolution. In: Bartz-Beielstein, T., Branke, J., Filipič, B., Smith, J. (eds.) PPSN 2014. LNCS, vol. 8672, pp. 465–475. Springer, Cham (2014). https://doi.org/10.1007/978-3-319-10762-2_46

33. Vanneschi, L., Castelli, M., Manzoni, L.: The K landscapes: a tunably difficult benchmark for genetic programming. In: Proceedings of the 13th Annual Conference on Genetic and Evolutionary Computation, pp. 1467–1474. ACM (2011)

34. Whigham, P.A., Dick, G., Maclaurin, J.: On the mapping of genotype to phenotype in evolutionary algorithms. Genet. Program. Evolvable Mach. **18**, 353–361 (2017)

35. Whigham, P.A., Dick, G., Maclaurin, J., Owen, C.A.: Examining the best of both worlds of grammatical evolution. In: Proceedings of the 2015 Annual Conference on Genetic and Evolutionary Computation, pp. 1111–1118. ACM (2015)

36. Whigham, P.A., et al.: Grammatically-based genetic programming. In: Proceedings of the Workshop on Genetic Programming: From Theory to Real-World Applications, vol. 16, pp. 33–41 (1995)

37. White, D.R., Mcdermott, J., Castelli, M., Manzoni, L., Goldman, B.W., Kronberger, G., Jaśkowski, W., O'Reilly, U.M., Luke, S.: Better GP benchmarks: community survey results and proposals. Genet. Program. Evolvable Mach. **14**(1), 3–29 (2013)

38. Wong, P.-K., Wong, M.-L., Leung, K.-S.: Hierarchical knowledge in self-improving grammar-based genetic programming. In: Handl, J., Hart, E., Lewis, P.R., López-Ibáñez, M., Ochoa, G., Paechter, B. (eds.) PPSN 2016. LNCS, vol. 9921, pp. 270–280. Springer, Cham (2016). https://doi.org/10.1007/978-3-319-45823-6_25

Multi-level Grammar Genetic Programming for Scheduling in Heterogeneous Networks

Takfarinas Saber[1]([⊠])(iD), David Fagan[1], David Lynch[1], Stepan Kucera[2], Holger Claussen[2], and Michael O'Neill[1]

[1] Natural Computing Research and Applications Group, School of Business, University College Dublin, Dublin, Ireland
{takfarinas.saber,david.fagan,m.oneill}@ucd.ie, david.lynch@ucdconnect.ie
[2] Bell Laboratories, Nokia, Dublin, Ireland
{stepan.kucera,holger.claussen}@nokia-bell-labs.com

Abstract. Co-ordination of Inter-Cell Interference through scheduling enables telecommunication companies to better exploit their Heterogeneous Networks. However, it requires from these entities to implement an effective scheduling algorithm. The state-of-the-art for the scheduling in Heterogeneous Networks is a Grammar-Guided Genetic Programming algorithm which evolves, from a given grammar, an expression that maps to the scheduling of transmissions. We evaluate in our work the possibility of improving the results obtained by the state-of-the-art using a layered grammar approach. We show that starting with a small restricted grammar and introducing the full functionality after 10 generations outperforms the state-of-the-art, even when varying the algorithm used to generate the initial population and the maximum initial tree depth.

Keywords: Telecommunication · Heterogeneous Network
Scheduling · Grammar-Guided Genetic Programming
Multi-level grammar

1 Introduction

We have seen in the last decade a proliferation in the use of mobile phones worldwide to reach 4.47 billion users in 2017 and this number is expected to exceed the 5 billion barrier by 2019 [1]. Companies attempt to attract new costumers through price cuts and the introduction of new technologies, like the soon-to-come 5G networks. Due to the heterogeneity and growing size of the networks, there is a large and increasing need to optimise their performance [2].

In traditional single cellular networks, Macro Cells (MCs) are employed to cover all User Equipments (UEs) such as phones, tablets, and any other device equipped with a broadband adapter. However, with the explosion of connected devices, MCs struggle to cope with the load. Therefore, they have to be supplemented with local and less powerful Small Cells (SCs), creating a two-tiered

© Springer International Publishing AG, part of Springer Nature 2018
M. Castelli et al. (Eds.): EuroGP 2018, LNCS 10781, pp. 118–134, 2018.
https://doi.org/10.1007/978-3-319-77553-1_8

configuration called Heterogeneous Networks (HetNets). SCs are deployed in areas with traffic hot-spots to attract the near-by UEs, which offloads the MCs and mitigates their performance deficit. Despite being beneficial from cost and performance points of view, SCs share the same bandwidth as the MCs, thus, making them more susceptible to interference. To mitigate possible interference, the 3rd Generation Partnership Project (3GPP [3]) provisioned an enhanced Inter-Cell Interference Coordination (eICIC) mechanism i.e., Almost Blank Subframes (ABSs). The ABSs force MCs to mute periodically for a certain duration, allowing SCs to communicate with their UEs without interference from the nearby MCs.

Several challenges are induced by the configuration of these HetNets and require real-time optimisation. In our work, we particularly address the definition of the ABS and the scheduling of the UEs when communicating with their respective SC. The current state-of-the-art for optimising this problem in a real-time fashion (millisecond timescale) is using a Grammar-Guided Genetic Programming algorithm (G3P [4]). However, the designed algorithm starts with a randomly generated initial population and uses a unique and thorough grammar from the beginning to the end of the optimisation.

To improve the performance of evolutionary algorithms, some works in the literature use greedy techniques to generate good individuals as an initial population (e.g., [5,6]), while others promote an incremental introduction of the domain knowledge to the optimisation algorithm (for instance, McKay et al. [7] use a developmental strategy of the grammar in Genetic Programming, whereas NEAT [8] augments the typologies of neural networks.

Our work evaluates the advantage of using a succession of grammars during the evolution with incremental granularities instead of a single one. The idea is based on: (i) starting with a grammar that contains fewer terminals with the aim of guiding the optimisation towards individuals with 'ideal' forms, and (ii) introducing a larger and more thorough grammar after some generations with the aim of increasing the search space and thus improving the quality of the individuals further. We create a hybridisation of different G3P algorithms where the first ones are used to direct the search towards interesting individuals and the last one to probes the whole search space, similarly to [6,9,10]. Our work is organised around and aims at answering this main Research Question (mRQ): Is it good to use different grammar levels?

It has been shown by Nicolau [11] that the way the initial population is generated affects drastically the performance of grammar-based genetic programming algorithms. Therefore, we evaluate the way our approach is affected by the modification of two parameters related to the initial population, in two secondary Research Questions: (sRQ1) the algorithm used to generate the initial population, and (sRQ2) the maximum initial tree depth.

The rest of this paper is organised as follows: Sect. 2 defines the problem of scheduling in Heterogeneous Networks. Section 3 presents a short study of the works done on the problem. Section 4 describes the state-of-the-art algorithm G3P for the scheduling in HetNets, in addition to our multi-level grammar approach. In particular, we present the different grammars and the mapping of an expression to a schedule. Section 5 describes the dataset, the setup and the significance test. Section 6 aims at answering the aforementioned research questions and shows results of the experiments. Section 7 concludes our study and proposes some future directions that we would like to explore.

2   Problem Definition

Let us consider a Heterogeneous Network $\mathcal{N}$ with a set of Macro Cells $\mathcal{M}$ and Small Cells $\mathcal{S}$. We also consider a set of User Equipements $\mathcal{U}$ and that every UE $u_i \in \mathcal{U}$ receives a signal σ_i^j from the cell $c_j \in \mathcal{M} \cup \mathcal{S}$.

2.1   Attaching UEs

UEs are known to attach greedily to the cell from which they receive the strongest signal. Since SCs have low power, the number of UEs that attach to them is limited. To cope with this issue, the 3GPP provisioned a bias mechanism i.e., Range Expansion Bias (REB) enabling SCs to attach a larger number of UEs beyond the area where their signal is higher than the near-by MCs. Therefore, the signal σ_i^j of every cell $c_j \in \mathcal{M} \cup \mathcal{S}$ to a UE $u_i \in \mathcal{U}$ is biased by an REB β_j, with $\beta_j = 0$ for every $c_j \in \mathcal{M}$. Every UE u_i is attached to a cell $c_j \in \mathcal{M} \cup \mathcal{S}$:

$$c_j = \arg\max_{k=1}^{|\mathcal{M} \cup \mathcal{S}|}(\sigma_i^k + \beta_k) \tag{1}$$

Definition 1. *Expanded Region E_j of an SC $c_j \in \mathcal{S}$ is the area where UEs would attach to c_j, but would not attach to it without the using the bias β_j. We say that a UE u_i is in the expanded region E_j an SC $c_j \in \mathcal{S}$ if and only if:*

$$c_j = \arg\max_{k=1}^{|\mathcal{M} \cup \mathcal{S}|}(\sigma_i^k + \beta_k) \quad \wedge \quad c_j \neq \arg\max_{k=1}^{|\mathcal{M} \cup \mathcal{S}|}(\sigma_i^k) \tag{2}$$

Fig. 1 shows an example that summarises the aforementioned concepts.

2.2   Almost Blank Subframes

Using the same communication channel between MCs and SCs exasperates the interference at the expanded regions. The 3GPP framework defines a time domain (i.e., a frame $\mathcal{F}$) containing 40 subframes (SF) with a 1ms time interval for each subframe. The interference can be mitigated by muting the transmission of the MCs at some of the subframes using the ABS mechanism. Therefore, allowing near-by SCs to communicate with UEs in their expanded region with low interference. By muting the MCs during some SFs, UEs at the expanded regions experience a large reduction in interference. However, UEs attached to MCs cannot communicate with their respective cells during that time frame.

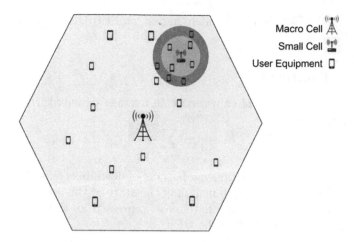

Fig. 1. Example of a Heterogeneous Network with one Macro Cell, one Small Cell, and 20 User Equipements. The grey hexagon corresponds to the area where UEs will attach to the MC in absence of any SC. The blue area is the region where the signal from the SC is stronger than the one coming from the MC. In red is the expanded region where UEs attach to the SC thanks to the bias.

2.3 Scheduling

The downlink rate R_i^f of a UE u_i quantifies the amount of data that can be transferred in the SF S_f. R_i^f is described by Shannon's formula [12] as depending on: (i) the bandwidth B, (ii) the number of UEs communicating at the same SF S_f, and (iii) the Signal to Interference and Noise Ratios (SINRs):

$$R_i^f = \frac{B}{N_f} \times log_2(1 + SINR_i^f) \tag{3}$$

UEs attached to MCs experience high SINR making their downlink always high. Therefore, scheduling UEs attached to MCs is trivial as they can all be allocated to all the SFs when the MCs are active (i.e., not muted). However, UEs attached to SCs experience a relatively low signal (SCs are low powered devices) and are subject to high interference from MCs (during their active SFs).

While the bandwidth is expensive and scarce and thus, hard to improve, both the $SINR$ and the number of communicating UEs N could be improved. The $SINR_i^f$ can be improved by muting MCs at the given SF S_f. However, exaggerating this process would lead to a substantial reduction in the overall downlink rate of UEs attached to MCs (which may be more numerous) as they would not receive any data in the mean time. Similarly, reducing the number of UEs communicating simultaneously and only communicating with fewer of them would improve the downlink for the active ones, but would mean that dismissed UEs will not be receiving any transmission. All these aspects make the management of transmissions not trivial, requiring an autonomic scheduling system that would specify: (i) SFs at which MCs are muted, and (ii) UEs communicating at any given SF.

2.4   Fitness Function

On average, every UE $u_i \in \mathcal{U}$ experiences an average downlink $\bar{R}_i$ over all the SFs of the same frame.

$$\bar{R}_i = \frac{1}{|\mathcal{F}|} \sum_{S_f \in \mathcal{F}} R_i^f \qquad (4)$$

HetNets typically aim at optimising fairness of downlinks experienced by the different users in the network [13] including the state-of-the-art work [4] we are comparing to in this paper. This fairness is expressed as the sum of average downlink logs (i.e., $log(\bar{R}_i)$) of all the UEs:

$$Fairness = \sum_{u_i \in \mathcal{U}} log\left(\bar{R}_i\right) \qquad (5)$$

Maximising the logs of average downlinks sets a high penalty when having UEs with low average downlinks, while at the same time does not provide a large reward when having UEs with excessively high average downlinks. In this work, we aim at maximising the fairness in Eq. 5 as the fitness function.

3   Previous Works

Most works in the literature to address the scheduling of transmission in HetNets put forward algorithms designed by expert agents. The most employed strategy is to partition the UEs that are attached to SCs into two different groups [14] based on the SFs they are scheduled to communicate at: (i) ABS-SFs; SFs in which MCs are muted or (ii) Non-ABS SFs; SFs in which MCs are active. Jiang and Lei [15] model the problem as a two-player bargaining game between ABS and Non-ABS SFs to attract the UEs to transmit within their time intervals. Lopez et al. [16] aim at balancing the downlinks for the UEs in both groups to equalise each other.

Autonomic solutions for the scheduling problem in HetNets are only proposed in the recent years. Lynch et al. [4] use a Grammar-Guided Genetic Programming algorithm [17] to evolve the schedules in a reinforcement learning fashion. Their algorithm has been proven to outperform the state-of-the-art systems designed by the experts across multiple metrics. The authors also showed by running a genetic algorithm for a longer period (not real-time) a large potential for performance improvement, and this is one of the major motivations for our work. The authors use a single full grammar during the entire evolution process. In our work, we propose feeding the algorithm with a smaller and more compact grammar, before extending it during the evolution (after some generations).

There has been much research into grammars. Many different approaches have been proposed and investigated from probabilistic grammars, where each production updates the probability of the production being allowed to happen again, to developmental evaluation [7] which evolves the grammar during the evolution. A comprehensive survey of these methods is presented by Hemberg [18]. With all these approaches the idea of layered learning is key. The goal is to learn during evolution and then use that knowledge to bootstrap to the next solution. In a similar vein, the approach proposed in this paper looks to establish a strong corpus of individuals that are then allowed to explore the much wider search space of the unrestricted grammar.

4   Multi-Level Grammar-Guided Genetic Programming

In this section, we describe both the state-of-the-art algorithm for scheduling in HetNets (i.e., G3P) and our proposed approach (i.e., multi-level grammar).

4.1   State-of-the-Art: Grammar-Guided Genetic Programming

The state-of-the-art for the scheduling in HetNets is a Grammar-Guided Genetic Programming [4] algorithm which uses a unique grammar in a Backus-Naur Form (BNF) to incorporate domain knowledge:

```
<expr> ::= <reg> | <reg> | <reg> | <Terminal>
<reg> ::= <expr><op><expr> | <expr><op><expr> | <expr><op><expr> | <expr><op><expr> |
          <non-linear>(<expr>) | <non-linear>(<expr>)
<op> ::= + | - | * | / (protected)
<non-linear> ::= sin | log (protected) | sqrt (protected) | step
<Terminal> ::= <sign><const> | <statistic>
<sign> ::= - | +
<const> ::= 0.0 | 0.1 | 0.2 | 0.3 | 0.4 | 0.5 | 0.6 | 0.7 | 0.8 | 0.9 | 1.0
<statistic> ::= downlink | num_variable | num_att | airtime | congestion |
                avg_downlink_frame | max_downlink_frame | min_downlink_frame |
                avg_downlink_SF | max_downlink_SF | min_downlink_SF |
                avg_downlink_cell | max_downlink_cell | min_downlink_cell
```

While most of the rules in this grammar are common to the GP world and easy to understand, <statistic> contains terminals that are from the network domain and we refer the reader to [4] for their formal definition.

The state-of-the-art algorithm is G3P: an adaptation of a grammar-based form of GP [19] as implemented in the PonyGE 2 framework [20]. G3P is used to evolve an expression that maps the SINR related statistics and attachment information to a binary decision for each UE per SF: whether to schedule the UE to communicate at the given SF or not. The authors use Algorithm 1 (please refer to [4] for a more detailed version) to do this mapping, before evaluating the fitness function with the resulting schedule.

4.2   Our Approach: Multi-Level Grammar

In addition to the full and more thorough grammar (i.e., F) defined by the state-of-the-art, we define two other grammars by only updating the list of available

input : E: Expression
output: M: Schedule Matrix
for $c_j \in S$ **do**
$\quad$ $M[j] \leftarrow zeros(|\mathcal{F}| \times |\mathcal{U}|)$ // `define a transmission schedule matrix`
$\quad$ **for** $S_f \in \mathcal{F}$ **do**
$\quad\quad$ **for** $u_i \in \mathcal{U}$ **do**
$\quad\quad\quad$ $interest \leftarrow evaluate(E, M, i, f)$ // `evaluate expression for` u_i
$\quad\quad\quad\quad$ `in f with current Schedule`
$\quad\quad\quad$ **if** $interest > 0$ and $SINR_i^f \geq 1$ **then**
$\quad\quad\quad\quad$ | $M_j[j][i][f] \leftarrow 1$ // `set as 'scheduled'`
$\quad\quad\quad$ **end**
$\quad\quad$ **end**
$\quad$ **end**
end
return N;

Algorithm 1: Mapping of an expression to a transmission schedule.

terminals. We have created two incremental grammars: (i) S: small, and (ii) M: medium, such that S is included in M and M is included in F.

The small grammar is defined by modifying <const> and <statistic>. The number of terminals is reduced to the strict minimum by only keeping a small subset of constants and what seems to be the most important statistics. The downlink is what we would like to optimise. Whereas maximising the value of min_downlink_frame would improve the smallest downlinks. Therefore, improving it would have a better impact on the fitness function. We set in S:

```
<const> ::= 0.0 | 0.5 | 1.0
<statistic> ::= downlink | min_downlink_frame
```

The medium grammar is also defined by modifying <const> and <statistic>. We add 6 terminals to the medium grammar: 4 constants (2 signs × 2 constants) and 2 statistics (i.e., max_downlink_frame and min_downlink_cell) that are also related to the downlink, in addition to the terminals from the grammar S. We set <const> and <statistic> in M as follows:

```
<const> ::= 0.0 | 0.3 | 0.5 | 0.8 | 1.0
<statistic> ::= downlink | min_downlink_frame | max_downlink_frame | min_downlink_cell
```

After defining these grammars (i.e., S, M and F), we adapted the state-of-the-art algorithm G3P to take one grammar at the start of the experiment and dynamically modify the grammar to a more complex one (e.g., from S to M, M to F, or S to F). All individuals obtained using a given grammar are seeded as an initial population [21] to G3P using the following grammar. We do not require any modification in the representation of the individuals when updating the grammar as they are represented in a tree form and the grammars are included within each other. This means that an individual has both the same representation and the same interpretation (in terms of schedule), before and after changing the grammar.

Although modifying the values of some parameters would have probably been ideal when introducing a new grammar (e.g., increasing the mutation rate for few generations facilitates the introduction of new terminals), we chose to not modify any parameter. We make this choice in order to mitigate any implication from changing the values of these parameters and only leave one varying element at the time (i.e., the grammar).

5   Experiment Design

We describe in this section the dataset and the setup used in our experiment, in addition to the test used to assess the significance of our results.

5.1   Dataset

We simulate in our work three HetNets following the same process as described in [4] and all of them serving the same geographical area that encompasses $3.61 \, \text{km}^2$ of Dublin city centre. All the HetNets contain 21 MCs spread in a hexagonal pattern. However, they differ in the number of SCs they contain. The first HetNet is the least dense with 21 SCs (1 SC per MC on average). The second HetNet is denser than the first one with 63 SCs (3 SCs per MC on average). The third and last HetNet is the densest among them with 105 SCs (5 SCs per MC on average). Additionally, we consider that a total of 1250 UEs are in the considered geographical area and are attached to one of the MCs or SCs.

5.2   Setup

We use the state-of-the-art algorithm G3P provided by the authors [4]. To validate our approach, we design different grammar configurations (see Sect. 6). We compare the best fitness function obtained when using each of the configurations in G3P instead of the full grammar. We set the population size to 100 and allow the algorithm to run for 100 generations. Furthermore, we use the Ramped Half-Half (RHH [22]) algorithm to generate the initial population with a maximum tree depth of 20. We use the sub-tree crossover with a probability 0.5, and undergo a sub-tree mutation to 60% of the population, while point mutating the remaining 40%. We set all the other parameters as described in [4]. Moreover, we repeat every experiment 30 times to minimise the effect of randomness.

5.3   Significance

In order to validate the significance of our comparisons, we perform a statistical test using a non-parametric test: the two-tailed Mann-Whitney U test (MWU). In every experiment, MWU takes in the different performance values (best fitness function values) obtained by two algorithms from each run (i.e., 30). MWU returns the p-value that one of the algorithms obtains different values than the other. We consider tests significant when the p-value is below 0.05.

6   Evaluation

We aim in this section to answer the research questions that were formalised in Sect. 1 experimentally.

6.1   mRQ: Is It Good to Use Different Grammar Levels?

In order to show the relevance of combining different grammars, we compare 7 grammar configurations on the three instances (21 SCs, 63 SCs and 105 SCs). We designed 6 different grammar configurations in addition to the default scenario F (one full grammar from beginning to end):

- S5M10F: start with S and introduce M and F at generations 5 and 10.
- S10M20F: start with S and introduce M and F at generations 10 and 20.
- S5M20F: start with S and introduce M and F at generations 5 and 20.
- S1F: start with S and introduce F at generation 1 (after generation 0).
- S5F: start with S and introduce F at generation 5.
- S10F: start with S and introduce F at generation 10.

We set parameters of G3P to the same values over all the grammar configurations as described in Sect. 5.

Figure 2 shows the evolution per generation of the best fitness on each instance, obtained by G3P when using the different grammar configurations (results are averaged over 30 runs).

We notice from Fig. 2 that G3P improves the best fitness function for all instances (constantly improves the baseline i.e., the smallest recorded values: 231.764, 319.588 and 616.874 for 21 SCs, 63 SCs and 105 SCs respectively) regardless of the grammar configuration it is used with. We also notice that the number of generations (i.e., 100) is not enough to achieve a full convergence of the algorithms and increasing this parameter would allow achieving a better performance –but would increase the execution time though.

We see that using the full grammar only (i.e., F) achieves the best results on the 21 SCs instance (outperforming the second best grammar S5F with 1.69% on average). However, its performance worsens significantly on the two other instances (i.e., 63 SCs and 105 SCs) where S10F achieves the best results (S10F achieves 7.54% and 9.95% better results than F on average on 63 SCs and 105 SCs respectively). S5F also achieves similar results as S10F. Although S5F does not reach the same quality of results as S10F on 63 SCs and 105 SCs, it slightly outperforms it on 21 SCs. Despite S1F being based on the same principle as S5F and S10F (only using the small and the full grammar), it does not achieve good results. This is mostly due to the fact that using the small grammar for only one generation only affects the individuals in the initial population. S1F generates individuals with phenotypes composed of a smaller set of terminals and does not aim at converging towards 'ideal' individuals. This acts almost as a handicap for the evolution as it does not provide either the greediness to converge faster or the variety to explore the search space. Whereas, S5F and S10F optimise the

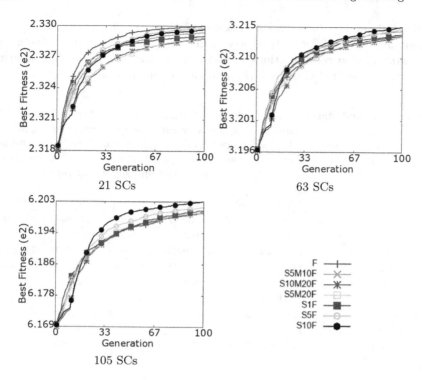

Fig. 2. Average over 30 runs of the evolution of the best fitness obtained by G3P on the different instances using various grammar configurations.

initial population further using the same small set of terminals, thus exploiting this small number of terminals for a better convergence.

Surprisingly, algorithms which use a succession of three grammar levels do not achieve as good results as those only using two grammars, with the exception of S5M10F. This is even more surprising as S10M20F achieves the worst performance in almost all instances. This goes against the intuition that if we have a grammar configuration (in this case S10M20F) similar to another one which achieves good results (e.g., S10F), the former is likely to achieve a good performance as well. If we look closely at the improvement curves of S10F and S10M20F at about 10 generations, they all seem to converge (or at least improve slowly). The introduction of the new grammar (i.e., a full grammar in the case of S10F) enables it to introduce new terminals to the evolution and improve drastically its performance. Whereas the introduction of the medium grammar allows S10M20F a much more limited improvement. In case of S10M20F, we notice a second convergence/stagnation around 20 generations before the full grammar gets introduced. However, given the limited number of generations, G3P with S10M20F could not reach the improvement of the other ones. We believe, however, that given a larger number of generations, this configuration could improve its performance.

Table 1 shows the mean and standard deviation over 30 runs of the best fitness function on the different instances, achieved by G3P when using the aforementioned grammar configurations. It also includes the p-value when comparing every approach against G3P with the grammar configuration F.

Table 1. Mean and standard deviation over 30 runs of the best fitness obtained by G3P when using different grammars. In addition, we include the p-value (using MWU) in comparison to the results obtained against G3P with grammar F. Note that we put '-' when computing the p-value for F against F as it is always 0.5 and thus not worth including in the results. We put in bold the best mean and significant p-values.

Instance	Function	F	S5M10F	S10M20F	S5M20F	S1F	S5F	S10F
21 SCs	Mean	**233.025**	232.963	232.903	232.915	232.927	233.004	232.991
	Stdev	0.038	0.088	0.082	0.090	0.072	0.065	0.128
	p-value	-	**1.89E-04**	**3.56E-09**	**4.92E-08**	**5.78E-08**	1.56E-01	3.81E-01
63 SCs	Mean	321.341	321.415	321.330	321.328	321.367	321.439	**321.473**
	Stdev	0.172	0.132	0.162	0.142	0.135	0.224	0.188
	p-value	-	**3.51E-02**	3.42E-01	2.55E-01	3.10E-01	**4.52E-04**	**7.59E-04**
105 SCs	Mean	619.950	619.953	619.990	619.972	620.022	620.104	**620.256**
	Stdev	0.261	0.273	0.225	0.176	0.269	0.321	0.299
	p-value	-	4.09E-01	2.46E-01	3.81E-01	1.03E-01	**3.19E-03**	**6.68E-06**

Table 1 confirms what has been noticed in Fig. 2 and shows that F achieves the best results on 21 SCs on average, whereas S10F achieves the best results on average on both 63 SCs and 105 SCs. It also shows that the standard deviation is rather large with regards to the difference in means. However, the p-value confirms that results obtained with S10F and S5F are significantly better than those obtained with F on both 63 SCs and 105 SCs. Whereas, the results obtained with F on 21 SCs are not significantly better than those of S10F and S5F.

The answer to mRQ is: Yes. It is good to use different grammar levels in most cases. More particularly in our case, starting with the small grammar and introducing the full one after 10 generations is the best grammar configuration.

6.2   sRQ1: How Are the Results Affected by the Algorithm Used to Generate the Initial Population?

We have shown from the previous research question that using S10F as a grammar configuration for G3P achieves significantly better results than using F in most cases. However, we would like to check whether these results are dependent on the way we generate the initial population. Therefore, we compare both grammar configurations F and S10F by varying the algorithms used to generate the initial population and fixing the other parameters to the same values. We use four different initialisation algorithms: (i) RHH: Ramped Half-Half [22], (ii) PIG: Position-Independent Grow [23], (iii) UT: Uniform Tree [23], and (iv) UG: Uniform Genome [23].

Figure 3 shows the evolution per generation of the best fitness on each instance, obtained by G3P when using either F or S10F with different initialisation algorithms (results are averaged over 30 runs).

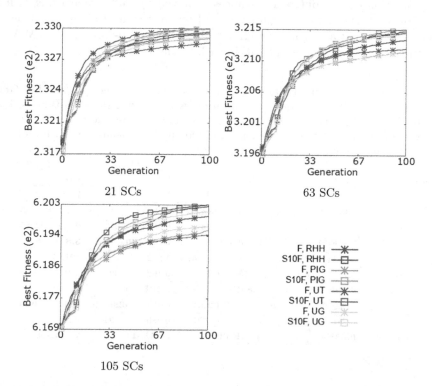

Fig. 3. Average over 30 runs of the evolution of the best fitness obtained by G3P on the different instances using either the full grammar (i.e., F) or the two-level grammar (i.e., S10F), when generating the initial populations with various initialisation algorithms (i.e., RHH, PIG, UT and UG).

We see from Fig. 3 that G3P successfully improves the fitness function over the 100 generations regardless of the grammar configuration and the algorithm used for the initialisation. We also see that the results do not fully converge within the 100 generations and that increasing this parameter is likely to improve the results.

In terms of performance, we clearly see that G3P achieves the best results when using RHH to generate the initial population on most instances, except 63 SCs. This validates the default setting chosen for the state-of-the-art algorithm [4]. We also see that with the exception of RHH on 21 SCs, using S10F allows getting better results than F. This is an important indicator that using the two-level grammar is better than the single grammar regardless of the algorithm used to generate the initial population. We even notice that S10F outperforms F on all instances and achieves better results when using initialisation algorithms different from RHH (S10F achieves 4.93%, 17.95%, 17.64% and 12.73% better results on average than F when using respectively RHH, PIG, UT and UG).

Table 2 shows the mean and standard deviation over 30 runs of the best fitness function on the different instances achieved by G3P when using either F or S10F as a grammar configuration, and when varying the algorithm used to generate the initial population. It also includes the p-value (using MWU) between the results with F and S10F in each scenario.

Table 2. Mean and standard deviation over 30 runs of the best fitness obtained by G3P using either F or S10F as grammar configuration, when varying the algorithm used to generate the initial population. In addition, we include the p-value (using MWU) to test the significance of the results. We put in bold the best mean between F and S10 F and significant p-values.

Instance	Function	RHH		PIG		UT		UG	
		F	S10F	F	S10F	F	S10F	F	S10F
21 SCs	Mean	**233.025**	232.991	232.926	**233.027**	232.882	**232.976**	232.936	**233.019**
	Stdev	0.038	0.128	0.057	0.042	0.064	0.073	0.081	0.053
	p-value	3.81E-01		**2.16E-08**		**5.09E-06**		**3.83E-05**	
63 SCs	Mean	321.341	**321.473**	321.161	**321.501**	321.210	**321.450**	321.173	**321.442**
	Stdev	0.172	0.188	0.113	0.096	0.102	0.093	0.109	0.118
	p-value	**7.59E-04**		**1.84E-11**		**2.10E-10**		**7.05E-10**	
105 SCs	Mean	619.950	**620.256**	619.567	**620.201**	619.448	**620.210**	619.676	**620.071**
	Stdev	0.261	0.299	0.277	0.256	0.285	0.194	0.269	0.307
	p-value	**6.68E-06**		**4.24E-09**		**3.35E-11**		**2.99E-05**	

Table 2 confirms that using S10F always leads to better results than F regardless of the algorithm used to generate the initial population (except with RHH on 21 SCs). We also notice a relatively large standard deviation in comparison to the difference in mean values. However, the standard deviation is similar to both grammar configurations. Despite the large standard deviations, using S10F allows achieving significantly better results with respect to the MWU test. Furthermore, we see that the unique case where F achieves better mean results than S10F is not significant.

The answer to sRQ1 is: The two-level grammar S10F allows achieving even better performance in comparison to the full grammar when used with different initialisation algorithms.

6.3   sRQ2: How Are the Results Affected by the Maximum Initial Tree Depth Used to Generate the Initial Population?

We have confirmed in the previous research questions that we achieve a better performance when using the two-level grammar S10F against when using only the full one (i.e., F). We have also confirmed that the results are not biased by the choice of the initialisation algorithm, as they are better with all the of them (except in case of RHH with 21 SCs).

In this part, we attempt to confirm whether the quality of results obtained using the two-level grammar is not negatively impacted by the maximum depth of the initial trees. We, therefore, run G3P using both F and S10F, with various maximum initial tree depths (i.e., 5, 10, 20 and 30), while setting the other parameters to their default values (more particularly, the algorithm used to generate the initial population is set to RHH).

Figure 4 shows the evolution per generation of the best fitness on each instance, obtained by G3P when using either F or S10F with different maximum depths for the initial trees (averaged over 30 runs).

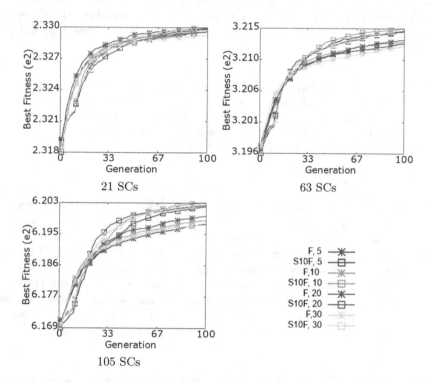

Fig. 4. Average over 30 runs of the evolution of the best fitness obtained by G3P on the different instances using either the full grammar (i.e., F) or the two-level grammar (i.e., S10F), when generating the initial populations with various maximum initial tree depths (i.e., 5, 10, 20 and 30).

We see from Fig. 4 that G3P improves the fitness function with all the maximum initial tree depths without fully converging within the 100 generations. We see that using the depth 20 achieves the best results only on the instance 21 SCs, whereas the maximum depth 10 achieves the best results on the other ones. Similarly to what has been noticed when varying the initialisation algorithm,

S10F outperforms F in all the cases (instances × maximum initial tree depths) except in 21 SCs with a depth of 20. This clearly shows that using the two-level grammar S10F is better than the full one and that it is not biased by the maximum initial tree depth. In addition to the fact that the value of maximum initial tree depth affects the final results, it also seems to affect the difference between the two grammars (i.e., S10F outperforms F more when using a maximum depth of 30 with 12.93% improvement than when using a maximum depth of 20 with 4.93% improvement on average).

Table 3 shows the mean and standard deviation over 30 runs of the best fitness function by G3P with either F or S10F as a grammar configuration while varying the maximum initial tree depth. It also includes the p-value (using MWU) between results with F and S10F in each scenario. As with Table 2, Table 3 confirms that using S10F always leads to significantly (i.e., based on MWU) better results than F regardless of the maximum initial tree depth (except with depth 20 on 21 SCs).

Table 3. Mean and standard deviation over 30 runs of the best fitness obtained by G3P using either F or S10F as grammar configuration, when generating the initial populations with various maximum initial tree depths (i.e., 5, 10, 20 and 30). We also include the p-value (using MWU) to test the significance of the results. We put in bold the best mean between F and S10F and significant p-values.

Instance	Function	5		10		20		30	
		F	S10F	F	S10F	F	S10F	F	S10F
21 SCs	Mean	232.991	**233.018**	232.993	**233.035**	**233.025**	232.991	232.954	**233.033**
	Stdev	0.044	0.060	0.050	0.035	0.038	0.128	0.055	0.051
	p-value	**3.62E-02**		**2.54E-03**		3.81E-01		**7.15E-06**	
63 SCs	Mean	321.288	**321.485**	321.310	**321.521**	321.341	**321.473**	321.255	**321.491**
	Stdev	0.085	0.100	0.080	0.086	0.172	0.188	0.067	0.090
	p-value	**1.75E-09**		**4.88E-10**		**7.59E-04**		**2.49E-11**	
105 SCs	Mean	619.731	**620.213**	619.833	**620.324**	619.950	**620.256**	619.750	**620.266**
	Stdev	0.194	0.221	0.176	0.141	0.261	0.299	0.214	0.208
	p-value	**4.24E-09**		**6.64E-11**		**6.68E-06**		**9.28E-10**	

The answer to sRQ2 is: The two-level grammar S10F allows achieving even better performance in comparison to the full grammar when used with different maximum initial tree depths.

7   Conclusion

We studied the use of different levels of grammars as a mean to improve the quality of the schedules in HetNets obtained by the G3P algorithm. Our approach consists of starting the optimisation with a short grammar which contains only the most important terminals in order to direct the search towards 'ideal' individuals. Then, to introduce a more thorough grammar during the evolution.

We showed that starting with the small grammar and introducing the full one after 10 generations, allows us to outperform the standard configuration which only uses one full grammar with up to 10% on average. We also showed that our approach is better in most cases regardless of the initialisation algorithm and maximum initial tree depth used to generate the initial populations.

In the future, we would like to analyse the sensitivity of the approach towards the number and the quality of the terminals in the small grammar. We also would like to investigate whether adding a local search to the two-level grammar, creating a three-step method [6], would be beneficial. Furthermore, we would like to study the way the approach is affected when the grammar is modified at positions other than terminals. Moreover, the quality of results obtained on the scheduling in HetNets motivates us to study the performance of our two-level grammar approach on problems from other domains.

Acknowledgement. This research is based upon works supported by the Science Foundation Ireland under Grant No. 13/IA/1850.

References

1. Number of mobile phone users worldwide from 2013 to 2019 (in billions). www.statista.com/statistics/274774/forecast-of-mobile-phone-users-worldwide
2. Andrews, J.G., Buzzi, S., Choi, W., Hanly, S.V., Lozano, A., Soong, A.C., Zhang, J.C.: What will 5G be? IEEE J. Sel. Areas Commun. **32**(6), 1065–1082 (2014)
3. 3GPP: The 3rd generation partnership project. www.3gpp.org
4. Lynch, D., Fenton, M., Kucera, S., Claussen, H., O'Neill, M.: Scheduling in heterogeneous networks using grammar-based genetic programming. In: EuroGP, pp. 83–98 (2016)
5. Saber, T., Marques-Silva, J., Thorburn, J., Ventresque, A.: Exact and hybrid solutions for the multi-objective vm reassignment problem. Int. J. Artif. Intell. Tools **26**, 1760004 (2017)
6. Saber, T., Ventresque, A., Gandibleux, X., Murphy, L.: Genepi: A multi-objective machine reassignment algorithm for data centres. In: HM, pp. 115–129 (2014)
7. McKay, R.I.B., Hoang, T.H., Essam, D.L., Nguyen, X.H.: Developmental evaluation in genetic programming: the preliminary results. In: Collet, P., Tomassini, M., Ebner, M., Gustafson, S., Ekárt, A. (eds.) EuroGP 2006. LNCS, vol. 3905, pp. 280–289. Springer, Heidelberg (2006). https://doi.org/10.1007/11729976_25
8. Stanley, K.O., Miikkulainen, R.: Evolving neural networks through augmenting topologies. Evol. Comput. **10**(2), 99–127 (2002)
9. Saber, T., Ventresque, A., Brandic, I., Thorburn, J., Murphy, L.: Towards a multi-objective VM reassignment for large decentralised data centres. In: UCC, pp. 65–74 (2015)
10. Saber, T., Thorburn, J., Murphy, L., Ventresque, A.: VM reassignment in hybrid clouds for large decentralised companies: a multi-objective challenge. Future Gener. Comput. Syst. **79**, 751–764 (2018)
11. Nicolau, M.: Understanding grammatical evolution: initialisation. Genet. Program Evolvable Mach. **18**, 467–507 (2017)
12. Shannon, C.E.: Communication in the presence of noise. IRE **37**(1), 10–21 (1949)

13. Weber, A., Stanze, O.: Scheduling strategies for HetNets using eICIC. In: ICC, pp. 6787–6791 (2012)
14. Fagan, D., Fenton, M., Lynch, D., Kucera, S., Claussen, H., O'Neill, M.: Deep learning through evolution: a hybrid approach to scheduling in a dynamic environment. In: IJCNN, pp. 775–782 (2017)
15. Jiang, L., Lei, M.: Resource allocation for eICIC scheme in heterogeneous networks. In: PIMRC. pp. 448–453 (2012)
16. Lopez-Perez, D., Claussen, H.: Duty cycles and load balancing in HetNets with eICIC almost blank subframes. In: PIMRC, pp. 173–178 (2013)
17. Dempsey, I., O'Neill, M., Brabazon, A.: Foundations in Grammatical Evolution for Dynamic Environments, vol. 194. Springer, Heidelberg (2009). https://doi.org/10.1007/978-3-642-00314-1
18. Hemberg, E.A.P.: An exploration of grammars in grammatical evolution. Ph.D. thesis, University College Dublin (2010)
19. McKay, R.I., Hoai, N.X., Whigham, P.A., Shan, Y., O'Neill, M.: Grammar-based genetic programming: a survey. Genet. Program Evolvable Mach. 11, 365–396 (2010)
20. Fenton, M., McDermott, J., Fagan, D., Forstenlechner, S., Hemberg, E., O'Neill, M.: PonyGE2: grammatical evolution in python. In: GECCO, pp. 1194–1201 (2017)
21. Saber, T., Brevet, D., Botterweck, G., Ventresque, A.: Is seeding a good strategy in multi-objective feature selection when feature models evolve? Information and Software Technology (2017)
22. Ryan, C., Azad, R.M.A.: Sensible initialisation in grammatical evolution. In: GECCO, pp. 142–145 (2003)
23. Fagan, D., Fenton, M., O'Neill, M.: Exploring position independent initialisation in grammatical evolution. In: CEC, pp. 5060–5067 (2016)

Scaling Tangled Program Graphs
to Visual Reinforcement Learning
in ViZDoom

Robert J. Smith$^{(\boxtimes)}$ and Malcolm I. Heywood

Dalhousie University, Halifax, NS, Canada
{rsmith,mheywood}@cs.dal.ca

Abstract. A tangled program graph framework (TPG) was recently proposed as an emergent process for decomposing tasks and simultaneously composing solutions by organizing code into graphs of teams of programs. The initial evaluation assessed the ability of TPG to discover agents capable of playing Atari game titles under the Arcade Learning Environment. This is an example of 'visual' reinforcement learning, i.e. agents are evolved directly from the frame buffer without recourse to hand designed features. TPG was able to evolve solutions competitive with state-of-the-art deep reinforcement learning solutions, but at a fraction of the complexity. One simplification assumed was that the visual input could be down sampled from a 210×160 resolution to 42×32. In this work, we consider the challenging 3D first person shooter environment of ViZDoom and require that agents be evolved at the original visual resolution of 320×240 pixels. In addition, we address issues for developing agents capable of operating in multi-task ViZDoom environments *simultaneously*. The resulting TPG solutions retain all the emergent properties of the original work as well as the computational efficiency. Moreover, solutions appear to generalize across multiple task scenarios, whereas equivalent solutions from deep reinforcement learning have focused on single task scenarios alone.

Keywords: Visual reinforcement learning · Emergent modularity
First person shooter · Partially observable

1 Introduction

State-of-the-art reinforcement learning (RL) algorithms increasingly emphasize the encoding of important properties from visual inputs, typically screen capture [17]. Several benchmarks have appeared including (but not limited to) arcade gaming titles [2], racing games [16] and first person shooter (FPS) environments [10]. Arcade formulations tend to emphasize the ability to play tens of titles using the same learning algorithm, but assume a relatively low resolution input and frequently supply complete information of game state. Conversely, the 'first person' perspective of FPS games provide a very rich environment – multiple

© Springer International Publishing AG, part of Springer Nature 2018
M. Castelli et al. (Eds.): EuroGP 2018, LNCS 10781, pp. 135–150, 2018.
https://doi.org/10.1007/978-3-319-77553-1_9

rooms, opponents, objects for capture/ avoidance – in which to investigate learning algorithms under partial observability of state, much higher dimensionality and a corresponding increase in the variation of content.

In this work we are interested in assessing the utility of a recently proposed Tangled Program Graph (TPG) approach [8,9] to the ViZDoom FPS environment. TPG was previously demonstrated to be capable of discovering emergent solutions to reinforcement learning problems based on visual input alone which were orders of magnitude simpler than solutions from deep learning [8,9]. Our motivation in this work is to answer to what degree the previous TPG results were dependent on a relatively low resolution state capture[1] and assess the capacity to perform task transfer between different typically partially observable task scenarios described in the ViZDoom environment. This is of merit as: (1) visual task domains are increasingly assuming very high resolution inputs that will potentially result in a loss of objects when down sampled too much, and (2) transferring skills between different source training scenarios can potentially lead to stronger performance in a target scenario, such as ViZDoom 'deathmatches' (e.g. [13,21]).

In this work we show that, by incorporating some additional task agnostic learning heuristics, TPG is able to simultaneously provide solutions to a total of ten custom task scenarios, eight of which come with the ViZDoom distribution and two are specifically introduced in this work (Sect. 4). We believe that this demonstrates that TPG is capable of scaling to much larger state spaces, while still retaining the capacity to discover solutions that are very efficient to execute without special hardware support.

2   Related Research

Gaming environments are increasingly being employed as suitably difficult tasks for demonstrating artificial general intelligence [22]. In this work, we are particularly interested in scenarios that focus on the ability of the machine learning algorithm alone, as opposed to hybrid formulations that augment machine learning with search, e.g. deep learning with Monte Carlo Tree Search [4]. From this perspective, the result of Mnih et al. was particularly noteworthy in that it demonstrated an effective strategy for playing a cross section of games using state information defined by screen capture alone [17], i.e. without recourse to any features crafted by humans. Moreover, the resulting performance in some game titles was better than/competitive with that of a human player. Since then, multiple deep reinforcement learning frameworks have been proposed and benchmarked on multiple console game titles (e.g. [11,12]), as have neuro-evolutionary approaches [3,5] and schemes based on some form of GP [6,8,9].

More recently, researchers have began to focus on gaming tasks outside of arcade console environments, where the case of first person shooter (FPS)

---

[1] 1344 pixels in the down sampled visual interface of [8,9] versus 76800 pixels in the TPG deployment demonstrated for ViZDoom.

environments is of particular relevance to this work. Specifically, FPS environments introduce three dimensional state information and partial observability. This means that strategies need discovering for navigating multiple rooms while avoiding specific objects/opponents and engaging with other types of opponents/objects. In short, FPS environments represent a different set of challenges from that experienced in console style games.

The ViZDoom framework was developed in 2016 in order to explicitly facilitate the efficient evaluation of 3D vision based machine learning algorithms on a FPS environment [10]. Machine learning agents therefore view the game world through the frame buffer and learn how to play on the basis of visual stimuli alone. ViZDoom gives researchers control over the size of the screen buffer, access to the depth buffer, as well as the ability to customize the learning process using WAD ("Where's All the Data?") files. These files were originally defined by John Carmack while building the original version of Doom and are thus interchangeable, giving ViZDoom a breadth of options for providing AI agents with learning environments ranging from the official Doom levels to custom scenarios.

ViZDoom has been used by researchers in two generic settings, a competition setting in which the principle objective is to survive some form of 'deathmatch' with other agents. Particular examples of this include deploying the best deep learning RL approach from other general game playing challenges [19], or modifying a general deep learning RL approach to include properties specifically useful for VizDoom, e.g. some form of prior task decomposition [13,21].

A second use of ViZDoom has been through the use of custom scenarios representing underlying tasks of value for developing better general purpose visual learning agents. For example, [1] use a health pack gathering scenario in which the objective is to avoid mines and collect health packs, whereas [18] assume a single scenario involving learning to aim against a static target. One interesting observation from Alvernaz and Togelius was that the deep learning element might produce a representation that failed to recognize hazards [1].

3   Tangled Program Graphs

The tangled program graph (TPG) represents a generalization of GP teaming into the *emergent* discovery of graphs of teams of programs [8,9]. To do so, a definition of GP is assumed in which each program provides a single (real valued) scalar output, b_i, which is interpreted as the 'bid'. Each program also has a single discrete atomic action, a_i, selected from the set of task specific atomic actions $\mathcal{A}$. Two populations (P and H) representing programs and teams respectively are then coevolved in a symbiotic relationship, or symbolic bid-based GP (SBB) [14,15]. Fitness is only defined explicitly at the team population. Each member of the team population identifies a unique subset of programs from the program population (the same program may appear in multiple teams). A variable length representation is assumed for members of each population, hence team complement evolves in the team population and program complement evolves in the program population.

SBB is independent of the specific representation assumed for a program. In this work we adopt a linear GP representation (as did the original work [14,15]), thus instructions take the form of operations on registers of the general form: $R[x] = R[x] < op > R[y]$, where x and y are register references and $< op >$ is an 'opcode'. Likewise we also adopt the eight opcodes common to Lichodzijewski's work: $\{+, -, \times, \div, ln, cos, cos, IF\}$, where the conditional is interpreted as IF $(R[x] < R[y])$ THEN $(R[x] = -R[x])$. Each instruction also has a mode bit which enables $R[y]$ to either index a register or an input attribute (in this case the visual screen buffer). The $R[y]$ references are always to registers of which there are a total of 8 in this work.

The output of a team is established by executing each program (from the same team) on the current state of the environment, $s(t)$, and identifying a single 'winning' bid, i.e. the program with maximum output [14,15]. Let this be p^*. Program p^* has won the right to suggest its action, call this $a(t)$. Under a reinforcement learning task, action $a(t)$ defines the action of the agent at time step t. Such an action potentially changes the state of the environment and the process repeats until some end condition is encountered (e.g. terminal state of the game, maximum number of training interactions). Fitness of a team can then be assessed using an appropriate measure of game score (Sect. 4). The process repeating over all teams, or a training 'generation'.

After each generation the teams are ranked (Step 2(b)v, Algorithm 1) and the worst performing H_{gap} individuals deleted (Step 2(b)vi). Any programs from the program population lacking a reference from at least one team are also then deleted (Step 2(b)vii). H_{gap} parents are then identified, and an equivalent number of child teams produced (care of a set of variation operators) to replace the deleted teams (Step 2(b)ii). One set of variation operators operate at the level of teams [14,15]: add or delete a reference to a program (P_a, P_d), modify a program currently in the team (P_m), or modify the action of a program within the team (P_a). Likewise a second set of variation operators operate at the level of programs [14,15] (only called upon when P_m test true), in which case the effected program is first cloned before applying variation operators to the cloned program (delete or add an instruction (P_{del}, P_{add}), mutate an instruction (P_{mut}) or swap two instructions within the same program (P_{swp}).

SBB provides a framework for evolving teams of arbitrary complement, i.e. without having to specify the number of programs per team [14,15]. Moreover, only the size of the team population needs explicitly defining (H_{size}). The size of the program population floats as programs either fail to be referenced by a team (delete program from P) or a program is cloned (add program to P).

TPG assumes SBB as the starting point and introduces an additional variation operator, P_α, which establishes the *type* of action change when P_a tests true. Thus, for P_α false, the cloned program's action, a_i is selected from the set of task specific atomic actions $a_i \in \mathcal{A}$, whereas for P_α true, the new action is a pointer to any team in the team population of generation g, or $a_i \in H(g)$. Once this happens we have a graph in which each 'node' is a team and 'arcs' are programs. Evaluation of a graph follows the same process as per SBB, and commences with the root team (all individuals in the initial team population are root nodes), w.r.t. state $s(t)$ of the environment. However, if the action of the winning program is a pointer to another team, then the new team is also evaluated on state, $s(t)$. As teams in the graph are evaluated they are 'marked'. If following execution of a team the winning program's action selects a team that was previously visited (on state $s(t)$), then the runner up program's bid is (recursively) assumed as the action until either an unvisited team or an atomic action is selected (each team must have at least one atomic action). As soon as the winning program's action takes the form of an atomic action, evaluation of the graph (for state $s(t)$) is complete and any 'marked' teams are reset.

Algorithm 1 . TPG algorithm (Sect. 3) with incremental record keeping (Sect. 4. The term 'agent' is used to denote a TPG individual. Note that only teams representing root nodes constitute an agent for which fitness evaluation is performed.

1. Initialize team population, $H(g = 0)$, and program population, $P(g)$.
2. For $(b = 0; b < \text{Bags}; b = b + 1, g = g + 1)$
    (a) $S(b) = S$ ▷ Initialize set of task scenarios
    (b) For all $(t \in \text{Scenarios})$
        i. $s(b) = rnd(S(b))$ ▷ Select a task scenario
        ii. $H(g) = H(g) \cup create(H_{gap})$
        iii. For all $(agent \in H(g))$ and $(k \in \text{Episodes})$
            A. $task = rnd(s(b))$ ▷ Initialize task scenario instance
            B. $evaluate(agent, task)$ ▷ Eq. (1)
        iv. Update OverallFitness over scenarios $\{t, ..., t - R\}$ ▷ Eqs. (2) and (3)
        v. Rank $(agents \in H(g))$
        vi. Prune lowest ranked H_{gap} agents from $H(g)$
        vii. Prune all $p \in P(g)$ without an agent
        viii. $S(b) = S(b) - s(b)$ ▷ Remove evaluated task scenario

---

The number of root teams varies during evolution. That is to say, if a root team receives a pointer from another team (care of the variation operators), then that graph has been incorporated as a subgraph into another root team's graph. Naturally, the application of P_a, P_α can also result in a subgraph being decoupled from its parent, resulting in the reappearance of a root team (a root has no incoming arcs). The resulting complexification of single team root nodes into graphs is therefore an emergent process in which the complexity of the relation between teams, team complement and program complexity are all driven by the underlying complexity of the task [8, 9].[2]

---

[2] For an illustration of the incremental construction of TPG individuals see the presentation slides of Stephen Kelly from EuroGP'17 http://stephenkelly.ca/research_files/skelly-mheywood-eurogp-2017.pdf.

4   Scaling TPG to Large Visual State Spaces

In the following we summarize the additional heuristics investigated to deploy TPG under the ViZDoom environment. Specifically, TPG was originally deployed under the Atari arcade learning environment, [2], with a down sampled visual state space, $s(t)$, from the original 210×160 pixels to 42×32 pixels.[3] ViZDoom, on the other hand, represents a 3D environment capable of presenting a much richer range of encounters than under individual arcade console titles [10]. With this in mind we introduce five factors for specifically scaling TPG up to the challenge of evolving agents capable of solving challenges present in ViZDoom.

4.1   Representation of Features

The default structure for screen buffer data in ViZDoom is a series of nested arrays of integers, where the outer array represents the height of the buffer (240 rows of pixels) and the inner arrays represent the RGB integer values of pixels in a given row (320 pixels per row) [10]. In order to provide SBB with a 'flat' set of environment registers, the three RGB values were concatenated into a single 32-bit integer format, thus defining input as an array of 76,800 integers to represent each frame received from the screen buffer. This is distinct from deep learning approaches in which each of the RGB channels is subject to convolution filtering. Moreover, the computational cost of retaining independent channels with deep reinforcement learning might require some form of cropping or down sampling from the resolution of the original input.[4] In effect, GP 'sees' a single channel consisting of pixel values over a much larger dynamic range (due to the concatenation of the three RGB channels into a single integer value) as opposed to retaining the representation of each pixel in three colour channels over a lower dynamic range.

4.2   Task Scenarios

As the complexity of task domains increases, it might be beneficial to learn from specific training scenarios or source tasks. Such approaches have proved effective in a wide range of settings including ViZDoom competitions [21] and the evolution of soccer skills [20].

The ViZDoom distribution has 8 example scenarios: *Basic, Deadly Corridor, Defend the Centre, Defend the Line, Health Gathering, My Way Home, Predict Position,* and *Take Cover*.[5] Naturally each scenario represents a different skill that might be beneficial for a general purpose policy capable of 'surviving' within the ViZDoom environment. In addition to the scenarios included with the

---

[3] Conversely, the deep reinforcement learning approach of [17] cropped the original visual space to 84×84 pixels.

[4] For example, [1] assume a 120×160 visual input and [13] assume 60×108.

[5] https://github.com/mwydmuch/ViZDoom/tree/master/scenarios.

package, two further scenarios were constructed in order to promote a particular type of movement called circle-strafing. Circle-strafing is when a player continuously focuses their aim on a target in the game while sidestepping (strafing). If an agent does both of these things, they will naturally walk in a circle around their target, potentially making it more difficult for the target to hit the agent (because it is moving) while decreasing the health of the target.

Circling scenario 1 places the player on one side of a raised ring (6 o'clock). The ring itself causes a small amount of damage to the player periodically. The floor inside and outside of the ring causes a moderate amount of damage if a player stands on it for too long. The player is tasked with picking up a health kit on the other side of the ring (12 o'clock). Whenever a health kit is acquired, another health kit spawns on the opposite side of the ring. This scenario is meant to help the player specifically perform circular movements, though not necessarily with strafing involved. There are three atomic actions made available $A = \{F, TL, TR\}$ (see Table 1) and the rewards are $+1$ per time unit alive, -100 on agent death, and there is a scenario time limit of 2100 time units.

Circling scenario 2 is a continuation of the first. The ring remains the same, except health kits spawn sequentially at every 90 degree section of the ring rather than on opposite sides. An enemy spawns in the middle of the ring. The player is tasked with moving to acquire the health kit, but has the additional requirement of shooting the enemy in the centre of the room in order to make the next health kit spawn. Note that the turning speed in ViZDoom is static when using a key (rather than a mouse) and will not allow the player to run to a health kit, turn, shoot the enemy, and then return to moving. In order to be successful, the player will need to learn to circle-strafe around the ring in order to stay alive. There are six atomic actions made available $A = \{F, TL, TR, SL, SR, A\}$ (see Table 1) and the rewards are $+1$ per time unit alive, -100 on agent death, and there is a scenario time limit of 2100 time units.

Note that in all cases, there is no special 'task scenario' flag provided as an additional input. The agent has to infer the nature of the task from the visual clues in the environment itself.

4.3 Graduated Action Space

Programs can potentially assume any atomic action, $a_i \in A$. However, different learning scenarios might only explicitly require a subset of the atomic actions. Given that at initialization the programs are initialized randomly, limiting programs to a subset of the total set of atomic actions might be beneficial. Thus, instead of allowing all actions to be used from the beginning of the first generation, we gradually add new actions to the existing pool as they are required. Thus, the first scenario will determine the starting actions of the population A_0. When the next scenario, k, is introduced, any additional actions, A', specific to the scenario are added, or $A_k = A_{k-1} + A'$. Thus, new programs will be able to sample from an incrementally expanding set of atomic actions. As the additional scenarios are attempted, new actions are added until all scenarios have been encountered. This allows TPG to gradually integrate the new actions into the existing program graphs.

4.4   Random Bags

In order to reduce unintended sequencing biases into the order with which scenarios are introduced during evolution, we randomize their order. Thus, each scenario is presented for a set of episodes. The collection of scenarios is placed into a *bag*, where they are shuffled randomly to produce a scenario permutation (Step 2(b)i, Algorithm 1). Each agent experiences the task scenarios in the permutation order specific to the bag (Step 2(b)iii). Once a bag is empty (all of the tasks have been performed) a new bag (ordering of scenarios) is created.

Table 1. Parameterization of VizDoom environment and training scenarios. Total number of training encounters is Bags × Scenarios × Episodes

Parameter	Value
Dimension of visual input space	$320 \times 240 = 78,800$
Colour Settings	RGB24
Atomic action set ($\mathcal{A}$)	Move forward (F), Move backward (B), Turn Left (TL), Turn Right (TR), Strafe Left (SL), Strafe Right (SR), Shoot/Attack (A)
Record Keeping (R)	2
Bags	6
Scenarios	10
Episodes	20

4.5   Fitness Through Incremental Record Keeping

One of the most difficult aspects of learning multiple tasks from a single continuous run is forgetfulness, e.g. [11]. Forgetfulness implies that as agents experience different task scenarios, they specialize on their most recent task, and slowly forget the nuances of the tasks that were encountered earlier. In short, if agents are unable to discover how to leverage what they have previously learnt from earlier tasks into the new task, they risk loosing important properties.

Ideally, we would evaluate agent performance on all tasks at each generation. This might lead to a multi-objective evaluation through Pareto dominance, e.g. [7,9]. However, in this case we have additional factors to consider, including the cost of maintaining an absolute measure of fitness over the 10 visual reinforcement task scenarios for the entire population at any point in time, i.e. each task scenario requires evaluation over multiple task initializations ('Episodes' parameter, Table 1).

In essence, we are attempting to strike a balance between an absolute performance function (as measured across all task scenarios) and measuring performance on the single current task scenario. Thus, an individual is not considered fit merely for their ability to solve the current task, but based on their ability to solve the most recently attempted R tasks. Once an agent's current task fitness is calculated, we then go back and retrieve fitness scores for the agent in the previous R tasks. Naturally, the H_{gap} agents created in the current task evaluation (Step 2(b)ii, Algorithm 1) lack appropriate fitness evaluations for the R previous tasks. In these cases, additional evaluations are performed to establish fitness on the past R tasks. Thus, at each generation up to $H_{size} - H_{gap}$ agents are evaluated on the current task, and up to H_{gap} agents on the current and R previous tasks.[6]

Fitness scores are calculated in the following way:

1. Estimate the raw fitness for each agent across a set of episodes for the current task scenario (Sect. 4.2). Thus, the raw fitness of agent i on task t is:

$$RawFitness(i, t) = \sum_{k=1}^{Episodes} score_k(i, t) \tag{1}$$

2. Normalize all fitness values by dividing every agent's by the maximum:[7]

$$NormFitness(i, t) = \frac{RawFitness(i, t)}{max_k(RawFitness(k, t))} \tag{2}$$

3. Estimate overall fitness of agent i in terms of the performance over the last R tasks

$$OverallFitness(i) = \sum_{k=0}^{R} NormFitness(i, t - k) \tag{3}$$

Note that Eq. (3) is limited to the evaluated scenarios in the case of a 'code start' (first bag with less than R scenarios evaluated).

Once the overall fitness is calculated for all agents, they are ranked in descending order and the worst H_{gap} agents removed. In short, we are attempting to balance the cost of evaluation across all task scenarios versus reduced accuracy introduced by only estimating fitness over a subset of tasks. Moreover, in continuously shuffling the order of scenario evaluation we hope to mitigate the introduction of other biases. Unlike the previous instance of multi-task learning with TPG [9], we anticipate some continuity between the tasks encountered, but

---

[6] Note that H_{size} is the number of teams present and reflects the number of agents (root teams) at initialization. However, as teams are subsumed into graphs, the number of agents (root teams) will decrease. Likewise, application of the variation operator could switch an action from a team pointer to an atomic action, breaking a graph into two smaller graphs, resulting in an increase in the number of agents.

[7] Any negative normalized fitness values are treated as 0, thus producing a number in the range [0, 1].

this same continuity also potentially makes it more difficult to resolve which task an agent is facing. Moreover, in this work we are facing a total of 10 tasks as opposed to a maximum of 3 tasks in [9].

5   Results

TPG is based on the symbiotic bid-based (SBB) teaming framework from Lichodzijewski et al. [15], for which there are several code bases. In this work, we assumed the open source Java code base of SBBJ (https://web.cs.dal.ca/~rsmith/_sbb_fcube/) and added additional functionality to support TPG as summarized in Sect. 3.

Training sessions were divided into permutations of scenarios, with each scenario appearing for a number of episodes, or a bag (Sect. 4.4), Table 1. In total, 6 bags (entire rounds through all the task scenarios) are conducted and results summarized for 12 independent runs. All runs were performed on desktop computing platforms without recourse to parallel hardware. Table 2 summarizes the TPG algorithm parameters assumed in this work.

After all the episodes from the same task scenario conclude Eq. (1) can be estimated. This reward is normalized relative to the highest scoring agent in order to determine the fitness of agents, Eq. (2). This normalization is important due to the mis-alignment of reward scales across the various scenarios (see for example the difference between column values in Table 3). Once all teams are given a fitness value for the current task scenario, then the normalized fitness values from the previous R completed scenarios are retrieved for each agent or Eq. (3). For the purposes of this experiment, $R = 2$, (Table 1), implying that the three most recently attempted scenarios are considered in fitness calculation.

Table 2. Parameterization of tangled program graph algorithm.

Teams		Learners	
Parameter	Value	Parameter	Value
Team Population Size (H_{size})	450	Max. Instructions	1024
Team Gap (H_{gap})	50% of Root Teams	Prob. Delete Instr. (P_{del})	0.5
Max. Prog. per Team (ω)	9	Prob. Add Instr. (P_{add})	0.5
P_d, P_a	0.7	Prob. Mutate Instr. (P_{mut})	1.0
P_m	0.2	Prob. Swap Instr. (P_{swp})	1.0
P_n	0.1		

The overall fitness score then allows all agents (root teams) to be ranked according to their overall utility across multiple scenario types. Since the assortment of scenarios is randomly bagged, there is no predictability of previous records in practice. Thus, given enough generations, each agent should be evaluated across all permutations of scenarios bound by $R + 1$.

5.1 Human Results

The human results were gathered from a full test run by the researchers. Each scenario was played 10 times with the same mechanical restrictions presented during a normal training scenario. The values in Table 3 reflect a moderate level of familiarity with the ViZDoom control scheme and a high amount of familiarity with first person shooters in general. The labels in Table 3 have the following interpretation: (1) *Singe task* reflects the best single agent on a specific task scenario (larger scores are better). Hence, this need not be the same agent on each task. (2) *Multi-agent* reflects the best agent across all task scenarios. It is interesting to note that the 'single best agents' also actually returned best performance on more than one game. Moreover, the Multi-agent actually had better median performance for 3 of the 10 tasks.

Table 3. Best multi-task agent and best single-task agent vs. Human. In all cases larger scores are better. Performance reflects game score over 200 test games for each scenario for each agent (1st, 2nd, 3rd quartile performance). Human score reflects a single best play by an experienced player. $\{^a,^b,^c,^d\}$ identify each single-task agent and the subset of tasks on which they were best.

Task scenario	Multi-task agent			Single-task agent			Human
	Q1	Median	Q3	Q1	Median	Q3	
Basic	59	63	68	74	76.57^a	79	76
Deadly Corridor	272.8	393.5	508.8	258.5	470.7^d	653.08	2280
Defend the Centre	9	11	12	8	10^b	12	16
Defend the Line	5	9	11	9	11^b	13	25
Health Gathering	595.75	642	778.25	1006	1021^c	1032	1065
My Way Home	0.81	0.86	0.88	0.813	0.834^d	0.869	0.982
Predict Position	-0.071	0.88	0.951	-0.055	0.892^a	0.919	0.965
Take Cover	125.75	172	215.25	147.25	180^a	216.25	1445
Health Circle	57	73	88	82	110.5^c	143.25	848
Circle Strafing	18	24	35	17.75	23^c	29	632

Some of the human results shown in Table 3 are the result of discovering how to exploit the specifics of the task environment. For example, the Take Cover scenario has a simple dominant strategy where you slowly sidestep all the way to one side and then sprint back to the other side to avoid all projectiles. Similarly, the Deadly Corridor scenario is seemingly difficult for learning agents, but generally simple for human players to complete. The score of 2280 in this scenario is reasonably close to the maximum for Deadly Corridor. This also applies to the result in Health Gathering, which requires a player to continuously collect health kits on the floor in order to not die. Since this scenario is timed, understanding the bounds of the problem ahead of time provide prior knowledge that a human player is able to use to their advantage. In short, humans are still most effective at playing task scenarios of the VizDoom environment.

Relative to contemporary research using deep learning or hybrid deep learning and neuroevolution, we note that results only appear to exist for the 'Basic' and 'Health Gathering' task scenarios. Moreover, such results represent the performance of the agent trained on a single task. Caveats aside, Table 4 compares scores for each agent under the two tasks for which we could find comparative results. Note also, that it is also likely that different numbers of testing events are assumed in each case (for TPG we used 200 task initializations).

Table 4. Comparison with deep learning results under 'Basic' and 'Health Gathering' task scenarios. Skip Count reflects the frequency with which an agent is required to make decisions.

Framework	Skip Count	Basic task Average score (Std. Dev.)	Health Gathering task Average score (Std. Dev.)
Deep learning [10]	0	51.5($\pm$74.9)	unknown
Deep learning [10]	4	82.2($\pm$9.4)	$\approx$ 1,300
Hybrid [1]	Unknown	Unknown	657.1($\pm$397.1)
TPG Multi-task	0	63.3($\pm$4.91)	680.8($\pm$97.7)
TPG Single-task	0	76.6($\pm$4.5)	1020.9($\pm$18.7)

Kempka et al. recognized that increasing the Skip Count from zero (requires the agent to make a decision at each frame) to an optimal value of 4 skipped frames had a significant positive impact on the quality of deep learning agent policies [10]. All TPG runs were performed with a Skip Count of zero. We note that the

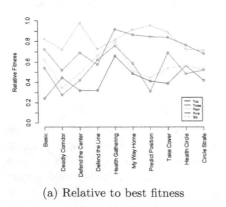
(a) Relative to best fitness

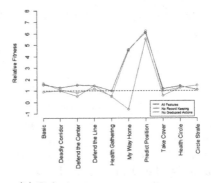
(b) Relative to average game score

Fig. 1. Score of the top team across an increasing number of scenario sessions (bags) using (a) normalization of single best agent, and (b) normalization relative to average game score.

comparator solutions are only significantly better than TPG when the non-zero Skip Count was employed. Future work could revisit this parameterization.

5.2   Population and Team Results

Incremental population champions were saved at the conclusion of each bag, the skills of which are shown in Fig. 1(a). This graph shows the normalized scores of that bag's champion when compared to the highest scoring individual of each

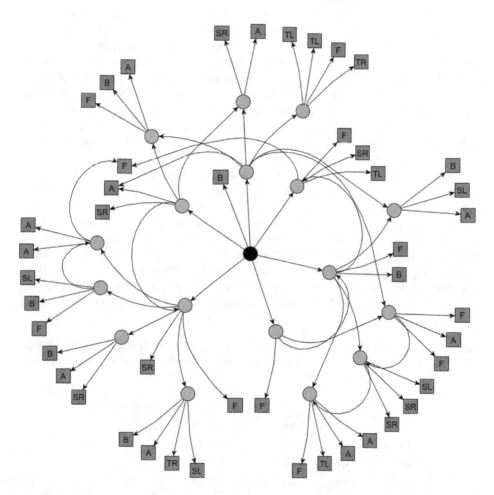

Fig. 2. TPG agent with highest average score over all tasks. The root node (black) defines the starting point for agent policy evaluation. The process of action selection follows the path defined by the arc of each node's winning bid until an atomic action is chosen. Because the path through the TPG graph is based on sequential node-wise bidding in which no node can be revisited, the search for an atomic action represents a low computational cost.

category. It should be noted that the champion is not necessarily the same root team at each level or task scenario.

As the number of bags increases, we see that the range of scenario scores are slowly moving toward unity, but do so relatively uniformly. For the first three bags only the Health Gathering task saw any significant improvement and thus the results show very low rewards. In the later bags, individual teams were capable of successfully completing those scenarios and drew their relative position higher. Since this experimental design rewarded better overall capabilities rather than localized specialization, this is an outcome we would expect to see as the TPG agents become more sophisticated throughout the learning process.

Figure 1(b) illustrates the impact of removing key scaling factors from the approach as outlined in Sect. 4. That is to say, 12 independent runs were also performed with a single missing scaling factor, after which we again summarize performance relative to the best agent from each task scenario. Specifically, without incremental record keeping during fitness evaluation (Sect. 4.5), it is not possible to maintain a consistent front of development across all task scenarios. Specifically, the 'MyWayHome' task scenario seemed to fail to improve at all without incremental records. Removing graduated actions did not seem to have a significant positive or negative effect.

The structure of the champion TPG agent is illustrated in Fig. 2. This individual corresponds a 'Multi-task agent' with performance as summarized in Tables 3 and 4. Each node represents a team (of programs) and each arc indicates the action of the program, i.e. pointer to either another node (a different team of programs) or an atomic action. There are a total of 18 teams, organizing a total of 72 programs which index a total of 33,708 unique pixels from the original visual input (or 43.9%). However, only a fraction of these programs (pixels) are executed (indexed) when TPG determines what action to suggest for any given state, Table 5. Thus, in order to make any particular decision between 14% (min)

Table 5. TPG Model complexity for Multi-task agent. Num. Teams/Instr/Pixels represent the average number of Teams (graph nodes), Instructions executed (including introns), Pixels indexed per decision

Task scenario	Num. Teams	Num. Instr	Num. Pixels
Basic	3.18	17,238	10,951
Deadly Corridor	3.53	18,038	11,847
Defend the Centre	2.86	16,081	10,692
Defend the Line	3.34	16,709	11,637
Health Gathering	3.41	16,524	10,685
My Way Home	3.91	17,423	11,225
Predict Position	3.37	16,952	11,749
Take Cover	2.76	15,746	11,488
Health Circle	3.34	17,785	10,653
Circle Strafing	3.47	17,537	11,866

to 15.5% (max) of the input space is actually used. Likewise, only 2 to 4 teams are evaluated per decision, implying that although the TPG solution might be 'complex', the path from root to action is still short.

6   Conclusion

We demonstrate the scaling up of tangled program graphs (TPG) to a set of visual reinforcement learning tasks described in the ViZDoom environment. Specifically, TPG agents are evolved from the entire content of the $320 \times 240 = 76,800$ pixel state space. Several mechanism are investigated for achieving this scaling of which basing the fitness function on recent previous task scenarios as well as the current task scenario appears to be the most significant.

TPG is still able to discover solutions that only index a fraction of the input space and only executes a subset of the programs in order to make each decision. At the same time, scaling across multiple task scenarios is demonstrated, where this is achieved through the ability to organize over 70 programs hierarchically using the discovery of appropriate graph connectivity. This capacity for emergent task decomposition/modularity provides a unique approach to discovering solutions under visual reinforcement learning problems. It is also fundamental to providing solutions that can be discovered and deployed without specialized hardware support (as is generally the case for deep learning). Comparison with contemporary deep learning results (for a subset of the task scenarios) indicates that the quality of the resulting policies does not appear to be compromised.

Acknowledgments. This research was supported by NSERC grant CRDJ 499792.

References

1. Alvernaz, S., Togelius, J.: Autoencoder-augmented neuroevolution for visual Doom playing. In: IEEE Conference on Computational Intelligence and Games, pp. 1–8 (2017)
2. Bellemare, M.G., Naddaf, Y., Veness, J., Bowling, M.: The arcade learning environment: an evaluation platform for general agents. J. Artif. Intell. Res. **47**, 253–279 (2012)
3. Braylan, A., Hollenbeck, M., Meyerson, E., Miikkulainen, R.: Reuse of neural modules for general video game playing. In: Proceedings of the AAAI Conference on Artificial Intelligence, pp. 353–359 (2016)
4. Guo, X., Singh, S., Lewis, R., Lee, H.: Deep learning for reward design to improve Monte Carlo Tree Search in ATARI games. In: Proceedings of the AAAI Conference on Artificial Intelligence, pp. 1519–1525 (2016)
5. Hausknecht, M., Lehman, J., Miikkulainen, R., Stone, P.: A neuroevolution approach to general Atari game playing. IEEE Trans. Comput. Intell. AI Games **6**(4), 355–366 (2014)
6. Jia, B., Ebner, M.: Evolving game state features from raw pixels. In: McDermott, J., Castelli, M., Sekanina, L., Haasdijk, E., García-Sánchez, P. (eds.) EuroGP 2017. LNCS, vol. 10196, pp. 52–63. Springer, Cham (2017). https://doi.org/10.1007/978-3-319-55696-3_4

7. Kelly, S., Heywood, M.I.: Knowledge transfer from Keepaway soccer to half-field offense through program symbiosis: Building simple programs for a complex task. In: ACM Genetic and Evolutionary Computation Conference, pp. 1143–1150 (2015)
8. Kelly, S., Heywood, M.I.: Emergent tangled graph representations for atari game playing agents. In: McDermott, J., Castelli, M., Sekanina, L., Haasdijk, E., García-Sánchez, P. (eds.) EuroGP 2017. LNCS, vol. 10196, pp. 64–79. Springer, Cham (2017). https://doi.org/10.1007/978-3-319-55696-3_5
9. Kelly, S., Heywood, M.I.: Multi-task learning in Atari video games with emergent tangled program graphs. In: ACM Genetic and Evolutionary Computation Conference, pp. 195–202 (2017)
10. Kempka, M., Wydmuch, M., Runc, G., Toczek, J., Jaśkowski, W.: ViZDoom: a doom-based AI research platform for visual reinforcement learning. In: IEEE Conference on Computational Intelligence and Games, pp. 1–8 (2016)
11. Kirkpatrick, J., Pascanu, R., Rabinowitz, N.C., Veness, J., Desjardins, G., Rusu, A.A., Milan, K., Quan, J., Ramalho, T., Grabska-Barwinska, A., Hassabis, D., Clopath, C., Kumaran, D., Hadsell, R.: Overcoming catastrophic forgetting in neural networks. CoRR abs/1612.00796 (2016)
12. Kunanusont, K., Lucas, S.M., Pérez-Liébana, D.: General video game AI: learning from screen capture. In: IEEE Conference on Computational Intelligence and Games, pp. 2078–2085 (2017)
13. Lample, G., Chaplot, D.S.: Playing FPS games with deep reinforcement learning. In: Proceedings of the AAAI Conference on Artificial Intelligence, pp. 2140–2146 (2017)
14. Lichodzijewski, P.: A symbiotic bid-based framework for problem decomposition using Genetic Programming. Ph.D. thesis, Faculty of Computer Science, Dalhousie University (2011)
15. Lichodzijewski, P., Heywood, M.I.: Symbiosis, complexification and simplicity under GP. In: Proceedings of the ACM Genetic and Evolutionary Computation Conference, pp. 853–860 (2010)
16. Loiacono, D., Lanzi, P., Togelius, J., Onieva, E., Pelta, D., Butz, M., Lonneker, T., Cardamone, L., Perez, D., Sáez, Y., Preuss, M., Quadflieg, J.: The 2009 simulated car racing championship. IEEE Trans. Comput. Intell. AI Games 2(2), 131–147 (2010)
17. Mnih, V., Kavukcuoglu, K., Silver, D., Rusu, A.A., Veness, J., Bellemare, M.G., Graves, A., Riedmiller, M., Fidjeland, A.K., Ostrovski, G., Petersen, S., Beattie, C., Sadik, A., Antonoglou, I., King, H., Kumaran, D., Wierstra, D., Legg, S., Hassabis, D.: Human-level control through deep reinforcement learning. Nature 518(7540), 529–533 (2015)
18. Poulsen, A.P., Thorhauge, M., Funch, M.H., Risi, S.: DLNE: a hybridization of deep learning and neuroevolution for visual control. In: IEEE Conference on Computational Intelligence and Games, pp. 1–8 (2017)
19. Ratcliffe, D.S., Devlin, S., Kruschwitz, U., Citi, L.: Clyde: a deep reinforcement learning DOOM playing agent. In: AAAI Workshop on What's Next for AI in Games, pp. 983–990 (2017)
20. Whiteson, S., Kohl, N., Miikkulainen, R., Stone, P.: Evolving keepaway soccer players through task decomposition. Mach. Learn. 59(1), 5–30 (2005)
21. Wu, Y., Tian, Y.: Training agent for first-person shooter game with actor-critic curriculum learning. In: International Conference on Learning Representations, pp. 1–10 (2017)
22. Yannakakis, G.N., Togelius, J.: A panorama of artificial and computational intelligence in games. IEEE Trans. Comput. Intell. AI Games 7(4), 317–335 (2015)

Towards in Vivo Genetic Programming: Evolving Boolean Networks to Determine Cell States

Nadia S. Taou and Michael A. Lones[✉]

School of Mathematical and Computer Sciences,
Heriot-Watt University, Edinburgh, UK
{nt2,m.lones}@hw.ac.uk

Abstract. Within the genetic programming community, there has been growing interest in the use of computational representations motivated by gene regulatory networks (GRNs). It is thought that these representations capture useful biological properties, such as evolvability and robustness, and thereby support the evolution of complex computational behaviours. However, computational evolution of GRNs also opens up opportunities to go in the opposite direction: designing programs that could one day be implemented in biological cells. In this paper, we explore the ability of evolutionary algorithms to design Boolean networks, abstract models of GRNs suitable for refining into synthetic biology implementations, and show how they can be used to control cell states within a range of executable models of biological systems.

Keywords: Gene regulatory networks · Boolean networks · Control
Evolutionary algorithms

1 Introduction

Gene regulatory networks (GRNs) are biochemical systems that process information and generate complex responses within biological organisms. In effect, they are the "genetic programs" of biological cells. Because of this, GRNs have long been a source of inspiration to the genetic programming community, with the first papers using representations motivated by GRNs published around the turn of the millenium [1,2]. Recently this interest has started to blossom, with a number of different research groups looking at how GRN-based approaches can be used to solve computational problems within computers [3]. The motivations for this are various: to increase the evolvability of genetic programming [2], to increase the robustness of executional systems [4], to support the evolution of complex computational behaviours [5], and to increase the compactness or efficiency of computation [6]. However, regardless of the motivation, the focus of this research has been pretty singular: evolving GRN models that will be executed *in silico*.

© Springer International Publishing AG, part of Springer Nature 2018
M. Castelli et al. (Eds.): EuroGP 2018, LNCS 10781, pp. 151–165, 2018.
https://doi.org/10.1007/978-3-319-77553-1_10

In the last decade, there has been considerable progress in the field of synthetic biology [7]. An important activity within synthetic biology is the design and assembly of novel biochemical pathways that can be deployed within an existing cell in order to change its behaviour. Typically this is aimed at implementing medical interventions in a way that is more precise and/or effective than conventional therapeutic methods. At the moment, these synthetic circuits usually take the form of conventional digital designs (i.e. feed-forward logic circuits), which can be coupled to the existing gene regulatory pathways through the control of particular transcription factors. These logic circuits are themselves implemented using proteins and nucleic acids, and can be deployed into cells using various mechanisms, including customised viruses [7,8]. It will likely soon reach the point where a synthetic circuit can be delivered to a particular cell within the human body.

Whilst the GP community is becoming more interested in computational representations found in biological cells, the synthetic biology community mainly focuses on using computational representations found in silicon systems. Given that we have found computational models of GRNs to exhibit desirable properties (e.g. compact expressiveness, intrinsic fault tolerance) that are not necessarily found in more conventional models of computation, this might seem like an odd situation. In particular, why not use the existing design principles of biological cells and build synthetic GRNs rather than Boolean logic circuits? One answer to this question is that most synthetic biologists (and indeed most people) do not understand the design principles of GRNs, and hence can not design synthetic GRNs that have particular computational behaviours. This would appear to open up an opportunity for a community that can design GRNs that have particular computational behaviours, and this is almost exactly what we have been doing in recent years in the GP community, albeit with a fairly ad hoc group of GRN models and target behaviours. In essence, we suggest there is significant scope for using GP (and related approaches) to evolve programs that could be run *in vivo*—that is, within biological cells—and hence close the circle from biological inspiration to computational optimisation and back again to biology.

In this paper, we present work on using evolutionary algorithms (EAs) to design Boolean networks (BNs) that have a particular biological function. BNs are a class of abstract GRN models that are known to have steady-state equivalence to more detailed quantitative models [9], such as systems of differential equations, and have been successfully used to model various biological networks [10]. Since they are composed of Boolean functions, they also lend themselves well to implementation using existing synthetic biology techniques [8], so in principle evolved models could be refined into biological implementations. This is not necessarily the case with the more complex models currently used in the GP community, and hence why we focus on this relatively abstract class of GRN models.

An important problem in biology is the control of cell state [11]. This plays a role in many diseases: for instance, many cancers are thought to be caused by a cell transitioning to an abnormal cell state [12], and a potential cure would involve guiding its transition back to a normal cell state. Transitions in cell state are governed by a cell's GRN, and it is necessary to intervene in the GRN's natural dynamics in order to change this. At present, such interventions are typically achieved by using a drug to target a single gene, forcing it to be permanently on or off. However, this is a rather blunt form of intervention and unlikely to be sufficient for causing large-scale changes within a cell's behaviour. More effective forms of intervention would involve the coordinated targeting of multiple genes in a particular temporal pattern. In effect, this requires a control strategy, and consequently much of the work in this area has focused on using conventional control techniques to generate appropriate patterns of intervention [13,14]. However, this is an NP hard problem [15], meaning that in practice exact analytical solutions can only be found for small systems. This in itself suggests a potential role for metaheuristics such as EAs, which are often used to address problems where analytical solutions are infeasible.

In previous work [16,17], we have shown that EAs can be used to design BNs that can carry out control (i.e. generate a series of control interventions) when coupled to a target BN. The target Boolean networks, in this case, were randomly sampled from the space of all BNs of a given size, giving an estimate of the general ability of evolved BNs to influence the dynamics of other BNs. In the work reported in this paper, by comparison, we focus on the control of actual models of biological regulatory networks, and show that evolved BNs are able to govern these systems so that they transition to a specific attractor corresponding to a particular cell state. In this sense, they are much closer to being viable "genetic programs".

Section 2 presents a brief introduction to BNs and describes the BN models of biological networks that serve as control targets in this work. Section 3 describes the experimental methodology. Sections 4 and 5 present results and discussion, and Sect. 6 concludes.

2   Boolean Networks

A Boolean network (BN) is a discrete-time non-linear dynamical system represented as a directed graph $G(V, E)$ composed of nodes, or vertices, V and edges E [18,19]. The time evolution of a BN is expressed by a set of Boolean functions f_i, $i = 1, 2, 3, ...$ Each BN node has a binary state s which is updated synchronously according to its Boolean function and the states of the k input nodes that are connected to it. Formally, $s(t + 1) = f_i(s(t))$, where s is a set of network states, N is the number of nodes, $s \in \{0, 1\}^N$, $t = 0, 1, 2, 3, 4, ...$ is the discrete time, and $f_i : \{0, 1\}^N \rightarrow \{0, 1\}$. Since a BN is deterministic $s(t + 1)$ is only determined by $s(t)$. The possible number of Boolean functions is 2^{2^k}, and the state space is finite and equal to 2^N. Since the state space is finite, states must eventually be repeated, leading to temporal structures called attractors. When used to model GRNs, these attractors can be interpreted as the stable states (or cell types) of a cell [20].

2.1   Boolean Models of Biological Networks

To evaluate the ability of our control method in realistic biological situations, we selected five BNs from the literature that model well-known genetic regulatory systems [21–24]. Several factors motivated this choice. First, we wanted to look at the effect of network size, and the selected BNs vary in size from 10 nodes to 40 nodes. Second, it is important to show that the method works on systems with different state space structures. To address this, we chose BNs with different numbers of stable states, since this gives some indication of the complexity of the dynamics: the selected BNs have between 3 and 13 stable states, all of which are point attractors (i.e. a single repeating expression state). Finally, the chosen models are biologically diverse, capturing a range of different biological processes (morphogenesis, signalling and cell cycle regulation) that occur in a range of different species (single-celled organisms, plants, and animals). Each is briefly described in this section.

T Cell Receptor Signalling Pathway. T cells are a subgroup of white blood cells that play a crucial role in the adaptive immune response, helping to protect the host against different pathogens such as virus and bacteria. The inappropriate activation of a T cell can lead to various autoimmune diseases. T cell receptor (TCR) is a membrane protein found on the surface of T cells which contributes to their activation by recognising antigen. A BN of the TCR signalling pathway is described in [22] and is depicted in Fig. 1a. It comprises 40 genes and has 8 point attractors, corresponding to different activation and proliferation cell states. See [22] for details of Boolean functions.

T Helper Cell Differentiation Network. T helper cells, commonly called Th cells, are a type of T cell that plays a critical and key role in the adaptive immune system, where they help the immune activities of other immune cells such as B cell antibodies, plasma cells and cytotoxic T cells. T helper cells differentiate into one of the largest subcategories of cells, for example TFH, Th1, Th2, Th3, Th9 and Th17, which produce and release several types of T cells cytokines to regulate immune responses. A BN model of Th cell differentiation was developed in [21]. This model, depicted in Fig. 1b, captures the activities of 23 genes and has three point attractors, corresponding to different Th cell types. See [21] for details of Boolean functions.

Flower Morphogenesis in Arabidopsis Thaliana. Morphogenesis, the development of an organism's form through the process of cell differentiation, is an important component of multicellular organisms, and often plays a role in disease development. The most widely studied models of morphogenesis concern flower development in plants, and particularly within the model species *arabidopsis thaliana*, a small flowering plant. Flower morphogenesis occurs during the entire life cycle from groups of undifferentiated cells known as meristems. These develop into various different cell types in order to form the organs of a

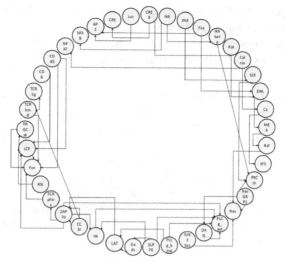

(a) T cell receptor signalling

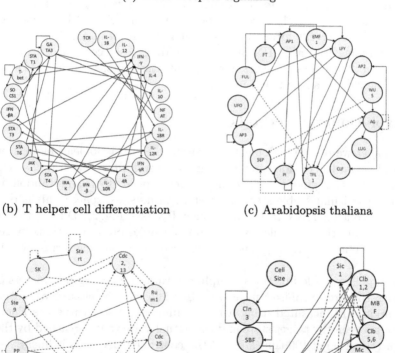

(b) T helper cell differentiation   (c) Arabidopsis thaliana

(d) Fission yeast cell cycle   (e) Budding yeast cell cycle

Fig. 1. Biological models used as case studies in this work.

flower, for example sepals, petals, stamens and carpels. A BN model of flower morphogenesis in arabidobis thaliana is described in [24]. It comprises 15 genes and has 10 point attractors, each corresponding to a different cell type. See Fig. 1c. Details of Boolean functions can be found in [24,25].

Fission Yeast Cell Cycle Regulation. Fission yeast is the common name of *schizosaccharomyces pombe*, a unicellular eukaryote whose cells are rod-shaped and divide by medial fission. It is a well known system used to study cell growth and division, mainly because of their simple shape and their place within the eukaryotic lineage. The fission yeast cell cycle is the sequence of events that occur in a cell leading to duplication of all its components and its division into two almost identical daughter cells. A BN model of fission yeast cell cycle regulation is given in [23]. It is formed by 10 genes and has 13 point attractors, corresponding to different stable cell states within the cell cycle. See Fig. 1d. Details of Boolean functions can be found in [23].

Budding Yeast Cell Cycle Regulation. Budding yeast is another species of yeast that has been widely used to study the eukaryotic cell life cycle. As the name implies, new cells form as a bud that grows from an existing cell, rather than undergoing fission. A BN model of budding yeast cell cycle regulation is described in [23]. It has 12 genes and 7 point attractors. See Fig. 1e. Details of Boolean functions can be found in [26].

3   Evolutionary Methods

In this paper we optimise Boolean networks to control Boolean models of real biological networks. We focus on applying a control intervention (i.e. a series of perturbations) that guides a trajectory of a controlled BN from a random initial state to a particular stable state (attractor) in its state space. This is done by coupling a controller BN to the controlled BN (see Fig. 2). During the course of its execution, the controller BN generates a series of interventions by setting the states of one or more target nodes (referred to as coupling terms) within the controlled BN.

The topology, node functions, coupling terms, and timing parameters of the controller BN are optimised using an EA. The effectiveness of a controller's interventions are measured using a fitness function that returns the Euclidean distance between the target state and the actual state that is reached by the end of a control period of 100 time steps of the controlled BN. This is linearly scaled to the interval $[0, 1]$, where a fitness of 1.0 indicates that the target state was reached. The fitness distribution over 20 runs is used to give an estimate of the ability of the EA to find a controller BN that can control a specified controlled BN so that it reaches a specified stable state. This is repeated for each stable state of each target network.

A controller BN has two evolved timing parameters, each within the range $[1, 50]$. The first timing parameter determines the number of time steps the controller BN will perform for each time step of the controlled BN, i.e. the relative speed of the controller. The second timing parameter indicates how frequently the controller BN is executed, in terms of the number of time steps of the controlled network, i.e. how often it intervenes.

A controller RBN is represented as an array of nodes, each comprising a Boolean function number between 0 and 2^{2^k}, an initial state, and a set of input nodes, where each input is indicated by its position within the array. For these experiments, k is fixed at 2 (the edge-of-chaos regime, where computational behaviours are hypothesised to be maximal [27]) and the controller has a fixed length of 15 nodes. In previous work, we have found this length to offer a fair trade-off between expressiveness and search space size [16]. The solution chromosome also contains the timing parameters and the coupling terms. See Fig. 3. A generational evolutionary algorithm is executed for 100 generations each run, with a population size of 500, tournament selection ($n = 3$), uniform crossover ($p_c = 0.15$), point mutation ($p_m = 0.06$) and elitism ($n = 1$).

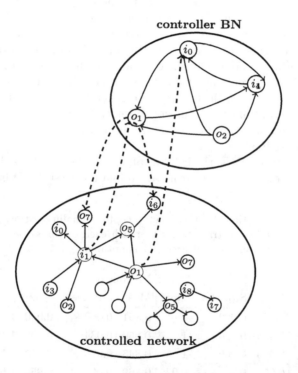

Fig. 2. Evolved controller Boolean network (representing a synthetic GRN) coupled to a controlled Boolean network (representing a native biological regulatory network). Coupling between the two networks is implemented by copying the expression states from designated nodes in the controller to designated nodes in the controlled network (depicted as dotted arrows in this diagram) at specified intervals.

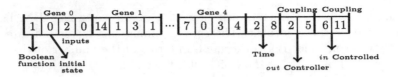

Fig. 3. Example of a Boolean network's genetic representation. Since $k = 2$, functions are numbered between 0 and 15. The timing and coupling terms indicate that this network is iterated twice each time it is executed, it is executed every 8 steps of the controlled network, its control outputs (interventions) are copied to nodes 2 and 5 of the controlled network, and its feedback (in) inputs from the controlled network are copied from nodes 6 and 11.

4   Results

Tables 1, 2, 3, 4 and 5 present summary statistics for the fitness distributions of both the natural (i.e. how close it gets to the target state in the absence of control) and controlled dynamics of each case study BN for each target stable state. Table 6 summarises these results, showing the mean fitness achieved and the number of target states reached (with and without control) within each biological network. Without control, only a small number of these attractors were reached (4/32). Even when they were reached, the standard deviations in fitness (i.e. in distance from the target) were generally large. This indicates that, for a particular evolutionary run, most randomly sampled initial states will not be within the basin of attraction of the target attractor, and hence the control problems are non-trivial.

In all the case study BNs, the EA was able to find controllers that can target the majority of the steady states from a random initial state. Where the target

Table 1. Fitness distributions for the T cell receptor signaling pathway control problem, indicating the ability of trajectories to reach each of the system's stable states both with and without control. A fitness of 1 is optimal.

Attractors	Control			No control			
	Mean	Std. Dev.	Max.	Mean	Std. Dev.	Max.	p-value
1	0.996	0.009	1	0.851	0.060	0.950	1.278×10^{-08}
2	0.975	0.026	1	0.850	0.034	0.900	1.596×10^{-08}
3	0.996	0.009	1	0.843	0.075	0.975	2.745×10^{-08}
4	0.975	0	0.975	0.869	0.066	0.950	3.664×10^{-09}
5	0.996	0.009	1	0.861	0.066	0.950	9.115×10^{-09}
6	0.969	0.010	0.975	0.917	0.055	0.975	1.49×10^{-05}
7	0.975	0	0.975	0.868	0.048	0.950	5.66×10^{-09}
8	1	0	1	0.844	0.051	0.950	3.073×10^{-09}
General mean	0.985	0.008	0.990	0.863	0.057	0.950	1.872×10^{-06}

Table 2. T-helper cell differentiation

Attractors	Control			No control			
	Mean	Std. Dev.	Max.	Mean	Std. Dev.	Max.	p-value
1	0.972	0.065	1	0.553	0.067	0.652	1.094×10^{-08}
2	1	0	1	0.601	0.179	0.826	3.823×10^{-09}
3	0.867	0.0446	0.913	0.510	0.161	0.826	5.285×10^{-08}
General mean	0.946	0.036	0.971	0.554	0.135	0.768	2.253×10^{-08}

Table 3. Arabidopsis thaliana flower morphogenesis

Attractors	Control			No control			
	Mean	Std. Dev.	Max.	Mean	Std. Dev.	Max.	p-value
1	1	0	1	0.863	0.137	1	1.094×10^{-08}
2	0.926	0.030	0.933	0.561	0.124	0.800	2.75×10^{-09}
3	0.989	0.033	1	0.635	0.100	0.733	5.693×10^{-09}
4	0.933	0	0.933	0.800	0.049	0.866	2.726×10^{-09}
5	1	0	1	0.835	0.144	0.933	2.549×10^{-09}
6	0.933	0	0.933	0.217	0.150	0.800	3.027×10^{-09}
7	1	0	1	0.919	0.042	1	3.3×10^{-08}
8	1	0	1	0.624	0.074	0.733	3.062×10^{-09}
9	1	0	1	0.382	0.184	0.933	4.479×10^{-09}
10	0.996	0.015	1	0.256	0.059	0.333	4.45×10^{-09}
General mean	0.977	0.0078	0.979	0.609	0.110	0.831	7.263×10^{-09}

was reached, the standard deviation between runs tended to be low, meaning that most runs are able to find BNs with optimal, or at least near-optimal, control strategies: the maximum likelihood estimation is 1.0 when BN controllers are successfully found and 0.9 otherwise. In all cases, the evolved controllers guided the system closer to the target states than could be achieved in the absence of control (see p-values in tables).

Some target BNs appear to be harder to control than others. The arabidopsis thaliana and T cell receptor signalling networks both have three steady states which were not reachable by the evolved controllers; although, in both cases, the systems could be controlled to states not far from the target state. However, there does not appear to be a simple relationship between the difficulty of the control task and the number of attractors: for example, the fission yeast BN, which has the most attractors, was the easiest to control. There is, however, a mild negative correlation (-0.23) between network size and control fitness, and indeed the largest network (T cell receptor signalling) was one of the hardest to control.

Table 4. Fission yeast cell cycle

Attractors	Control			No control			p-value
	Mean	Std. Dev.	Max.	Mean	Std. Dev.	Max.	
1	1	0	1	0.442	0.285	1	3.916×10^{-08}
2	1	0	1	0.321	0.171	0.900	2.25×10^{-09}
3	0.921	0.042	1	0.594	0.102	0.700	6.823×10^{-09}
4	0.994	0.022	1	0.447	0.219	0.900	4.107×10^{-09}
5	0.994	0.022	1	0.505	0.246	0.900	5.482×10^{-09}
6	1	0	1	0.573	0.133	0.900	1.921×10^{-09}
7	1	0	1	0.484	0.121	0.800	2.377×10^{-09}
8	0.900	0	0.900	0.763	0.095	0.900	2.088×10^{-09}
9	1	0	1	0.600	0.124	0.800	12.483×10^{-06}
10	0.984	0.0373	1	0.405	0.154	0.900	2.457×10^{-09}
11	0.921	0.041	1	0.552	0.134	0.800	1.274×10^{-08}
12	1	0	1	0.382	0.184	0.933	8.583×10^{-09}
13	0.994	0.022	1	0.536	0.134	0.800	3.873×10^{-09}
General mean	0.997	0.014	0.992	0.508	0.161	0.864	1.980×10^{-07}

Table 5. Budding yeast cell cycle

Attractors	Control			No control			p-value
	Mean	Std. Dev.	Max.	Mean	Std. Dev.	Max.	
1	1	0	1	0.543	0.165	0.666	2.788×10^{-09}
2	1	0	1	0.627	0.321	0.916	2.088×10^{-09}
3	1	0	1	0.442	0.416	0.916	2.25×10^{-09}
4	1	0	1	0.500	0.328	0.833	2.544×10^{-09}
5	1	0	1	0.605	0.393	0.916	2.859×10^{-09}
6	0.916	0	0.916	0.521	0.249	0.750	2.335×10^{-09}
7	1	0	1	0.434	0.479	1	1.036×10^{-05}
General mean	0.988	0	0.988	0.524	0.335	0.855	1.482×10^{-06}

Although the majority of target states could be reached, the evolved controllers were not able to reach all target states. Further research is required to understand exactly why this is the case, though we can speculate it is likely due to at least two reasons. First, in some target networks, the majority of random states may fall far from the basin of attraction of a particular stable state, making the problem intrinsically hard when an arbitrary initial state is chosen. Second, transitions between states in BNs are typically not between adjacent states, meaning that in many cases there will not be valid transitions from states which differ by a single bit from the target: this is likely to lead to deceptive local

Table 6. Summary of the results, showing the mean fitness (1 is optimal) across all runs, and the number of attractors reached, for each case study BN both when under the control of an evolved BN and when following its natural dynamics (no control) from a random initial state.

Network name	Size	Mean fitness		Attractors reached		
		Control	No control	Total	Control	No control
Fission yeast cell cycle	10	0.997	0.508	13	12	1
Budding yeast cell cycle	12	0.988	0.524	7	6	1
Arabidopsis thaliana	15	0.977	0.609	10	7	2
T helper cell differentiation	23	0.946	0.554	3	2	0
T cell receptor signalling	40	0.985	0.863	8	5	0

optima in the state space. If this is the case, there may be scope for using diversity preservation techniques (e.g. crowding, fitness sharing) to navigate around local optima during optimisation. This is something we plan to look at in future work. Nevertheless, the results are promising, and demonstrate that even basic evolutionary algorithms can solve state space targeting problems, and can do so in a way that does not require *a priori* understanding of the structure of the state space.

5 Discussion

The results of this study suggest that it is possible to evolve synthetic GRN models that have specific, biologically-relevant, behaviours. This is not the first time that EAs have been used to design and optimise GRNs for use within a synthetic biology context. For example, a number of different research groups have previously used EAs to design GRN models that have simple dynamical behaviours such as oscillation and bistability [28–30]. Nevertheless, unlike these earlier studies, the synthetic GRN models evolved in this work have behaviours that could reasonably be described as computational or programmatic, since controllers are essentially programs that carry out decisions based on their inputs.

This work is very much motivated by previous work in the GP community where GRN models have been used to carry out computation. Control, in particular, has been a recurring application in this nascent research field, with GRN models evolved to solve control tasks in robotics [6], computer gaming [4], and chaotic systems [5], to name but a few. This seems natural, since control is one of the principal behaviours carried out by biological GRNs. However, evolved GRNs have also been used to solve more diverse tasks (e.g. image compression [31]) and theoretical studies have shown that GRN models such as BNs are computationally universal [3], so in principle evolved synthetic GRNs could be used to carry out a much broader range of computational tasks, and perhaps even be used as the basis of general-purpose cellular computers. There is also no reason to limit the scope of this research to GRNs. BNs, for example, can be used to model other important biological networks, such as intracellular signalling networks.

In this work, we intentionally used a standard evolutionary algorithm and a linear solution representation (essentially a genetic algorithm) in order to keep things simple. In practice, there is plenty of scope for using more advanced approaches. Notably, there has been a lot of work on evolving network structures, and much of this would be directly applicable especially if we aim to evolve larger, more complex networks. Recent work in applying NEAT to GRN models [32], and applying Cartesian GP to recurrent networks [33] seem particularly relevant.

This is still early work, and there remains significant work to be done to show that this is a viable approach to designing synthetic GRNs. Initially, we plan to study the evolved controllers in order to gain insight into the nature (and diversity) of the computational behaviours that are carried out when solving these control tasks. However, we also need to take into account biological constraints when evolving controllers: for instance, restricting coupling terms to biologically accessible targets (since currently any node in the target network can be used for coupling), and focusing on biologically-meaningful initial conditions rather than randomly sampling starting states.

Whilst this will help to build confidence that evolved controllers are doing something useful and viable, we also need to demonstrate that the evolved controllers are robust. There are several aspects to this. First, there is the generality of a controller's behaviour; for instance, can a single controller tolerate different initial conditions? Second, there are the differences between simulation and reality. Research in evolutionary robotics has shown that this kind of "reality gap" can restrict the generality of evolved controllers. We know from existing research [4,34] that GRN models are less susceptible to this problem, given their natural robustness. However, we also need to consider that the simulation environment used in this work is quite different to biological reality. For instance, biological cells are stochastic environments, both in terms of what occurs and when things occur. We might address this, for example, by using probabilistic BN models. There is also the question of how much confidence we have in the executable model used to evaluate a BN model, since this will determine how much confidence we have in the evolved model. However, this is a more general issue of biological modelling, and there has been significant progress in developing reliable executable models of biological systems [35].

Refining evolved BN programs into actual synthetic biology realisations would involve a number of extra challenges. For instance, in this work we evolved timing parameters to allow the controller and controlled systems to operate over different timescales. This may also be possible to do within synthetic biology implementations, e.g. using RNA interference rather than transcription factors to speed up the controller's logic, but it would not be trivial. Another issue might be limitations placed on the controller's size or topology due to the difficulty of avoiding cross-talk within synthetic biology circuits.

6   Conclusions

The control of a cell's state is an important problem: it is instrumental for controlling many disease processes, yet in practice it is very difficult to find a series of interventions that will guide a cell between two different states. In this paper, we describe a novel approach to solving this problem which involves optimising a synthetic gene regulatory network which is then used to generate a pattern of interventions based on the state of a target cell. The approach is evaluated using computational simulation, representing the gene regulatory network as a Boolean network, and the target cells as executable Boolean models. An evolutionary algorithm is then used to carry out optimisation of the Boolean network. In the majority of the case studies we looked at, the evolutionary algorithm was able to find Boolean networks that could successfully guide the target cell model from a randomly sampled initial state to a biologically-meaningful cell state.

The choice of Boolean networks is not arbitrary. The fact that they are constructed from Boolean logic gates means that there is a potential pathway from model to biological implementation through the use of existing synthetic biology principles. The choice of an evolutionary algorithm is also not arbitrary, and is motivated by a larger body of work in the field of genetic programming which is concerned with using evolutionary algorithms to design programmatic behaviours. In recent years, the genetic programming community has increasingly made use of models of gene regulatory networks to represent evolving computation. In addition to being intrinsically evolvable, these representations have also proved able at expressing complex computational behaviours that are robust yet compact. However, to our knowledge, this is the first time that evolutionary algorithms have been used to design actual "genetic programs".

From this perspective, it is interesting to note that synthetic biology focuses on implementing feed-forward logic circuits and traditional models of computation within biological cells, rather than using native biological design principles. This is in contrast to the opposing direction of travel in the genetic programming community. There are various reasons for this, but a significant factor is the difficulty of designing gene regulatory networks, which are based around principles of non-linear dynamical systems rather than well understood digital design principles. However, the ability of evolutionary algorithms to optimise these structures suggests that the genetic programming community could play an important role in designing programs that will one day run *in vivo* within biological cells.

References

1. Reil, T.: Dynamics of gene expression in an artificial genome — implications for biological and artificial ontogeny. In: Floreano, D., Nicoud, J.-D., Mondada, F. (eds.) ECAL 1999. LNCS (LNAI), vol. 1674, pp. 457–466. Springer, Heidelberg (1999). https://doi.org/10.1007/3-540-48304-7_63

2. Banzhaf, W.: Artificial regulatory networks and genetic programming. In: Riolo, R., Worzel, B. (eds.) Genetic Programming Theory and Practice, Genetic Programming Series, vol. 6, pp. 43–61. Springer, Boston (2003). https://doi.org/10.1007/978-1-4419-8983-3_4

3. Lones, M.A.: Computing with artificial gene regulatory networks. In: Iba, H., Noman, N. (eds.) Evolutionary Algorithms in Gene Regulatory Network Research, pp. 398–424. Wiley (2016). https://doi.org/10.1002/9781119079453.ch15

4. Sanchez, S., Cussat-Blanc, S.: Gene regulated car driving: using a gene regulatory network to drive a virtual car. Genet. Program Evolvable Mach. **15**(4), 477–511 (2014). https://doi.org/10.1007/s10710-014-9228-y

5. Lones, M.A., Turner, A.P., Fuente, L.A., Stepney, S., Caves, L.S.D., Tyrrell, A.M.: Biochemical connectionism. Nat. Comput. **12**(4), 453–472 (2013). https://doi.org/10.1007/s11047-013-9400-y

6. Trefzer, M.A., Kuyucu, T., Miller, J.F., Tyrrell, A.M.: Evolution and analysis of a robot controller based on a gene regulatory network. In: Tempesti, G., Tyrrell, A.M., Miller, J.F. (eds.) ICES 2010. LNCS, vol. 6274, pp. 61–72. Springer, Heidelberg (2010). https://doi.org/10.1007/978-3-642-15323-5_6

7. Lienert, F., Lohmueller, J.J., Garg, A., Silver, P.A.: Synthetic biology in mammalian cells: next generation research tools and therapeutics. Nat. Rev. Mol. Cell Biol. **15**(2), 95–107 (2014)

8. Singh, V.: Recent advances and opportunities in synthetic logic gates engineering in living cells. Syst. Synth. Biol. **8**(4), 271–282 (2014)

9. Veliz-Cuba, A., Arthur, J., Hochstetler, L., Klomps, V., Korpi, E.: On the relationship of steady states of continuous and discrete models arising from biology. Bull. Math. Biol. **74**(12), 2779–2792 (2012)

10. Saadatpour, A., Albert, R.: Boolean modeling of biological regulatory networks: a methodology tutorial. Methods **62**(1), 3–12 (2013)

11. Cury, J.E., Baldissera, F.L.: Systems biology, synthetic biology and control theory: a promising golden braid. Annu. Rev. Control **37**(1), 57–67 (2013)

12. Huang, S., Ernberg, I., Kauffman, S.: Cancer attractors: a systems view of tumors from a gene network dynamics and developmental perspective. Semin. Cell Dev. Biol. **20**, 869–876 (2009). Elsevier

13. Motter, A.E.: Networkcontrology: Chaos: an Interdisciplinary. J. Nonlinear Sci. **25**(9), 097621 (2015)

14. Fornasini, E., Valcher, M.E.: Recent developments in Boolean networks control. J. Control Decis. **3**(1), 1–18 (2016)

15. Akutsu, T., Hayashida, M., Ching, W.K., Ng, M.K.: Control of Boolean networks: hardness results and algorithms for tree structured networks. J. Theoret. Biol. **244**(4), 670–679 (2007)

16. Taou, N.S., Corne, D.W., Lones, M.A.: Evolving Boolean networks for biological control: state space targeting in scale free Boolean networks. In: 2016 IEEE Conference on Computational Intelligence in Bioinformatics and Computational Biology (CIBCB), pp. 1–6 (2016). https://doi.org/10.1109/CIBCB.2016.7758125

17. Taou, N.S., Corne, D.W., Lones, M.A.: Towards intelligent biological control: controlling Boolean networks with Boolean networks. In: European Conference on the Applications of Evolutionary Computation, pp. 351–362 (2016). https://doi.org/10.1007/978-3-319-31204-0_23

18. Kauffman, S.A.: The Origins of Order: Self Organization and Selection in Evolution. Oxford University Press, New York (1993)

19. Drossel, B.: Random Boolean networks. In: Schuster, H.G. (ed.) Reviews of Nonlinear Dynamics and Complexity, pp. 69–110 (2008). https://doi.org/10.1002/9783527626359.ch3

20. Huang, S., Eichler, G., Bar-Yam, Y., Ingber, D.E.: Cell fates as high-dimensional attractor states of a complex gene regulatory network. Phys. Rev. Lett. **94**(12), 128701 (2005)

21. Mendoza, L., Xenarios, I.: A method for the generation of standardized qualitative dynamical systems of regulatory networks. Theor. Biol. Med. Model. **3**(1), 13 (2006)

22. Klamt, S., Saez-Rodriguez, J., Lindquist, J.A., Simeoni, L., Gilles, E.D.: A methodology for the structural and functional analysis of signaling and regulatory networks. BMC Bioinf. **7**(1), 56 (2006)

23. Davidich, M.I., Bornholdt, S.: Boolean network model predicts cell cycle sequence of fission yeast. PLoS ONE **3**(2), e1672 (2008)

24. Alvarez-Buylla, E.R., Chaos, Á., Aldana, M., Benítez, M., Cortes-Poza, Y., Espinosa-Soto, C., Hartasánchez, D.A., Lotto, R.B., Malkin, D., Santos, G.J.E., et al.: Floral morphogenesis: stochastic explorations of a gene network epigenetic landscape. PLoS ONE **3**(11), e3626 (2008)

25. Mendoza, L., Thieffry, D., Alvarez-Buylla, E.R.: Genetic control of flower morphogenesis in Arabidopsis thaliana: a logical analysis. Bioinformatics **15**(7), 593–606 (1999). Oxford, England

26. Li, F., Long, T., Lu, Y., Ouyang, Q., Tang, C.: The yeast cell-cycle network is robustly designed. Proc. Nat. Acad. Sci. U.S.A. **101**(14), 4781–4786 (2004)

27. Goudarzi, A., Teuscher, C., Gulbahce, N., Rohlf, T.: Emergent criticality through adaptive information processing in Boolean networks. Phys. Rev. Lett. **108**(12), 128702 (2012)

28. François, P., Hakim, V.: Design of genetic networks with specified functions by evolution in silico. Proc. Nat. Acad. Sci. U.S.A. **101**(2), 580–585 (2004)

29. Noman, N., Monjo, T., Moscato, P., Iba, H.: Evolving robust gene regulatory networks. PLoS ONE **10**(1), e0116258 (2015)

30. Garcia-Bernardo, J., Eppstein, M.J.: Evolving modular genetic regulatory networks with a recursive, top-down approach. Syst. Synth. Biol. **9**(4), 179–189 (2015)

31. Trefzer, M.A., Kuyucu, T., Miller, J.F., Tyrrell, A.M.: Image compression of natural images using artificial gene regulatory networks. In: Proceedings of the 12th Annual Conference on Genetic and Evolutionary Computation, GECCO 2010, pp. 595–602. ACM, New York (2010). https://doi.org/10.1145/1830483.1830593

32. Cussat-Blanc, S., Harrington, K., Pollack, J.: Gene regulatory network evolution through augmenting topologies. IEEE Trans. Evol. Comput. **19**(6), 823–837 (2015)

33. Turner, A.J., Miller, J.F.: Recurrent Cartesian genetic programming of artificial neural networks. Genet. Program Evolvable Mach. **18**(2), 185–212 (2017)

34. Roli, A., Manfroni, M., Pinciroli, C., Birattari, M.: On the design of Boolean network robots. In: Di Chio, C., Cagnoni, S., Cotta, C., Ebner, M., Ekárt, A., Esparcia-Alcázar, A.I., Merelo, J.J., Neri, F., Preuss, M., Richter, H., Togelius, J., Yannakakis, G.N. (eds.) EvoApplications 2011. LNCS, vol. 6624, pp. 43–52. Springer, Heidelberg (2011). https://doi.org/10.1007/978-3-642-20525-5_5

35. Timmis, J., Alden, K., Andrews, P., Clark, E., Nellis, A., Naylor, B., Coles, M., Kaye, P.: Building confidence in quantitative systems pharmacology models: an engineer's guide to exploring the rationale in model design and development. CPT Pharmacometrics Syst. Pharmacol. **6**(3), 156–167 (2017)

A Multiple Expression Alignment
Framework for Genetic Programming

Leonardo Vanneschi[(✉)][iD], Kristen Scott, and Mauro Castelli

NOVA IMS, Universidade Nova de Lisboa, 1070-312 Lisboa, Portugal
{lvanneschi,kscott,mcastelli}@novaims.unl.pt

Abstract. Alignment in the error space is a recent idea to exploit semantic awareness in genetic programming. In a previous contribution, the concepts of optimally aligned and optimally coplanar individuals were introduced, and it was shown that given optimally aligned, or optimally coplanar, individuals, it is possible to construct a globally optimal solution analytically. As a consequence, genetic programming methods, aimed at searching for optimally aligned, or optimally coplanar, individuals were introduced. In this paper, we critically discuss those methods, analyzing their major limitations and we propose new genetic programming systems aimed at overcoming those limitations. The presented experimental results, conducted on four real-life symbolic regression problems, show that the proposed algorithms outperform not only the existing methods based on the concept of alignment in the error space, but also geometric semantic genetic programming and standard genetic programming.

1 Introduction

In the last few years, the use of semantic awareness for improving Genetic Programming (GP) [1,2] and other heuristic methods [3] became popular. A survey discussing large part of the existing semantic approaches in GP can be found in [4]. In that survey, the existing work was categorized into three broad classes: approaches based on semantic diversity, on semantic locality and on semantic geometry. Among several other references, semantic diversity and semantic locality, and their relationship with the effectiveness of GP, were investigated in depth in [5,6]. On the other hand, the idea of studying semantic geometry revealed itself about a decade ago (see for instance [7,8]), and became a GP hot topic in 2013, when a new version of GP, called Geometric Semantic GP (GSGP) was introduced [9]. GSGP uses new operators, called geometric semantic operators (GSOs), instead of traditional crossover and mutation, and it owes part of its successes to the fact that GSOs induce a unimodal fitness landscape [10–12] for any supervised learning problem. In the last six years, a large number of contributions showed that GSGP is competitive with the state of the art in many

© Springer International Publishing AG, part of Springer Nature 2018
M. Castelli et al. (Eds.): EuroGP 2018, LNCS 10781, pp. 166–183, 2018.
https://doi.org/10.1007/978-3-319-77553-1_11

applicative domains (see for instance [13–15]). Few years after the introduction of GSGP, a new way of exploiting semantic awareness was presented in [16] and further developed in [17, 18]. The idea, which is also the focus of this paper, can be sketched as follows.

We define semantics of an individual as the vector of its output values on the training cases [9]. Hence, semantics can be represented as a point in a space that we call *semantic space*. In supervised learning, the target is also a point in the semantic space, but usually (unless the rare case where the target value is equal to zero for each training case) it does *not* correspond to the origin of the Cartesian system. Then, we translate each point in the semantic space by subtracting the target from it. In this way, for each individual, we obtain a new point, that we call *error vector*, and we call the corresponding space *error space*. The target, by construction, corresponds to the origin of the Cartesian system in the error space.

In [16], it was proven that, given sets of individuals with particular characteristics of alignment in the error space (called optimally aligned, and optimally coplanar, individuals), it is possible to analytically reconstruct a globally optimal solution (see Sect. 2.1). Keeping this in mind, it makes sense to develop GP systems whose objective is looking for optimally aligned, or optimally coplanar, individuals (instead of looking directly for an optimal solution, as in traditional GP). The first attempt at developing a system aimed at searching for optimally aligned, or optimally coplanar, individuals was presented in [16], where the ESAGP method was proposed. While ESAGP reported interesting results, it has the important limitation of constraining the alignment only in one particular direction in the error space, that is prefixed *a priori*. In order to overcome this limitation, a particular version of GP must be defined, that evolves individuals that are sets of programs, instead of just one program as in traditional GP. The first preliminary attempt was made in [17], where the POGP system was introduced. However, in [17] severe limitations of POGP, which make it unusable in practice, were reported.

The objective of this paper is to present new GP systems aimed at evolving sets of programs with the objective of generating optimally aligned individuals, and able to overcome all the limitations of POGP. The new systems (called Align, Nested_Align and Nested_Align_β) will be compared to standard GP, GSGP and ESAGP on four complex real-life symbolic regression problems.

The rest of the paper is structured as follows: in Sect. 2, we revise previous and related work, with particular focus on ESAGP and POGP, describing the known issues of POGP. Section 3 describes the proposed methods (Align, Nested_Align and Nested_Align_β), explaining how they overcome the previously discussed issues of POGP. In Sect. 4, we present our experimental study, in which the experimental settings and test problems are described and the obtained results discussed. Finally, Sect. 5 concludes the paper, also suggesting ideas for future research.

2   Previous and Related Work

2.1   Error Space Alignment GP

In [16], the concept of optimal alignment was introduced for the first time, together with a new GP method, called ESAGP (which stands for Error Space Alignment GP), that exploits it. Two individuals A and B are *optimally aligned* if a scalar constant k exists such that $e_A = k \cdot e_B$, where e_A and e_B are the error vectors of A and B respectively. From this definition, it is not difficult to see that two individuals are optimally aligned if the straight line joining their error vectors also intersects the origin in the error space. Analogously, and extending the idea to three dimensions, three individuals are *optimally coplanar* if the bi-dimensional plane in which their error vectors lie in the error space also intersects the origin. In [16], it is proven that given any pair of optimally aligned individuals A and B, it is possible to reconstruct a globally optimal solution P_{opt}. This solution is defined in Eq. (1):

$$P_{opt} = \frac{1}{1-k} * A - \frac{k}{1-k} * B \qquad (1)$$

Analogously, in [16], it is also proven that given any triplet of optimally coplanar individuals, it is possible to analytically construct a globally optimal solution (the reader is referred to [16] for the equation of the globally optimal solution in that case).

Keeping all this in mind, the ESAGP method introduced in [16] was composed by two GP systems: ESAGP-1, whose objective is looking for optimally aligned pairs of individuals, and ESAGP-2 whose objective is looking for triplets of optimally coplanar individuals. The biggest difference between these systems and traditional GP is that the search in ESAGP-1 and ESAGP-2 is not guided by the quality of the single solutions, but only on their alignment properties. Several possible ways of searching for alignments can be imagined. In ESAGP, one direction, called *attractor*, is fixed and all the individuals in the population are *pushed* towards an alignment with the attractor. In this way, ESAGP-1 and ESAGP-2 can maintain the traditional representation of solutions where each solution is represented by one program. The other face of the coin is that ESAGP-1 and ESAGP-2 strongly restrict what GP can do, forcing the alignment to necessarily happen in just one prefixed direction, i.e. the one of the attractor. The ESAGP systems were also studied in [18], where the operators used to reach the alignment with the attractor were GSOs. The authors of [18] report severe overfitting for this new ESAGP version. The objective of this paper is to relieve the constraint of ESAGP by defining a new GP system that is generally able to evolve *vectors* of programs (even though only vectors of size equal to 2 will be used in this paper). As already mentioned, a preliminary attempt is represented by POGP [17], described below.

2.2 Pair Optimization GP

In [17], Pair Optimization GP (POGP) was introduced. Limiting itself to the bi-dimensional case (i.e. to the case in which pairs of optimally aligned individuals are sought for), POGP extends ESAGP-1, releasing the limitation of forcing alignments in a prefixed direction. In POGP, individuals are pairs of programs, and fitness is the angle between the respective error vectors. From now on, for the sake of clarity, this type of individual (i.e. individuals characterized by more than one program) will be called *multi-individuals*. In [17], the following problems of POGP were reported: (*i*) generation of semantically identical, or very similar, expressions; (*ii*) k constant in Eq. (1) equal, or very close, to zero; (*iii*) generation of expressions with huge error values. These problems are discussed here:

Issue 1: generation of semantically identical, or very similar, expressions. A simple way for GP to find two expressions that are optimally aligned in the error space is to find two expressions that have exactly the same semantics (and consequently the same error vector). However, this causes a problem once we try to reconstruct the optimal solution as in Eq. (1). In fact, if the two expressions have the same error vector, the k value in Eq. (1) is equal to 1, which gives a denominator equal to zero. Experience tells us that GP tends very often to generate multi-individuals that have this kind of problem. Also, it is worth pointing out that even preventing GP from generating multi-individuals that have an identical sematics, GP may still push the evolution towards the generation of multi-individuals whose expressions have semantics that are very similar between each other. This leads to a k constant in Eq. (1) that, although not being exactly equal to 1, has a value that is very close to 1. As a consequence, the denominator in Eq. (1), although not being exactly equal to zero, may be very close to zero and thus the value calculated by Eq. (1) could be a huge number. This would force a GP system to deal with unbearably large numbers during all its execution, which may lead to several problems, including numeric overflow.

Issue 2: k constant in Eq. (1) equal, or very close, to zero. Looking at Eq. (1), one may notice that if k is equal to zero, then expression B is irrelevant and the reconstructed solution P_{opt} is equal to expression A. A similar problem also manifests itself when k is not exactly equal to zero, but very close to zero. In this last case, both expressions A and B contribute to P_{opt}, but the contribution of B may be so small to be considered as marginal, and P_{opt} would *de facto* be extremely similar to A. Experience tells us that, unless this issue is taken care of, the evolution would very often generate such situations. This basically turns a multi-individual alignment based system into traditional GP, in which only one of the expressions in the multi-individual matters. If we really want to study the effectiveness of multi-individual alignment based systems, we have to impede these kind of situations.

Issue 3: generation of expressions with huge error values. As previously mentioned, systems based on the concept of alignment in the error space could limit themselves to searching for expressions that are optimally aligned, without taking into account their performance (i.e. how close their semantics are to the target). However, experience tells us that, if we give GP the only task of finding aligned expressions, GP frequently tends to generate expressions whose semantics contain unbearably large numbers. Once again, this may lead to several problems, including numeric overflow, and a successful system should definitely prevent this from happening.

One fact that should be remarked is that none of the previous issues can be taken into account with simple conditions that prevent some precise situations from happening. For instance, one may consider solving Issue 1 by simply testing if the expressions in a multi-individual are semantically identical between each other, and rejecting the multi-individual if that happens. But, as already discussed, expressions that have very similar semantics between each other may also lead to problems. Furthermore, the idea of introducing a threshold ϵ to the semantic diversity of the expressions in a multi-individual, and rejecting all the multi-individuals for which the diversity is smaller than ϵ does not seem a brilliant solution. In fact, experience tells us that GP would tend to generate multi-individuals with a diversity equal, or very close to ϵ itself. Analogously, if we consider Issue 2, neither rejecting multi-individuals that have a k constant equal to zero, nor rejecting individuals that have an absolute value of k larger than a given threshold would solve the problem. Finally, considering Issue 3, also rejecting individuals that have the coordinates of the semantic vector larger than a given threshold δ_{max} would not solve the problem, since GP would tend to generate expressions in which the coordinates of the semantic vector are equal, or very close, to δ_{max} itself.

In such a situation, we believe that a promising way to effectively solve these issues is (besides defining the specific conditions mentioned above) to take the issues into account in the selection process, for instance giving more probability of being selected for mating to multi-individuals that have large semantic diversity between the expressions, values of k that are, as much as possible, far from zero and expressions whose semantics are, as much as possible, close to the target. These ideas are implemented in the proposed systems, which are described below.

3   Description of the Proposed Methods

In order to introduce the proposed methods in a compact way, we describe first the Nested_Align method, and then we discuss the other methods by simply pointing out the differences between them and Nested_Align.

3.1   Nested_Align

Here, we describe selection, mutation and population initialization of Nested_Align, keeping in mind that no crossover has been defined yet for this method.

Selection. Besides trying to optimize the performance of the multi-individuals, selection is the phase that takes into account the issues of the previous alignment-based methods discussed in Sect. 2.2. Nested_Align contains five selection criteria, that have been organized into a nested tournament. Let $\phi_1, \phi_2, ..., \phi_m$ be the expressions characterizing a multi-individual. Once again, it is worth pointing out that only the case $m = 2$ was taken into account in this paper. But the concept is general, and so it will be explained using m expressions. The selection criteria are:

- Criterion 1: diversity (calculated using the standard deviation) of the semantics of the expressions $\phi_1, \phi_2, ..., \phi_m$ (to be maximized).
- Criterion 2: the absolute value of the k constant that characterizes the reconstructed expression, as in Eq. (1) (to be maximized).
- Criterion 3: the sum of the errors of the single expressions $\phi_1, \phi_2, ..., \phi_m$ (to be minimized).
- Criterion 4: the angle between the error vectors of the expressions $\phi_1, \phi_2, ..., \phi_m$ (to be minimized).
- Criterion 5: the error of the reconstructed expression P_{opt} in Eq. (1) (to be minimized).

The nested tournament works as follows: an individual is selected if it is the winner of a tournament, that we call T_5, that is based on Criterion 5. All the participants in tournament T_5, instead of being individuals chosen at random as in the traditional tournament selection algorithm, are winners of previous tournaments (that we call tournaments of type T_4), which are based on Criterion 4. Analogously, for all $i = 4, 3, 2$, all participants in the tournaments of type T_i are winners of previous tournaments (that we will call tournaments of type T_{i-1}), based on Criterion $i - 1$. Finally, the participants in the tournaments of type T_1 (the kind of tournament that is based on Criterion 1) are individuals selected at random from the population. In this way, an individual, in order to be selected, has to undergo five selection *layers*, each of which is based on one of the five different chosen criteria. Motivations for the chosen criteria follow:

- Criterion 1 was introduced to counteract Issue 1 in Sect. 2.2. Maximizing the semantic diversity of the expressions in a multi-individual should naturally prevent GP from creating multi-individuals with identical semantics or semantics that are very similar between each other.
- Criterion 2 was introduced to counteract Issue 2 in Sect. 2.2. Maximizing the absolute value of constant k should naturally allow GP to generate multi-individuals for which k's value is neither equal nor close to zero.
- Criterion 3 was introduced to counteract Issue 3 in Sect. 2.2. If the expressions that characterize a multi-individual will have a "reasonable" error, then their semantics will be reasonably similar to the target, thus naturally avoiding the appearance of unbearably large numbers.
- Criterion 4 is a performance criterion: if the angle between the error vectors of the expressions $\phi_1, \phi_2, ..., \phi_m$ is equal to zero, then Eq. (1) allows us to reconstruct a perfect solution P_{opt}. Also, the smaller this angle, the smaller should

be the error of P_{opt}. Nevertheless, experience tells us that multi-individuals may exist with similar values of this angle, but very different values of the error of the reconstructed solution P_{opt}, due for example to individuals with a very large distance from the target. This fact made us conclude that Criterion 4 cannot be the only performance objective, and suggested to us to also introduce Criterion 5.

– Criterion 5 is a further performance criterion. Among multi-individuals with the same angle between the error vectors of the expressions $\phi_1, \phi_2, ..., \phi_m$, the preferred ones will be the ones for which the reconstructed solution P_{opt} has the smallest error.

Mutation. The mechanism we have implemented for applying mutation to a multi-individual is extremely simple: for each expression ϕ_i in a multi-individual, mutation is applied to ϕ_i with a given mutation probability p_m, where p_m is a parameter of the system. It is worth remarking that in our implementation all expressions ϕ_i of a multi-individual have the same probability of undergoing mutation, but this probability is applied independently to each one of them. So, some expressions could be mutated, and some other could remain unchanged. The type of mutation that is applied to expressions is Koza's standard subtree mutation [1].

To this "basic" mutation algorithm, we have also decided to add a mechanism of rejection, in order to help the selection process in counteracting the issues discussed in Sect. 2.2. Given a prefixed parameter that we call δ_k, if the multi-individual generated by mutation has a k constant included in the range $[1 - \delta_k, 1+\delta_k]$, or in the range $[-\delta_k, \delta_k]$, then the k constant is considered, respectively, too close to 1 or too close to 0 and the multi-individual is rejected. In this case, a new individual is selected for mutation, using again the nested tournament discussed above.

The combined effect of this rejection process and of the selection algorithm should strongly counteract the issues discussed in Sect. 2.2. In fact, when k is equal to 1, or equal to 0, or even close to 1 or 0 inside a given prefixed toleration radius δ_k, the multi-individual is not allowed to survive. For all the other multi-individuals, distance between k and 1 and between k and 0 are used as optimization objectives, to be maximized. This allows GP to evolve multi-individuals with k values that are "reasonably far" from 0 and 1.

Initialization. Nested_Align initializes a population of multi-individuals using multiple executions of the Ramped Half and Half algorithm [1]. More specifically, let n be the number of expressions in a multi-individual ($n = 2$ in our experiments), and let m be the size of the population that has to be initialized. Nested_Align runs n times the Ramped Half and Half algorithm, thus creating n "traditional" populations of trees $P_1, P_2, ..., P_n$, each of which containing m trees. Let $P = \{\Pi_1, \Pi_2, ..., \Pi_m\}$ be the population that Nested_Align has to initialize (where, for each $i = 1, 2, ..., m$, Π_i is an n-dimensional multi-individual). Then, for each $i = 1, 2, ..., m$ and for each $j = 1, 2, ..., n$, the jth tree of multi-individual Π_i is the jth tree in population P_i.

To this "basic" initialization algorithm, we have added an adjustment mechanism to make sure that the initial population does not contain multi-individuals with a k equal, or close, to 0 and 1. More in partcular, given a prefixed number of expressions α, that is a new parameter of the system, if the created multi-individual has a k value included in the range $[1 - \delta_k, 1 + \delta_k]$, or in the range $[-\delta_k, \delta_k]$ (where δ_k is the same parameter as the one used for implementing rejections of mutated individuals), then α randomly chosen expressions in the multi-individual are removed and replaced by as many new randomly generated expressions. Then the k value is calculated again, and the process is repeated until the multi-individual has a k value that stays outside the ranges $[1-\delta_k, 1+\delta_k]$ and $[-\delta_k, \delta_k]$. Only when this happens, the multi-individual is accepted inside the population. Given that only multi-individuals of two expressions are considered in this paper, in our experiments we have always used $\alpha = 1$.

3.2 Differences Between Align, Nested_Align_β and Nested_Align

Align. The difference between Align and Nested_Align is that Align does not use the nested tournament discussed above. Selection in Align is implemented by a traditional tournament algorithm, using as fitness the error of the reconstructed expression P_{opt} in Eq. (1). Mutation and initialization in Align work exactly as in Nested_Align. In this way, the only mechanism that Align has to counteract the issues described in Sect. 2.2 is to make sure that initialization and mutation only create multi-individuals with a k value outside the ranges $[1 - \delta_k, 1 + \delta_k]$ and $[-\delta_k, \delta_k]$. The motivation for the introduction of Align is that the nested tournament that characterizes Nested_Align may be complex and time-consuming. Comparing the performance of Nested_Align to the ones of Align, we will be able to evaluate to importance of the nested tournament and its impact on the performance of the system.

Nested_Align_β. This method integrates a multi-individual approach with a traditional single-expression GP approach. More precisely, the method begins as Nested_Align, but after β generations, the evolution is done by GSGP. In order to transform a population of multi-individuals into a population of traditional single-expression individuals, each individual is replaced by the reconstructed solution P_{opt} in Eq. (1). The rationale behind the introduction of Nested_Align_β is that alignment-based systems are known to have a very quick improvement in fitness in the first generations, which may sometimes cause overfitting of training data (the reader is referred to [16–20] for a discussion of the issue). Given that GSGP is instead known for being a slow optimization process, able to limit overfitting under certain circumstances (see [21]), the idea is transforming Nested_Align into GSGP, possibly before overfitting arises. Even though a deep study of parameter β is strongly in demand, only $\beta = 50$ was tested in this paper. For this reason, from now on, the name Nested_Align_50 will be used for this method.

4   Experimental Study

4.1   Experimental Settings and Test Problems

For each model, 30 runs were performed, each on a different randomly selected split of the dataset into training set (70%) and test set (30%). The parameters used are summarized in Table 1. Besides those parameters, the primitive operators were addition, subtraction, multiplication and division protected as in [1]. The terminal symbols included one variable for each feature in the dataset, plus the following numerical constants: $-1.0, -0.75, -0.5, -0.25, 0.25, 0.5, 0.75, 1.0$. Parent selection was done using tournaments of size 5, with the exception of the models which use nested selection (i.e. Nested_Align and, in the first 50 generations, Nested_Align_50), which used a tournament of size 10 for each layer of the nested selection. For standard GP subtree crossover and subtree mutation were used [1], where crossover rate was equal to 0.9 and mutation rate was equal to 0.1. For all the other studied methods, crossover rate was equal to zero (i.e. no crossover was performed during the evolution). While Align, Nested_Align and Nested_Align_50 do not have a crossover operator implemented yet, the motivation for not using crossover in GSGP can be found in [17], where it is clearly shown that GSGP using only mutation often overcomes GSGP using both crossover and mutation. The test problems that we have used in our experimental study are four symbolic regression real-life applications. All these problems have already been used in previous GP studies [17,21–24]. Table 2 reports, for each dataset, the number of features (variables) and the number of instances (observations). For a complete description of these datasets, the reader is referred to the references reported in the same table.

4.2   Experimental Results

The results we have obtained are reported in Figs. 1, 2, 3, 4 and 5. They are organized as follows: in Figs. 1 and 2, we report the results of the best error obtained on training data (more particularly, in Fig. 1 the proposed methods are compared to standard GP and GSGP, while in Fig. 2 they are compared to ESAGP-1 and

Table 1. GP parameters used in our experiments.

Parameter	Setting
Population size	100
Max. # of generations	200
Initialization	Ramped H-H
Max. depth for evolution	17
Max. depth for initialization	6
δ_k	0.02

Table 2. Description of the test problems. For each dataset, the number of features (independent variables) and the number of instances (observations) are reported.

Dataset	# Features	# Instances
Bioavailability [25]	241	206
PPB [25]	626	131
Toxicity [25]	626	234
Energy [24]	8	768

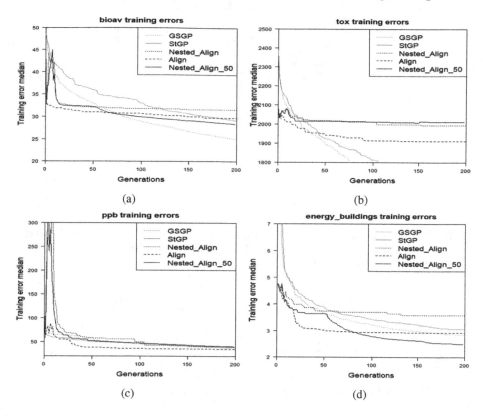

Fig. 1. Results on the *training set*. Comparison between the proposed methods (Nested_Align, Align and Nested_Align_50), standard GP and GSGP. Plot (a): Bioavailability; plot (b): Toxicity; plot (c): PPB; plot (d): Energy.

ESAGP-2); in Figs. 3 and 4, we report the results of the best training model on unseen test data (in Fig. 3 the proposed methods are compared to standard GP and GSGP, while in Fig. 4 they are compared to ESAGP-1 and ESAGP-2); in Fig. 5, we report the results relative to the size of the programs (calculated as the number tree nodes). In Figs. 1 and 3 (i.e. the ones where the proposed methods are compared to standard GP and GSGP), plot (a) reports the results obtained on the Bioavailability problem, plot (b) reports the ones obtained on the Toxicity problem, plot (c) on the PPB problem, and plot (d) on Energy. Concerning the ESAGP methods, we have taken the results directly from [16], for comparison. In that paper, only results relative to training and unseen error on the Bioavailability and Toxicity datasets were made available. For this reason, in Figs. 2 and 4, plot (a) reports the results obtained on the Bioavailability problem and plot (b) reports the ones obtained on the Toxicity problem, and those figures do not contain any other plot. Finally, Table 3 reports the results of the statistical tests performed on the obtained unseen errors.

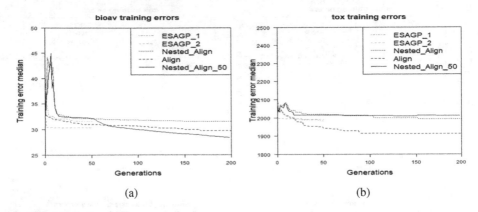

Fig. 2. Results on the *training set*. Comparison between the prososed methods (Nested_Align, Align and Nested_Align_50), ESAGP-1 and ESAGP-2. Plot (a): Bioavailability; plot (b): Toxicity;

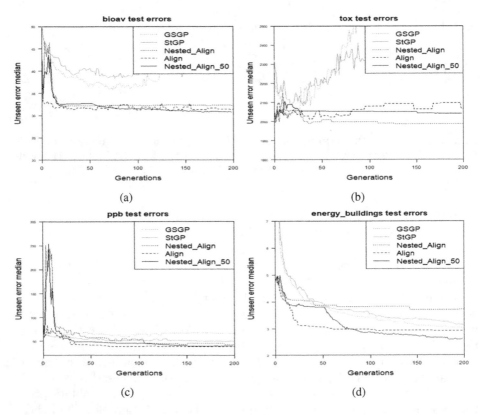

Fig. 3. Results on the *test set*. Comparison between the prososed methods (Nested_Align, Align and Nested_Align_50), standard GP and GSGP. Plot (a): Bioavailability; plot (b): Toxicity; plot (c): PPB; plot (d): Energy.

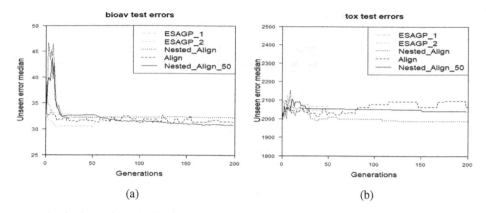

Fig. 4. Results on the *test set*. Comparison between the prososed methods (Nested_Align, Align and Nested_Align_50), ESAGP-1 and ESAGP-2. Plot (a): Bioavailability; plot (b): Toxicity;

Let us begin commenting the results on the training set. As Fig. 1 shows, on the training set Nested_Align_50 is the method that obtains the best results on one problem over four (Energy). On two of the other problems (Bioavailability and Toxicity) the method that was able to find the best results was GSGP. Finally, on the PPB dataset all the methods returned comparable results between each other, with a slight preference for Align. Remembering that, after 50 generations, Nested_Align_50 "turns into" GSGP, our interpretation of these results is that, in general, GSGP is an appropriate method for optimizing training data, which is not surprising, given that GSOs induce a unimodal fitness landscape. In particular, the "switch" between the Nested_Align algorithm and GSGP at generation 50 seems beneficial in much of the cases. This can be seen in the Bioavailability and Energy problems, where a rapid improvement of the curve of Nested_Align_50, looking like a sudden descending "step", is clearly visible at generation 50. So, given that in the last part of the runs Nested_Align_50 and GSGP are identical, Nested_Align_50 prevails if the initial phase in which Nested_Align was executed was beneficial. On the other hand, GSGP prevails if it was not. From the above discussed results, we can conclude that it is beneficial on one problem, while it is not on two others (and it is irrelevant in the fourth of the studied problems, where Nested_Align_50 and GSGP perform comparably). Concerning a comparison between the proposed methods and ESAGP-1 and ESAGP-2 (Fig. 2), two considerations have to be done: first of all, in [16] results were reported only until generation 50, and those are the only ESAGP-1 and ESAGP-2 results in our possession. Secondly, it is possible to "speculate" that both ESAGP-1 and ESAGP-2 are outperformed by other methods both on the Bioavailability and on the Toxicity datasets (more in particular, by Nested_Align_50 and Align on Bioavailability and by Align on Toxicity). In fact, even though we cannot be sure because we do not have the data of the last 150 generations, the curve of both the ESAGP methods, after a rapid decrease

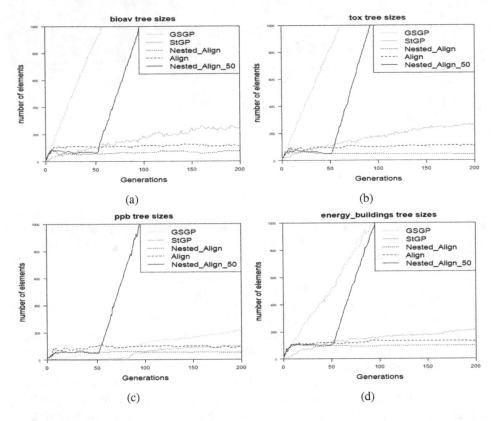

Fig. 5. Results concerning the *program size*. Comparison between the prososed methods (Nested_Align, Align and Nested_Align_50), standard GP and GSGP. Plot (a): Bioavailability; plot (b): Toxicity; plot (c): PPB; plot (d): Energy.

in the first 20 generations, seems to stabilize and to remain practically constant, approximately from generation 20 to generation 50.

Now, let us discuss the results on the test set, starting from Fig. 3. On the Bioavailability, PPB and Toxicity problems, the three proposed methods clearly outperform both GSGP and standard GP, with Nested_Align_50 that is slightly preferable compared to the other two methods on Bioavailability and Align on Toxicity. On the Energy problem, the method that performs better than all the others is Nested_Align_50, and, also on the test set, we can observe a clear fitness improvement, looking like a sudden descending "step", at generation 50, where the switch between Nested_Align and GSGP takes place. In conclusion, on the test set all the three methods that we have introduced in this paper show reasonable results, improving the ones of GSGP and standard GP. Among those methods, Nested_Align_50 seems the most preferable one, corroborating our intuition that Nested_Align learns fast in the beginning, while the switch to GSGP allows us to continue the learning while limiting overfitting. Concerning a comparison between the proposed methods and ESAGP-1 and ESAGP-2 (Fig. 4),

Table 3. p-values of the Wilcoxon rank-sum test on unseen data, under the alternative hypothesis that the samples do not have equal medians. Bold denotes statistically significant values.

Bioavailability

	STGP	Nested_Align	Nested_Align_50	Align	ESAGP-1
GSGP	0.133	**9.33E-05**	**1.02E-05**	**1.21E-04**	**3.18E-05**
STGP		**3.29E-04**	**1.81E-05**	**0.002**	**2.58E-04**
Nested_Align			0.289	0.438	0.624
Nested_Align_50				0.962	0.420
Align					0.646

Toxicity

	STGP	Nested_Align	Nested_Align_50	Align	ESAGP-1
GSGP	0.035	**1.13E-09**	**3.04E-07**	**1.88E-08**	**2.39E-09**
STGP		**3.26E-06**	**1.86E-04**	**5.98E-05**	**1.93E-05**
Nested_Align			0.511	0.307	0.246
Nested_Align_50				0.704	0.678
Align					0.986

PPB

	STGP	Nested_Align	Nested_Align_50	Align
GSGP	0.237	0.153	0.043	**0.001**
STGP		0.474	0.124	**6.98E-04**
Nested_Align			0.359	0.021
Nested_Align_50				0.099

Energy

	STGP	Nested_Align	Nested_Align_50	Align
GSGP	0.109	**1.24E-05**	**1.33E-05**	0.270
STGP		**8.75E-04**	**3.08E-06**	0.023
Nested_Align			**1.26E-08**	**6.03E-06**
Nested_Align_50				**1.01E-04**

what we can conclude using the data at our disposal is that both ESAGP-1 and ESAGP-2 are outperformed by Nested_Align_50 for the Bioavailability problem and by Nested_Align_50 and Nested_Align on the Toxicity problem. However, it is worth pointing out that, when discussing the results on the test set, having data only until generation 50 strongly limits our possible conclusions. In fact, we do not have any information that, later in the run, the ESAGP methods will not begin to overfit, as it happens to, for instance, to Align on the Toxicity problem. Actually, on the Toxicity problem, Align outperforms both ESAGP-1 and ESAGP-2 in the first 50 generations, and only later in the run the test error of Align starts increasing. In synthesis, we consider our conclusions (i.e. that the

ESAGP methods are outperformed by Nested_Align_50 for the Bioavailability problem and by Nested_Align_50 and Nested_Align on the Toxicity problem) the most "optimistic" scenario for the ESAGP methods. If we had the results until generation 200, the picture for ESAGP could be even worse.

We finally discuss Fig. 5, reporting the dimensions of the evolved programs, a very important criterion that has direct link with the models' interpretability [26]. GSGP and Nested_Align_50 generate much larger individuals compared to the other methods. This was expected, given that generating large individual is a known drawback of GSOs [9]. The fact that in the first 50 generations Nested_Align_50 does not use GSOs only partially limits the problem, simply delaying the code growth, that is, after generation 50, exactly as important as for GSGP. On the other hand, it is clearly visible that Align and Nested_Align are able to generate individuals that are smaller than the ones of standard GP. Furthermore, after a first initial phase in which the size of the individuals grow, we can see that Align and Nested_Align basically have no code growth (the curves of these two methods, after an initial phase of growth, are practically parallel to the horizontal axis). Last but not least, in all the studied problems the final models generated by Align and Nested_Align have around only 50 tree nodes.

All this considered, our conclusions are: if we are interested in performance and we can accept models that are "black boxes" (meaning with this, models that are too complicated to be interpreted and understood), then Nested_Align_50 seems the most appropriate of the proposed methods. On the other hand, if the readability of the model is an issue, then Align and Nested_Align are good compromises between performance and model simplicity.

To analyse the statistical significance of the results that we have obtained on unseen data, a set of tests has been performed. The Lilliefors test has shown that the data are not normally distributed and hence a rank-based statistic has been used. The Wilcoxon rank-sum test for pairwise data comparison with Bonferroni correction has been used, under the alternative hypothesis that the samples do not have equal medians at the end of the run, with a significance level $\alpha = 0.05$. The p-values are reported in Table 3, where statistically significant differences are highlighted with p-values in bold. As we can observe, on the Bioavailability and Toxicity datasets the differences between the proposed methods (Align, Nested_Align and Nested_Align_50) and the existing ones (standard GP, GSGP and ESAGP-1) are statistically significant, while the differences of the proposed methods between each other are not statistically significant. The same thing also holds for the Energy dataset, with the only exception of the Align method, whose results are not statistically different from the ones of GSGP and standard GP. The only dataset in which the statistical test gives us a different picture is PPB, where, among the proposed methods, Align is the only one that was able to return results that are statistically different from the ones of GSGP and standard GP.

5 Conclusions and Future Work

Three new genetic programming systems, called Align, Nested_Align and Nested_Align_50, based on the idea of alignment in the error space, were introduced in this paper. These new systems overcome some limitations of the previously existing alignment-based algorithms. On four real-life symbolic regression problems, the proposed systems have outperformed not only the state-of-the-art alignment-based methods, but also standard genetic programming and geometric semantic genetic programming. More specifically, Nested_Align_50 was the method that returned the best results, but Nested_Align_50 also generated very large programs. On the other hand, Align and Nested_Align, although returning results that are slightly worse compared to Nested_Align_50 in terms of accuracy, were able to evolve much smaller programs.

One of the most important limitations of this paper is that only alignments in two dimensions are considered. In other words, the proposed systems use individuals that are pairs of programs and they are only able to search for pairs of optimally aligned programs. Our current research is focused on extending the method to more then two dimensions. For instance, we are currently working on the development of systems that evolve individuals that are triplets of programs, aimed at finding triplets of optimally coplanar individuals. The subsequent step will be to further extend the method, possibly generalizing to any number of dimensions. The design of self-configuring methods, that automatically decide the most appropriate dimension, is one of the most ambitious goals of our current work.

References

1. Koza, J.R.: Genetic Programming: On the Programming of Computers by Means of Natural Selection. MIT Press, Cambridge (1992)
2. Poli, R., Langdon, W.B., Mcphee, N.F.: A Field Guide to Genetic Programming, March 2008
3. Castelli, M., Silva, S., Manzoni, L., Vanneschi, L.: Geometric selective harmony search. Inf. Sci. **279**, 468–482 (2014)
4. Vanneschi, L., Castelli, M., Silva, S.: A survey of semantic methods in genetic programming. Genet. Program Evolvable Mach. **15**(2), 195–214 (2014)
5. Nguyen, Q.U.: Examining semantic diversity and semantic locality of operators in genetic programming. Ph.D. thesis, University College Dublin (2011)
6. Castelli, M., Vanneschi, L., Silva, S.: Semantic search-based genetic programming and the effect of intron deletion. IEEE Trans. Cybern. **44**(1), 103–113 (2014)
7. Krawiec, K., Lichocki, P.: Approximating geometric crossover in semantic space. In: Proceedings of the 11th Annual Conference on Genetic and Evolutionary Computation, GECCO 2009, pp. 987–994. ACM, New York (2009)
8. Pawlak, T.P., Krawiec, K.: Competent geometric semantic genetic programming for symbolic regression and Boolean function synthesis. Evol. Comput. **15**(1), 1–28 (2017)

9. Moraglio, A., Krawiec, K., Johnson, C.G.: Geometric semantic genetic programming. In: Coello, C.A.C., Cutello, V., Deb, K., Forrest, S., Nicosia, G., Pavone, M. (eds.) PPSN 2012. LNCS, vol. 7491, pp. 21–31. Springer, Heidelberg (2012). https://doi.org/10.1007/978-3-642-32937-1_3

10. Vanneschi, L.: An introduction to geometric semantic genetic programming. In: Schütze, O., Trujillo, L., Legrand, P., Maldonado, Y. (eds.) NEO 2015. SCI, vol. 663, pp. 3–42. Springer, Cham (2017). https://doi.org/10.1007/978-3-319-44003-3_1

11. Verel, S., Collard, P., Tomassini, M., Vanneschi, L.: Fitness landscape of the cellular automata majority problem: view from the "olympus". Theor. Comput. Sci. **378**(1), 54–77 (2007)

12. Vanneschi, L., Tomassini, M., Collard, P., Vérel, S., Pirola, Y., Mauri, G.: A Comprehensive View of Fitness Landscapes with Neutrality and Fitness Clouds. In: Ebner, M., O'Neill, M., Ekárt, A., Vanneschi, L., Esparcia-Alcázar, A.I. (eds.) EuroGP 2007. LNCS, vol. 4445, pp. 241–250. Springer, Heidelberg (2007). https://doi.org/10.1007/978-3-540-71605-1_22

13. Castelli, M., Trujillo, L., Vanneschi, L., Popovič, A.: Prediction of energy performance of residential buildings: a genetic programming approach. Energ. Buildings **102**, 67–74 (2015)

14. Castelli, M., Castaldi, D., Giordani, I., Silva, S., Vanneschi, L., Archetti, F., Maccagnola, D.: An efficient implementation of geometric semantic genetic programming for anticoagulation level prediction in pharmacogenetics. In: Correia, L., Reis, L.P., Cascalho, J. (eds.) EPIA 2013. LNCS (LNAI), vol. 8154, pp. 78–89. Springer, Heidelberg (2013). https://doi.org/10.1007/978-3-642-40669-0_8

15. Castelli, M., Vanneschi, L., De Felice, M.: Forecasting short-term electricity consumption using a semantics-based genetic programming framework: the South Italy case. Energ. Econom. **47**, 37–41 (2015)

16. Ruberto, S., Vanneschi, L., Castelli, M., Silva, S.: ESAGP – a semantic GP framework based on alignment in the error space. In: Nicolau, M., Krawiec, K., Heywood, M.I., Castelli, M., García-Sánchez, P., Merelo, J.J., Rivas Santos, V.M., Sim, K. (eds.) EuroGP 2014. LNCS, vol. 8599, pp. 150–161. Springer, Heidelberg (2014). https://doi.org/10.1007/978-3-662-44303-3_13

17. Castelli, M., Vanneschi, L., Silva, S., Ruberto, S.: How to exploit alignment in the error space: two different GP models. In: Riolo, R., Worzel, W.P., Kotanchek, M. (eds.) Genetic Programming Theory and Practice XII. Genetic and Evolutionary Computation, Ann Arbor, pp. 133–148. Springer, Cham (2014). https://doi.org/10.1007/978-3-319-16030-6_8

18. Gonçalves, I., Silva, S., Fonseca, C.M., Castelli, M.: Arbitrarily close alignments in the error space: a geometric semantic genetic programming approach. In: Proceedings of the 2016 on Genetic and Evolutionary Computation Conference Companion, GECCO 2016 Companion, pp. 99–100. ACM, New York (2016)

19. Castelli, M., Manzoni, L., Silva, S., Vanneschi, L.: A comparison of the generalization ability of different genetic programming frameworks. In: 2010 IEEE Congress on Evolutionary Computation (CEC), pp. 1–8. IEEE (2010)

20. Castelli, M., Manzoni, L., Silva, S., Vanneschi, L.: A quantitative study of learning and generalization in genetic programming. In: Silva, S., Foster, J.A., Nicolau, M., Machado, P., Giacobini, M. (eds.) EuroGP 2011. LNCS, vol. 6621, pp. 25–36. Springer, Heidelberg (2011). https://doi.org/10.1007/978-3-642-20407-4_3

21. Vanneschi, L., Castelli, M., Manzoni, L., Silva, S.: A new implementation of geometric semantic GP and its application to problems in pharmacokinetics. In: Krawiec, K., Moraglio, A., Hu, T., Etaner-Uyar, A.Ş., Hu, B. (eds.) EuroGP 2013. LNCS, vol. 7831, pp. 205–216. Springer, Heidelberg (2013). https://doi.org/10. 1007/978-3-642-37207-0_18

22. Castelli, M., Vanneschi, L., Silva, S.: Prediction of the unified Parkinson's disease rating scale assessment using a genetic programming system with geometric semantic genetic operators. Expert Syst. Appl. **41**(10), 4608–4616 (2014)

23. Castelli, M., Vanneschi, L., Silva, S.: Prediction of high performance concrete strength using genetic programming with geometric semantic genetic operators. Expert Syst. Appl. **40**(17), 6856–6862 (2013)

24. Castelli, M., Trujillo, L., Vanneschi, L., Popovič, A.: Prediction of energy performance of residential buildings: a genetic programming approach. Energ. Buildings **102**, 67–74 (2015)

25. Archetti, F., Lanzeni, S., Messina, E., Vanneschi, L.: Genetic programming for computational pharmacokinetics in drug discovery and development. Genet. Program Evolvable Mach. **8**(4), 413–432 (2007)

26. Poli, R., McPhee, N.F., Vanneschi, L.: The impact of population size on code growth in GP: analysis and empirical validation. In: Proceedings of the 10th Annual Conference on Genetic and Evolutionary Computation, GECCO 2008, pp. 1275–1282. ACM, New York (2008)

Short Presentations

Multi-objective Evolution of Ultra-Fast General-Purpose Hash Functions

David Grochol$^{(\boxtimes)}$ and Lukas Sekanina

IT4Innovations Centre of Excellence, Faculty of Information Technology,
Brno University of Technology, Božetěchova 2, 612 66 Brno, Czech Republic
{igrochol,sekanina}@fit.vutbr.cz

Abstract. Hashing is an important function in many applications such as hash tables, caches and Bloom filters. In past, genetic programming was applied to evolve application-specific as well as general-purpose hash functions, where the main design target was the quality of hashing. As hash functions are frequently called in various time-critical applications, it is important to optimize their implementation with respect to the execution time. In this paper, linear genetic programming is combined with NSGA-II algorithm in order to obtain general-purpose, ultra-fast and high-quality hash functions. Evolved hash functions show highly competitive quality of hashing, but significantly reduced execution time in comparison with the state of the art hash functions available in literature.

1 Introduction

Hash functions are highly nonlinear functions assigning a relatively short numerical representation to an arbitrary data record of a predefined structure and size. Hash functions are frequently used in many applications of computer science and engineering such as hash tables, caches and Bloom filters. Hash functions are evaluated with respect to two fundamental properties: (i) quality of hashing – which can be defined in different ways (see Sect. 2.1) and (ii) complexity, which is highly correlated with the execution time. Some additional properties are crucial for the so-called cryptographic hash functions, but this paper only deals with *non-cryptographic* hash functions. As the design of a good hash function is tricky and requires a lot of insight and experience, evolutionary algorithms (genetic programming (GP) in particular) have been employed to accomplish this task.

The existing body of literature dealing with evolutionary design of hash functions is relatively rich; however, except paper [1] none of them is explicitly oriented to the optimization of the time of execution (latency or delay in other words) which becomes crucial in contemporary high end applications such as high speed network monitoring, big data indexing and finding duplicate records.

In the literature, the latency is usually considered as a constraint and the optimization goal is to maximize the quality of hashing. The hash function design problem is then formulated as a single objective design problem.

© Springer International Publishing AG, part of Springer Nature 2018
M. Castelli et al. (Eds.): EuroGP 2018, LNCS 10781, pp. 187–202, 2018.
https://doi.org/10.1007/978-3-319-77553-1_12

In some cases, hash functions are evolved as application-specific functions and evaluated in a very specific environment [1–4], providing thus much better solutions in particular applications than the so called *general-purpose hash functions*. For example, a multi-objective evolutionary design approach focusing not only on the quality of hashing, but also on the execution time has been proposed for network flow hashing [1]. In this case, evolved hash functions had a fixed-size input (96 bits) and consisted of a linear sequence of instructions which is executed just once to obtain the hash.

The goal of this paper is to present and evaluate a *multi-objective evolutionary approach* for the design of high-quality and ultra-fast *general-purpose* hash functions. The main difference with respect to [1] is that the resulting hash functions are capable of accepting multiple k-bit inputs (in order to be general-purpose ones) and the evaluation is performed on various principally different test sets such as randomly generated data, network flow records, passwords and Facebook and Twitter data. The proposed approach is based on *linear genetic programming* (LGP) combined with a multi-objective NSGA-II algorithm, where the objectives are the number of collisions (after embedding the hash function to a hash table) and the execution time. As measuring the real execution time on a particular machine is time consuming (during the evolution), the execution time is estimated according to the number and type of instructions used by a particular candidate hash function. In order to estimate this value for modern processors, a specialized procedure is developed which considers not only the complexity of instructions, but also their scheduling on SIMD architectures. Evolved hash functions are compared in terms of quality of hashing and execution time with 8 human-designed and 2 evolved general-purpose hash functions available in the literature.

The rest of the paper is organized as follows. Section 2 briefly introduces the principles of hash functions and previous work on evolving hash functions. The proposed multi-objective method is introduced in Sect. 3. Section 4 describes our results from the experiments performed in order to evaluate the proposed method and compare resulting hash functions with existing solutions. Conclusions are given in Sect. 5.

2 Related Work

In this section, the principles of hash functions are presented and evolutionary approaches developed to the design of hash functions are briefly surveyed.

2.1 Hash Functions

A *hash function* is a mathematical function h that maps an input binary string (of length k) to a binary string of fixed length (l), $h : 2^k \rightarrow 2^l$, where $k >> l$. The output value is called *hash value* or simply *hash* [5]. The definition of hash function implies the existence of collisions, i.e. $h(x) = h(y)$, where x, y are two input messages such that $x \neq y$. One of desirable properties of hash functions is that similar input vectors produce completely different outputs. This is called the avalanche effect.

The most important application of hash functions is the *hash table* [6]. Based on the key (the input to the hash function) a particular row (index) of the table is activated and data are read/stored from/to a memory slot with that index. In order to handle collisions (different data mapped to the same index), a separate chaining method, cuckoo hashing, coalesced hashing and other techniques have been developed. In the case of the separate chaining method, a list of records having the same hash is operated for each index of the table. A newly entered data record is then stored to the first empty item of the list connected to the particular index. If there is at most one occupied record at index i then the time complexity of lookup is $O(1)$; if n records exist then the complexity is $O(n)$ for the i-th index.

The quality of non-cryptographic hash functions is given in terms of the collision resistance (good hash functions generate a minimum number of collisions), avalanche effect, distribution of outputs, execution time and table load factor (for a given memory size). The hash function is typically called several times in order to obtain desired address because the memory addressing system can be designed as hierarchical, for example, in the cuckoo hashing scheme [7].

2.2 Hash Function Design

Non-cryptographic hash functions are mostly used in hash tables [6]. Other important applications are Bloom filters [8], geometric hashing [9], coherency sensitive hashing [10,11] etc. A common approach to the automatic hash function design is to apply a general construction procedure such as the Merkle-Damgård construction. The literature provides us with various implementations of general-purpose human-created hash functions including DJBHash [12], DEKHash [5], FVN (Fowler-Noll-Vo) [13], One At Time, Lookup3 [14], MurmurHash2, MurmurHash3 [15] and CityHash [16].

Evolutionary approaches have been primarily focused on the non-cryptographic hash function design and evolved with genetic algorithms [17], tree GP [18], linear GP [1], grammar evolution [19] and Cartesian GP [20]. They can further be divided according to the purpose, i.e. either application-specific hash functions [1,21] or general-purpose hash functions [18,22]. The difference lies in the input data size and the evaluation approach. The fitness function is usually based on measuring the avalanche effect [23,24] or the number of collisions [1,22].

3   Multi-objective Linear GP in Hash Function Design

As target hash functions are optimized with respect to the execution time, it is natural to represent them at the level of machine instructions. Hence, linear genetic programming in which candidate programs are represented as sequences of instructions for a register machine [25–27] is employed to evolve hash functions. In order to ensure a multi-objective design, LGP is connected with NSGA-II as introduced in [1]. This section deals with proposed representation and evaluation of candidate hash functions.

3.1   Candidate Program Processing

General-purpose hash functions are typically constructed using instructions such as logical functions (e.g. XOR, AND, OR), addition, multiplication and rotation. These instructions then define the instruction set for LGP. The initial population is generated randomly using these instructions. As the size of the input is arbitrary in the case of general-purpose hashing, it is necessary to partition the input stream into several blocks and process them sequentially. Since the loop responsible for reading the input is always present, it makes no sense to evolve it. We will evolve just the body of the loop. Figure 1 shows that a candidate hash function is called in each iteration to read a new block and combine it with intermediate results obtained from processing the previous blocks. Particularly in this case, 32 bits are copied from the input stream to register r[1] in each iteration. The resulting hash is produced to register r[0].

```
unsigned int candidateProgram (*input){
    r[0] = input[0];

    FOR (i = 1; i < length(input); i++){
        r[1] = input[i];

        <Candidate program>

    }
    return r[0] ⊕ (r[0] >> 32);
}
```

Fig. 1. Framework for candidate program evaluation. In this case, a 32 bit data input is read in each iteration.

3.2   Quality of Hashing

Inspired in [1], the quality of hashing is measured in terms of the number of collisions. Let K_i inputs (keys) be mapped into i-th memory slot by a candidate hash function h. Then the fitness $f(h)$ is defined as the weighted number of collisions:

$$f(h) = \sum_{i=1}^{s} g_i, \text{ where} \tag{1}$$

$$g_i = \begin{cases} 0 & \text{if } K_i \leq 1 \\ \sum_{j=2}^{K_i} j^2 & \text{if } K_i \geq 2 \end{cases} \tag{2}$$

where s is the number of memory slots. This function clearly penalizes candidate hash functions showing many collisions at one slot. The objective is to minimize $f(h)$.

Algorithm 1. Execution time estimation

Input: Candidate program p

Output: The number of used instructions

1 $c \leftarrow$ RotateCodeOutputRegisterLast(p);
2 used-instructions $= 0$;
3 previous-used-instructions $= 0$;
4 used-registers $\leftarrow$ Insert(output-register);
5 **while** *previous-used-instructions $==$ used-instructions* **do**
6   previous-used-instructions $=$ used-instructions;
7   used-instructions $= 0$;
8      $c_p \leftarrow c$;
9   **while** $\langle\ i \leftarrow getLastInstruction(c_p)\ \rangle$ **do**
10   **if** *DestinationRegister(i) $\in$ used-registers* **then**
11   used-registers $\leftarrow$ Insert(source-registers(i));
12   Increment(used-instructions);
13   remove instruction i from c_p;

14 **return** RotateBack(used-instructions);

3.3   Execution Time Estimation

As hash functions are very frequently called in some applications, it is important to optimize them with respect to the execution time. In order to capture features of modern processors supporting the Single Instruction Multiple Data (SIMD) paradigm, a method performing the execution time estimate takes into account not only the number of instructions and their type, but also their eventual parallel processing (which in principle reduces the execution time). In LGP, not all instructions of a candidate program contribute to the result. There are two types of redundant instructions. Firstly, the genotype may contain instructions whose output is not consumed by any other instruction (the so-called structural redundancy). Secondly, there could be instructions used in the phenotype, but not contributing to the resulting value. For example, if the code contains r[5] = r[1] + r[0]; r[5] = r[2] + r[0], the first instruction can be removed. The algorithm developed to estimate the execution time removes unused instructions in the first step and, in the second step, it identifies those instructions that can be executed in parallel.

Because we evolve the body of a loop and the evolved code is executed multiple times, we cannot use the same approach as [1] (i.e. analyzing the algorithm from the last to the first instruction and removing unused instructions) to estimate the execution time. The reason is that unused instructions of one iteration can be important in the next iteration. Hence, Algorithm 1, removing the unused instructions, has more steps. Firstly, the instructions of the candidate program have to be rotated to a state in which the output register of the hash function is at the last position of the program. The program is analyzed in rounds, until all used instructions are not marked. Then unused instructions can be removed. Finally, the resulting code has to be rotated back, because the next step performs instruction scheduling and the order of instructions is important (see Algorithm 1). Example is presented in Fig. 2.

We exploit the instruction level parallelism [28] enabling to process multiple data with a single instruction. Modern CPUs can typically process 256 bits at once which means that eight 64-bit operations can be executed in one instruction instead of executing 4 instructions sequentially. As introduced in [1], instruction

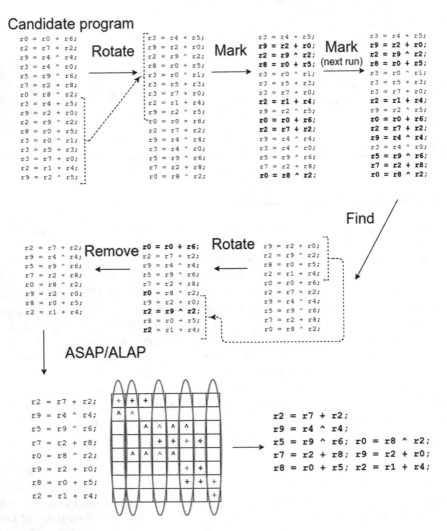

Fig. 2. Removal of unused instructions consists of rotating the candidate program to a configuration in which output register r0 is at position of the last instruction, identifying used instructions (in bold), removing unused instructions and rotating the code back. The optimized code is then scheduled for parallel execution. The final program consists of 5 steps in which 1, 1, 2, 2 and 2 instructions are executed in parallel.

```
unsigned int EvoHash1 (*input){          unsigned int EvoHash2 (*input){
   r[0] = input[0];                          r[0] = input[0];

   FOR (i = 1; i < length(input); i++){      FOR (i = 1; i < length(input); i++){
      r[1] = input[i];                          r[1] = input[i];
      r[8] = r[3] ⊕ r[1];                       r[4] = r[2] ⊕ r[5];
      r[5] = 0xA54FF53A;                         r[2] = r[1] + r[4];
      r[2] = r[5] + r[8];                        r[0] = r[0] + r[4];
      r[4] = r[1] * r[6];                     }
      r[0] = r[2] | r[2];                     return r0 ⊕ (r0 >> 32);
      r[3] = r[4] | r[2];                  }
   }
   return r0 ⊕ (r0 >> 32);
}
```

Fig. 3. Evolved hash functions that were selected from Pareto front in Fig. 4.

scheduling lies in determining when the instructions can be executed based on analyzing dependences among them. The ASAP (As Soon As Possible) and ALAP (As Late As Possible) routines are employed for this purpose. Figure 2 shows that in our example, the optimized 8-instruction program is finally executed in 5 steps in which 1, 1, 2, 2 and 2 instructions are executed in parallel.

3.4   Search Algorithm

A common version of LGP (with tournament selection, single-point crossover and mutation) is combined with NSGA-II [29]. According to [1], the maximum

Table 1. LGP parameters.

Parameter	Value
Population size	100
Crossover probability	90 %
Mutation probability	15 %
Program length	12
Registers count/type	8/64 b – int
Constants	{0x6a09e667, 0xbb67ae85, 0x3c6ef372, 0xa54ff53a, 0x510e527f, 0x9b05688c, 0x1f83d9ab, 0x5be0cd19, 0x428a2f98, 0x71374491}
Instruction set (weight)	{ADD (1), MUL (3), XOR (1), OR (1)}
Tournament size	4
Maximum number of generations	100
Crossover type	One-point

program size is limited to 12 instructions. The function set contains those operations that are typical for the hash function design (XOR, AND, OR, addition, multiplication and right rotation). As multiplication is more complex than the remaining instructions, its execution time is counted with weight 3 in the programs. Common hash functions contain various "magic" constants. We extracted those appearing in the initial phase of hash function SHA-2 [30] and included them to the set of constants available in LGP. The setup for LGP is summarized in Table 1. NSGA-II is employed to find the best trade-offs between the number of collisions (according to Eq. 2) and estimated execution time for a training set (see Sect. 4).

4   Experiments and Results

This section describes the data sets used for evaluation, experiments and their analysis in terms of quality of hashing and execution time. Results will be compared with hash functions from the literature.

4.1   Data Sets

In order to evaluate candidate hash functions on different types of problems, we used (i) randomly generated data and (ii) real-world data coming from network flows, user passwords, and Facebook and Twitter posts.

We randomly generated the training data set (using a random text generator) in such a way that it contains 200,000 vectors with a random size ranging from 16 to 1024 characters. The best-evolved hash functions and the hash functions taken from the literature were then compared using 9 different randomly generated test data sets (Dataset1–9) whose parameters are summarized in Table 2.

In the case of real-world data, data sets Netset1–3 are formed from identifiers of network flows (source and destination IP addresses, source and destination ports and transport protocol). The size of each input vector is 96 bits (see details in [1]). The Passwords data set contains 10 million user passwords. Every passwords consists of 5 to 16 characters. Finally, Facebook and Twitter data sets contain 1 million posts from selected social network groups. These posts are in English, German, Hungarian, Czech and Slovak languages.

4.2   Hash Functions Used for Comparison

Evolved hash functions will be compared with human-created hash function DJBHash, DEKHash, One At Time, Lookup3, FVNHash, Murmur2, Murmur3, CityHash and evolved hash functions available in the literature (GPHash [23,24] and EFHash [22]). A 32-bit hash table is used for testing all functions. A direct comparison with [1] is possible only for the specific data sets used in [1]. Application-specific hash functions (XORhash, NSGAHash1, NSGA-Hash2, NSGAHash3, NSGAHash4, NSGAHash5, NSGAHash6, NSGAHash7 [1]) operate with a 96-bit input and produce a 16 bit hash value. Evolved hash functions produce a 32 bit hash value. The XOR folding is used for reduction from 32 to 16 bits.

Table 2. Data sets.

Name	Number of vectors	Length [bytes]
Dataset1	100,000	64
Dataset2	100,000	128
Dataset3	100,000	256
Dataset4	100,000	512
Dataset5	100,000	1024
Dataset6	100,000	2048
Dataset7	1,000,000	16 − 4096
Dataset8	1,000,000	16 − 4096
Netset1	20,000	12
Netset2	50,000	12
Netset3	100,000	12
Passwords	10,000,000	5 − 16
Facebook	1,000,000	3 − 280
Twitter	1,000,000	3 − 5000

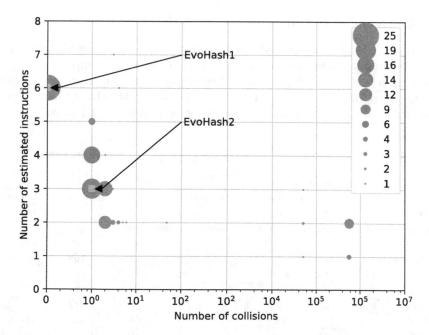

Fig. 4. Pareto fronts obtained from 100 independent runs of LGP. The size of the circle represents the number of identical solutions with the same properties. Selected hash functions (blue squares) are given in Fig. 3. (Color figure online)

4.3   Resulting Pareto Fronts

As we used the same parameters of LGP as [1], we do not report the impact of LGP parameters on the equality of evolution. The main focus is on a comparison of key parameters of evolved hash functions with existing hash functions.

We performed 100 independent runs of our multi-objective LGP and plotted in Fig. 4 parameters of all solutions appearing on the (100) final Pareto fronts. As many identical trade-offs were discovered in several (independent) runs, we plotted them using a circle whose diameter depends on the number of such cases. From all these designs, we selected two the most frequently occurring candidates (blue squares) and analyzed their properties in greater detail. EvoHash1 (see the C code in Fig. 3) produces zero collisions on the training data set, but includes relative many instructions. EvoHash2 (see the C code in Fig. 3) shows the best trade-off between the number of instructions and the number of collisions.

Since there are no clear outliers on Pareto fronts and the designs showing desired trade-offs are represented by larger circles (i.e. there are many good solutions), we can conclude that the proposed algorithm produces stable solutions. It can be seen in Fig. 4 that there are almost no solutions showing $10^1 - 10^4$ collisions. Our explanation for this behavior is that there are only a few discrete points for the second objective (the number of instructions) and these points are already covered by good solutions.

4.4   The Number of Collisions

The hash functions from the literature introduced in Sect. 4.2 were implemented in C programming language and compiled with the same compiler setting as evolved hash functions. All tests were then carried out with these implementations to ensure fair comparisons. The evaluation of all these hash functions was performed on an Intel Xeon E5-2620v3 processor running at 2.4 GHz.

Table 3. The number of collisions for randomly generated data sets.

Hash function	The number of collisions							
	DataSet1	DataSet2	DataSet3	DataSet4	DataSet5	DataSet6	DataSet7	DataSet8
DJBHash	0	3	0	1	1	3	132	116
DEKHash	60004	90000	90000	90000	90000	90000	122	118
FVNHash	0	4	1	1	1	0	115	122
One At Time	1	2	2	2	1	1	108	115
lookup3	1	0	0	2	1	2	122	111
Murmur2	1	1	1	0	3	3	125	126
Murmur3	2	0	2	1	1	3	114	111
CityHash	3	1	1	1	1	0	125	111
GPHash	1	1	1	1	0	0	115	102
EFHash	38137	53488	63353	64983	65119	65209	799933	799825
EvoHash1	2	2	2	1	1	1	133	116
EvoHash2	1	1	0	3	3	1	119	108

Table 3 gives the number of collisions for all randomly generated datasets for a 32 bit hash table. The best values are typed in **bold**; the second best values in ***bold-italic***. It can be seen that hash functions evolved by LGP produce a very similar number of collisions as other hash functions from the literature; except DEKHash and EFHash where many collisions are visible. From the point of view of the number of collisions, evolved hash functions are as good as the other hash functions. The same phenomenon can be observed for real-world data sets (see Tables 4 and 5).

4.5   The Execution Time and Performance

Tables 6, 7, 8 show the average execution time obtained from 50 independent runs of all hash functions on all data sets. The task is to compute a hash value for each vector of a given dataset. The evolved hash functions exhibit the shortest execution time in almost all cases. Similar parameters show Google's CityHash.

Table 4. The number of collisions for network data from [1].

Hash function	The number of collisions		
	NetSet1	NetSet2	NetSet3
DJBHash	2835	15113	48925
DEKHash	2926	15247	49017
FVNHash	2756	14957	48780
One At Time	2821	14988	**48636**
lookup3	2742	15009	48737
Murmur2	2800	15050	48749
Murmur3	2744	*14911*	48763
CityHash	2807	14990	*48647*
XORHash	2864	15011	48575
GPHash	2777	15052	48750
EFHash	5317	25266	63175
NSGAHash1	2923	15677	49336
NSGAHash2	2746	15170	48835
NSGAHash3	2689	15575	49292
NSGAHash4	2692	15010	48715
NSGAHash5	2759	14975	48749
NSGAHash6	*2650*	**14839**	48680
NSGAHash7	**2639**	14975	48650
EvoHash1	2849	15185	48652
EvoHash2	2821	14982	48695

Table 5. The number of collisions for real-world data sets.

Hash function	The number of collisions		
	Passwords	Facebook	Twitter
DJBHash	11663	247	137
DEKHash	14114	357	153
FVNHash	11845	115	115
One At Time	11590	105	138
lookup3	11567	119	107
Murmur2	11637	112	123
Murmur3	11589	103	*89*
CityHash	*11530*	122	122
GPHash	11634	117	113
EFHash	9983806	873270	824153
EvoHash1	11871	*23*	98
EvoHash2	**11469**	**10**	1

Evolved EvoHash2 is slightly faster (4%) than CityHash, but significantly faster (2x) than very popular Murmur hash 3.

Table 7 shows that the application-specific hash functions have a shorter execution time for the network data sets. But evolved hash functions are faster than the best conventional hash functions (CityHash, lookup3).

Finally, we compared all hash functions in terms of throughput that can be obtained by SMHasher [31]. This is a test suite designed to test performance properties of non-cryptographic hash functions. In the Bulk speed test (with

Table 6. The average execution time for randomly generated data sets.

Hash function	Execution time [ms]							
	DataSet1	DataSet2	DataSet3	DataSet4	DataSet5	DataSet6	DataSet7	DataSet8
DJBHash	19.56	32.914	45.311	72.31	126.081	231.675	2556.226	2554.123
DEKHash	12.907	19.352	28.141	46.975	81.419	156.839	1875.878	1872.019
FVNHash	17.354	31.694	48.371	83.761	155.702	294.259	3223.727	3220.844
One At Time	20.208	36.895	57.667	100.993	189.24	360.009	3918.302	3916.603
lookup3	12.867	22.685	28.403	42.581	72.585	125.851	1437.492	1433.961
Murmur2	12.06	20.332	25.718	36.065	60.202	102.426	1195.029	1190.402
Murmur3	12.863	21.622	27.796	40.367	68.557	119.167	1368.135	1363.745
CityHash	10.906	18.591	20.344	24.807	36.806	*54.535*	*683.363*	*679.325*
GPHash	25.497	47.418	80.294	147.286	283.533	550.774	5949.786	5948.746
EFHash	24.394	41.66	69.332	127.822	246.387	479.26	5237.982	5237.599
EvoHash1	*10.383*	*17.084*	*19.056*	*23.897*	*35.508*	55.838	685.604	681.327
EvoHash2	**10.385**	17.411	19.022	23.825	53.132	37.334	**659.185**	**656.647**

Table 7. The average execution time for network data from [1].

Hash function	Time [ms]		
	NetSet1	NetSet2	NetSet3
DJBHash	1.861	5.134	12.724
DEKHash	1.221	4.373	10.407
FVNHash	1.301	4.721	9.633
One At Time	1.769	5.290	12.352
lookup3	0.925	2.891	7.435
Murmur2	1.034	3.095	7.925
Murmur3	1.193	3.215	8.727
CityHash	0.960	2.625	7.407
XORHash	0.838	2.318	6.652
GPHash	1.865	4.671	12.558
EFHash	2.472	13.527	49.495
NSGAHash1	0.529	2.804	8.507
NSGAHash2	*0.527*	**2.072**	6.564
NSGAHash3	**0.514**	2.779	8.492
NSGAHash4	0.530	*2.073*	*6.219*
NSGAHash5	0.534	2.081	6.288
NSGAHash6	*0.527*	2.083	6.249
NSGAHash7	0.547	2.175	6.449
EvoHash1	0.802	2.569	7.455
EvoHash2	0.830	2.825	7.835

262144 byte keys), evolved hash functions EvoHash1 and EvoHash2 outperformed the remaining hash functions (Table 9).

Table 8. The average execution time for real-world data sets.

Hash function	Time [ms]		
	Passwords	Facebook	Twitter
DJBHash	5438.594	17.331	16.726
DEKHash	5067.882	13.240	13.119
FVNHash	5499.328	14.174	12.767
One At Time	6072.904	15.410	13.955
lookup3	4543.399	12.009	10.919
Murmur2	4464.339	11.723	10.774
Murmur3	4573.453	11.955	10.966
CityHash	4385.625	11.149	10.355
GPHash	6389.323	17.966	16.167
EFHash	5101.523	14.304	13.746
EvoHash1	**4268.402**	*10.895*	*9.996*
EvoHash2	*4277.341*	**10.832**	**9.954**

Table 9. Speed test according to SMHasher [31].

Bulk speed test – 262144-byte keys – MiB/sec

Hash function	Alignment							
	0	1	2	3	4	5	6	7
DJBHash	1268.27	1271.40	1271.40	1271.40	1271.40	1271.40	1271.40	1271.40
DEKHash	1906.95	1907.01	1907.02	1907.01	1907.00	1907.06	1907.06	1907.05
FVNHash	953.63	953.63	953.63	953.63	953.63	953.63	953.63	953.63
OneAtTime	634.20	634.12	634.12	634.15	634.14	634.12	634.15	634.14
lookup3	2750.08	2735.18	2735.27	2735.29	2749.80	2735.26	2735.20	2735.14
Murmur2	3813.36	3780.15	3780.15	3780.15	3813.46	3780.25	3780.25	3780.25
Murmur3	7476.99	7332.31	7335.21	7332.47	7333.44	7334.75	7332.51	7334.79
CityHash	*15450.42*	14386.41	14370.53	14389.85	14390.17	14372.77	14385.49	14400.47
GPHash	475.67	475.68	475.68	475.69	475.69	475.68	475.68	475.69
EFHash	543.60	543.59	543.59	543.58	543.60	543.58	543.59	543.59
EvoHash1	15121.84	*14661.90*	*14662.12*	*14663.13*	*14662.58*	*14662.96*	*14662.41*	*14662.68*
EvoHash2	**17578.29**	**16726.21**	**16726.44**	**16725.27**	**16730.33**	**16726.50**	**16727.08**	**16728.04**

5   Conclusions

In this paper, we proposed and evaluated a multi-objective evolutionary design approach in which LGP is combined with NSGA-II algorithm in order to obtain general-purpose, ultra-fast and high-quality hash functions. This proposal addressed current needs of IT industry which seeks for high quality, but ultra fast hash functions. The fitness function was based on (i) the number of collisions with penalization for candidate hash functions producing many collisions and (ii) the execution time.

The best evolved hash functions were compared with 10 hash functions from literature. In terms of quality, evolved hash functions produce almost the same number of collisions as other good hash functions. In terms of the execution time and performance, a hash function improving parameters of a high quality conventional solution (CityHash) was discovered.

Our future work will be devoted to improving the design framework (in terms of supporting other objectives and accelerating the design process) and detailed testing of evolved hash functions in other real-world applications.

Acknowledgments. This work was supported by the Czech science foundation project 16-08565S. The authors would like to thank Dr. Martin Zadnik for his valuable comments to this research.

References

1. Grochol, D., Sekanina, L.: Multiobjective evolution of hash functions for high speed networks. In: Proceedings of the 2017 IEEE Congress on Evolutionary Computation, pp. 1533–1540. IEEE Computer Society (2017)

2. Dobai, R., Korenek, J., Sekanina, L.: Adaptive development of hash functions in FPGA-based network routers. In: 2016 IEEE Symposium Series on Computational Intelligence, pp. 1–8. IEEE Computational Intelligence Society (2016)

3. Kidoň, M., Dobai, R.: Evolutionary design of hash functions for ip address hashing using genetic programming. In: 2017 IEEE Congress on Evolutionary Computation (CEC), pp. 1720–1727. IEEE (2017)

4. Kocsis, Z.A., Neumann, G., Swan, J., Epitropakis, M.G., Brownlee, A.E.I., Haraldsson, S.O., Bowles, E.: Repairing and optimizing hadoop *hashCode* implementations. In: Le Goues, C., Yoo, S. (eds.) SSBSE 2014. LNCS, vol. 8636, pp. 259–264. Springer, Cham (2014). https://doi.org/10.1007/978-3-319-09940-8_22

5. Knuth, D.E.: The Art of Computer Programming, vol. 3 (1973)

6. Maurer, W.D., Lewis, T.G.: Hash table methods. ACM Comput. Surv. (CSUR) **7**(1), 5–19 (1975)

7. Pagh, R., Rodler, F.F.: Cuckoo hashing. In: auf der Heide, F.M. (ed.) ESA 2001. LNCS, vol. 2161, pp. 121–133. Springer, Heidelberg (2001). https://doi.org/10.1007/3-540-44676-1_10

8. Song, H., Dharmapurikar, S., Turner, J., Lockwood, J.: Fast hash table lookup using extended bloom filter: an aid to network processing. SIGCOMM Comput. Commun. Rev. **35**(4), 181–192 (2005), https://doi.org/10.1145/1090191.1080114

9. Lamdan, Y., Wolfson, H.J.: Geometric hashing: a general and efficient model-based recognition scheme (1988)

10. Korman, S., Avidan, S.: Coherency sensitive hashing. IEEE Trans. Pattern Anal. Mach. Intell. **38**(6), 1099–1112 (2016)

11. Datar, M., Immorlica, N., Indyk, P., Mirrokni, V.S.: Locality-sensitive hashing scheme based on p-stable distributions. In: Proceedings of the Twentieth Annual Symposium on Computational Geometry, SCG 2004, pp. 253–262. ACM, New York (2004), https://doi.org/10.1145/997817.997857

12. Bernstein, D.J.: Mathematics and computer science, https://cr.yp.to/djb.html. Accessed 31 Jan 2016

13. Fowler, G., Vo, P., Noll, L.C.: FVN Hash, http://www.isthe.com/chongo/tech/comp/fnv/. Accessed 31 Jan 2016

14. Jenkins, B.: A hash function for hash table lookup, http://www.burtleburtle.net/bob/hash/doobs.html. Accessed 31 Jan 2016

15. Appleby, A.: Murmur hash functions, https://github.com/aappleby/smhasher. Accessed 31 Jan 2016

16. Pike, G., Alakuijala, J.: Introducing cityhash (2011)

17. Safdari, M., Joshi, R.: Evolving universal hash functions using genetic algorithms. In: International Conference on Future Computer and Communication, ICFCC 2009, pp. 84–87, April 2009

18. Estebanez, C., Saez, Y., Recio, G., Isasi, P.: Automatic design of noncryptographic hash functions using genetic programming. Comput. Intell. **30**(4), 798–831 (2014)

19. Berarducci, P., Jordan, D., Martin, D., Seitzer, J.: Gevosh: using grammatical evolution to generate hashing functions. In: MAICS, pp. 31–39 (2004)

20. Widiger, H., Salomon, R., Timmermann, D.: Packet classification with evolvable hardware hash functions – an intrinsic approach. In: Ijspeert, A.J., Masuzawa, T., Kusumoto, S. (eds.) BioADIT 2006. LNCS, vol. 3853, pp. 64–79. Springer, Heidelberg (2006). https://doi.org/10.1007/11613022_8

21. Kaufmann, P., Plessl, C., Platzner, M.: EvoCaches: application-specific adaptation of cache mappings. In: Adaptive Hardware and Systems (AHS), pp. 11–18. IEEE CS (2009)

22. Karasek, J., Burget, R., Morský, O.: Towards an automatic design of non-cryptographic hash function. In: 2011 34th International Conference on Telecommunications and Signal Processing (TSP), pp. 19–23. IEEE (2011)

23. Estébanez, C., Hernández-Castro, J.C., Ribagorda, A., Isasi, P.: Finding state-of-the-art non-cryptographic hashes with genetic programming. In: Runarsson, T.P., Beyer, H.-G., Burke, E., Merelo-Guervós, J.J., Whitley, L.D., Yao, X. (eds.) PPSN 2006. LNCS, vol. 4193, pp. 818–827. Springer, Heidelberg (2006). https://doi.org/10.1007/11844297_83

24. Estebanez, C., Hernandez-Castro, J.C., Ribagorda, A., Isasi, P.: Evolving hash functions by means of genetic programming. In: Proceedings of the 8th Annual Conference on Genetic and Evolutionary Computation, pp. 1861–1862. ACM (2006)

25. Brameier, M., Banzhaf, W.: Linear Genetic Programming. Springer, New York (2007). https://doi.org/10.1007/978-0-387-31030-5

26. Oltean, M., Grosan, C.: A comparison of several linear genetic programming techniques. Complex Syst. **14**(4), 285–314 (2003)

27. Wilson, G., Banzhaf, W.: A comparison of cartesian genetic programming and linear genetic programming. In: O'Neill, M., Vanneschi, L., Gustafson, S., Esparcia Alcázar, A.I., De Falco, I., Della Cioppa, A., Tarantino, E. (eds.) EuroGP 2008. LNCS, vol. 4971, pp. 182–193. Springer, Heidelberg (2008). https://doi.org/10.1007/978-3-540-78671-9_16

28. Wall, D.W.: Limits of Instruction-level Parallelism, vol. 19. ACM, New York (1991)

29. Deb, K., Agrawal, S., Pratap, A., Meyarivan, T.: A fast elitist non-dominated sorting genetic algorithm for multi-objective optimization: NSGA-II. In: Schoenauer, M., Deb, K., Rudolph, G., Yao, X., Lutton, E., Merelo, J.J., Schwefel, H.-P. (eds.) PPSN 2000. LNCS, vol. 1917, pp. 849–858. Springer, Heidelberg (2000). https://doi.org/10.1007/3-540-45356-3_83

30. NIST: Secure hashing, https://csrc.nist.gov/projects/hash-functions. Accessed 10 Oct 2017

31. Appleby, A.: Smhasher, https://github.com/aappleby/smhasher. Accessed 1 Nov 2017

A Comparative Study on Crossover in Cartesian Genetic Programming

Jakub Husa[1]($\boxtimes$) and Roman Kalkreuth[2]

[1] Faculty of Information Technology, Brno University of Technology,
Brno, Czech Republic
ihusa@fit.vutbr.cz

[2] Department of Computer Science, TU Dortmund University, Dortmund, Germany
Roman.Kalkreuth@tu-dortmund.de

Abstract. Cartesian Genetic Programming is often used with mutation as the sole genetic operator. Compared to the fundamental knowledge about the effect and use of mutation in CGP, the use of crossover has been less investigated and studied. In this paper, we present a comparative study of previously proposed crossover techniques for Cartesian Genetic Programming. This work also includes the proposal of a new crossover technique which swaps block of the CGP phenotype between two selected parents. The experiments of our study open a new perspective on comparative studies on crossover in CGP and its challenges. Our results show that it is possible for a crossover operator to outperform the standard $(1 + \lambda)$ strategy on a limited number of tasks. The question of finding a universal crossover operator in CGP remains open.

Keywords: Cartesian Genetic Programming · Crossover
Comparative study

1 Introduction

Genetic Programming (GP) as popularized by Koza [1–3] uses syntax trees as program representation. Cartesian Genetic Programming (CGP) as introduced by Miller et al. [4] offers a novel graph-based representation which in addition to standard GP problem domains, makes it easy to be applied to many graph-based applications such as electronic circuits, image processing, and neural networks. CGP is mainly used with mutation as the only genetic operator. The reason for this is that previous work on crossover in CGP has provided mixed results and comparative results about the use of crossover are missing.

Tree-based GP was originally introduced with a sub-tree crossover technique which swaps randomly chosen sub-branches of the parent trees to produce new offsprings. Koza considered crossover as the dominant genetic operator as a result of his experiments [2,3]. However, later research with more comprehensive and detailed experiments found that the beneficial effects of crossover cannot be generalized in GP [5–7].

© Springer International Publishing AG, part of Springer Nature 2018
M. Castelli et al. (Eds.): EuroGP 2018, LNCS 10781, pp. 203–219, 2018.
https://doi.org/10.1007/978-3-319-77553-1_13

In contrast to fundamental knowledge about crossover in tree-based GP, the state of knowledge in CGP appears to be comparatively weak. Furthermore, the potential and understanding of crossover in CGP seem to be an open and remaining question. In this paper, we present the results of a first comparative study on crossover in CGP which includes the comparison of formerly proposed crossover techniques. Furthermore, we introduce a new method of crossover for CGP, called Block crossover, which is also investigated in our study.

Section 2 of this paper describes CGP briefly and surveys previous work on crossover in CGP. This section also surveys former attempts of comparative crossover studies in tree-based GP and reviews its contribution to the understanding of GP. In Sect. 3 we introduce our new form of crossover for CGP. Section 4 is devoted to the experimental results of our study and the description of our experiments. In Sect. 5 we discuss the results of our experiments. Finally, Sect. 6 gives a conclusion and outlines future work.

2   Related Work

2.1   Cartesian Genetic Programming

In contrast to tree-based GP, CGP represents a genetic program via genotype-phenotype mapping as an indexed, acyclic and directed graph. Originally the structure of the graphs was a rectangular grid of N_r rows and N_c columns, but later work also focused on a representation with just one row. The genes in the genotype are grouped, and each group refers to a node of the graph, except the last group which represents the outputs of the phenotype. Each node is represented by two types of genes which index the function number in the GP function set and the node inputs. These nodes are called *function nodes* and execute functions on the input values. The number of input genes depends on the maximum arity N_a of the function set. The last group in the genotype represents the indexes of the nodes which lead to the outputs.

A backward search is used to decode the corresponding phenotype. The backward search starts from the outputs and processes the linked nodes in the genotype. In this way, only active nodes are processed during the evaluation. The number of inputs N_i, outputs N_o and the length of the genotype is fixed. Every candidate program is represented with $N_r * N_c * (N_a + 1) + N_o$ integers. Even when the length of the genotype is fixed for every candidate program, the length of the corresponding phenotype in CGP is variable which can be considered as a significant advantage of the CGP representation.

CGP traditionally operates with a $(1+\lambda)$ evolutionary algorithm (EA) in which λ is often chosen with a size of four. The new population in each generation consists of the best individual of the previous population and the λ created offspring. The breeding procedure is mostly done by a point mutation which creates offsprings by changing a small number of randomly selected genes from the parent genotype to a random value within the permissible range. One of the most important techniques is a special rule for the selection of the new parent. In the case when two or more individuals can serve as the parent, an individual which has not served as the parent in the previous generation will be selected as a new parent. This strategy is important because it ensures the diversity of the population and has been found highly beneficial for the search performance of CGP.

2.2   Previous Work on Crossover in CGP

Some of the first experiments on crossover in CGP included the investigation of four variations of crossover which were tested on the simple regression problem $x^2 + 2x + 1$. Clegg et al. [8] reported that all tested variations of crossover techniques influenced the convergence of CGP negatively. In comparison to the mutation-only CGP algorithm, the addition of the crossover techniques hindered the performance of CGP. The crossover techniques were applied to the standard integer-based representation of CGP.

For instance, the genetic material was recombined by swapping parts of the genotypes of the parent individuals or randomly exchanging selected nodes. Clegg et al. [8] stressed that merely swapping the integers (in whatever manner) on a genotypic level in CGP disrupts the performance.

This was the motivation for a new form of crossover which has been introduced by Clegg et al. [8] and is based on a real-valued representation. This variation of CGP represents the graph as a fixed length list of real-valued numbers in the interval [0,1]. The genes are decoded to the integer-based representation with the help of normalization values (e.g. the number of functions or maximum input range). The recombination of two genotypes is performed with a standard Arithmetic crossover operation which uses a random weighting factor and can also be found in the field of real-valued Genetic Algorithms. The experiments of Clegg et al. showed that the new representation in combination with crossover improves the convergence behavior of CGP. However, for the convergence behavior in the later generations, Clegg et al. showed that the use of crossover in real-valued CGP leads to disruptive effects on one of the two tested problems. The improved convergence of the Arithmetic crossover was evaluated in the domain of symbolic regression and has been found useful in this problem domain [8].

Slaný et al. [9] analyzed the fitness landscapes of functional-level CGP on image operator design problems. Slaný et al. analyzed single and multi-point crossover operators. It was demonstrated that the mutation operator and the single-point crossover operator generate the smoothest landscapes for the tested problems.

For a multi-chromosome approach to CGP, Walker et al. [10] investigated a multi-chromosome crossover operator which joins the best chromosome parts from all individuals. This crossover technique was found useful for problems with multiple outputs and independent fitness assignment.

A beneficial effect of crossover in CGP was obtained by the use of an implicit context representation for CGP in which recombination is useful for the Even Parity-3 problem [11].

CGP has been extended for the automatic definition and reuse of functions by Walker et al. [12] and Kaufmann et al. [13]. Kaufmann et al. adopted the module creation mechanisms for a cone- and age-based CGP crossover [13]. Cone-based crossover showed good results for functions with repetitive inner patterns, while age-based crossover excels for randomized inner structures.

Recently, a new form of crossover has been introduced by Kalkreuth et al. [14]. The subgraph crossover recombines random parts of the CGP phenotype of two former selected individuals. This crossover technique has been found beneficial for the performance of CGP on symbolic regression, Boolean functions, and image operator design problems.

To the best of our knowledge, the most recent work on crossover in CGP has been done by Kalkreuth et al. However, while some crossover operators for standard CGP have been introduced and investigated, comprehensive comparative studies are still missing. This has been the motivation for our work.

3   The Block Crossover

The Block crossover is a new method of crossover for standard CGP. The method is mainly inspired by the cone-based crossover of Kaufmann et al. [13] for Embedded CGP, which integrates selected modules of a donor genotype into a receptor genotype. Since Kaufmann et al. have been successful with this crossover technique for specific boolean functions, our motivation for the proposal of the Block crossover is to adapt this mechanism for standard CGP. The Block crossover is also inspired by the subgraph crossover which has been introduced by Kalkreuth [14]. Since CGP suffers from a lack of a diverse and effective set of crossover techniques, the introduction and investigation of new crossover technique is significant.

The Block crossover technique focuses on the one-dimensional representation of CGP where the number of rows is limited to one. Given a previously selected genotypes of two individuals serving as parents, the Block crossover generates a list of all blocks of nodes that meet the following criteria:

- The block contains a desired number of nodes.
- All nodes in the block are directly linked through their inputs or outputs.
- All nodes in the block are part of the genotype's active path.

In our implementation, we have chosen to use blocks consisting of three nodes. To fulfill the other criteria, we have constructed the blocks by evaluating the genotype's active path, and selecting active nodes who's inputs were two distinct nodes and not primary inputs of the genotype. The time complexity of this simple method is linear, and it is performed along with the standard evaluation of the genotype's active path that precedes its evaluation.

The Block crossover then randomly selects one block from each list and swaps them between the genotypes. The position of the nodes transferred as part of the block may change inside the new genotype. However, their mutual links are preserved and the function performed by the block stays the same. Therefore, the created offsprings retain the same active path but performs a new operation. If either parent contains no swap-able blocks, no crossover operation is performed and the offsprings are simply cloned from their parents. The crossover operation is then followed by the standard point mutation.

Figure 1 illustrates the crossover procedure. First the active paths are determined, and the swap-able blocks are stored in the lists M_1 and M_2. Then, two blocks N_1 and N_2 are chosen from their respective lists. In order to produce the first offspring O_1, the first parent P_1 is cloned, and the function nodes inside the selected block N_1, are replaced by nodes from block N_2. Nodes $(2, 5, 6)$ have been moved to position $(2, 3, 4)$, but by maintaining them in the same order within the one-row genotype, we can ensure their mutual connection, and their logical function stay the same. The second offspring O_2 is produced in the same way but the roles of the parents P_1 and P_2 are reversed.

4   Experiments

4.1   Experimental Setup

We have performed experiments on problems from the symbolic regression and Boolean function domains. To evaluate the search performance, we measured the best fitness value found after a predefined number of fitness function evaluations (*best-fitness-of-run*). For all problems, the fitness was to be minimized. Our comparison has focused on four crossover operators. Standard One-point crossover, Subgraph crossover, Arithmetic crossover and our newly proposed Block crossover.

The evolution used a generational model. The initial population was randomly generated. Parent genomes for the next generation are picked using two separate tournaments, which allow for the same individual to be picked multiple times. The parents and a crossover operator are used ot create two offsprings, which are then mutated. This process is repeated until a sufficient number of offsprings has been created. Next generation consists of offsprings and a certain percentage of the best individuals (elites) from the previous generation.

In addition, two more evolutionary setups were added for comparison. The None crossover uses the same evolutionary scheme, but the offsprings it creates are identical clones of their parents, leaving mutation as the only active genetic operator. The $(1 + \lambda)$ setup forgoes the above described setup and implements the traditional CGP algorithm.

Our experiments have focused on examining the following hypothesis.

Hypothesis 1. *The $(1 + \lambda)$ CGP algorithm performs better than the crossover operators in all domains.*

In order to test this hypothesis, we first performed two rounds of meta-evolutionary experiments in order to determine which evolutionary parameters were critical, so that the crossover operators can all use their optimal setting, and be compared in a fair way. The two most important parameters were then subject to a parameter sweep, and for every crossover operator the best performing combination of parameters has been selected for comparison. To classify the significance of our results, we have used the Mann-Whitney U Test, to compare the standard $(1 + \lambda)$-CGP with all other crossover operators.

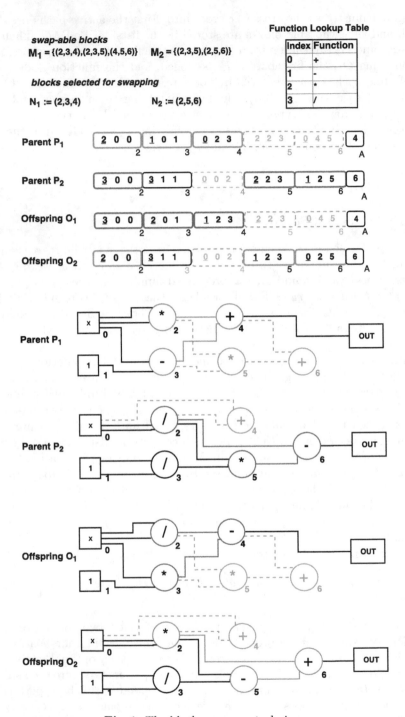

Fig. 1. The block crossover technique.

The implementation was done in Java, using the ECJ Evolutionary Computation Research System. All experiments were performed on a computing cluster with the following hardware configuration: 2 x Intel Xeon E5-2680v3 processor, 2.5 GHz, 12 cores; 128 GB RAM, 5.3 GB cache per core, DDR4@2133 MHz; InfiniBand FDR56 network connection.

4.2   Meta-evolution

For the meta-level, we used a basic canonical GA to tune five parameters we considered most important to the evolutionary process. Meta-evolution is very costly in terms of the computational effort necessary to find an optimal parameter setting. Furthermore, since GP benchmark problems can be very noisy in terms of finding the ideal solution, the evaluation of the evolved individuals is repeated multiple times, with fitness defined as the mean result.

During the first round of meta-evolution, all problems used the same setting, and the evolved parameters have been limited to discrete values, as seen in Table 1. During the second round, the granularity and range were modified to better fit each individual problem. Because the $(1 + \lambda)$ scheme does not use tournament selection nor elitism, the two parameters have been ignored during its meta-evolution.

Table 1. Configuration of the first round of the meta-evolutionary GA.

Property	Setting	Evolved parameter	Possible values
Maximum generations	50	Mutation rate	$0.01 - 0.20$
Population size	10	Elitism rate	$0, 0.05 - 0.50$
Mutation probability	0.5	Population size	$2 - 1024^\star$
Mutation type	Random walk	Genotype length	$2 - 1024^\star$
Tournament size	2	Tournament size	$2 - 1024^\star$
Number of trials	5		

$^\star \{2, 3, 4, 6, 8, 12, 16, 24, 32, 48, 64, 96, 128, 192, 256, 384, 512, 768, 1024\}$.

Results of the first round of meta-evolution have revealed that the tournament size parameter behaves wildly and does not converge to any specific value for any problem nor type of crossover. In some cases, it even significantly outgrew the population size. This caused the tournaments to include the entire population, resulting in a crossover of the best individual with itself, and wholly defeated the purpose of the crossover operator. To prevent this from happening, the tournament size has not been included in the second round of meta-evolution and its value has been fixed to four.

Table 2 shows the results of the second round of meta-evolution which were used to set up the ensuing parameter sweep. Because the computational effort required to perform a parameter sweep grows exponentially with the number of parameters, only the two most important parameters, mutation rate and population size, were included in the sweep.

Table 2. Results of the second round of the meta-evolutionary GA. The table shows the best-performing combination of the four tuned parameters.

Problem	Algorithm	Mutation rate	Elitism rate	Population size	Genotype length
Adder	$(1 + \lambda)$	0.01	–	4	512
	None	0.01	0.08	4	384
	Block	0.01	0.10	4	1536
	Subgraph	0.01	0.06	6	768
	One-point	0.02	0.24	6	96
	Arithmetic	0.025	0.26	6	96
Multiplier	$(1 + \lambda)$	0.05	–	3	24
	None	0.02	0.20	4	96
	Block	0.035	0.22	4	128
	Subgraph	0.04	0.04	4	64
	One-point	0.035	0.02	4	64
	Arithmetic	0.01	0.06	6	384
Bent	$(1 + \lambda)$	0.14	–	24	128
	None	0.09	0.20	24	512
	Block	0.045	0.22	3	128
	Subgraph	0.04	0.20	12	256
	One-point	0.10	0.24	12	256
	Arithmetic	0.05	0.20	6	256
Resilient	$(1 + \lambda)$	0.07	–	2	64
	None	0.07	0.20	32	2048
	Block	0.12	0.26	3	96
	Subgraph	0.09	0.26	3	96
	One-point	0.035	0.20	192	512
	Arithmetic	0.025	0.28	6	256
Koza-3	$(1 + \lambda)$	0.08	–	24	64
	None	0.15	0.10	64	64
	Block	0.19	0.22	96	32
	Subgraph	0.07	0.20	14	16
	One-point	0.09	0.08	16	24
	Arithmetic	0.12	0.28	12	32
Nguyen-4	$(1 + \lambda)$	0.07	–	24	192
	None	0.05	0.14	192	1024
	Block	0.11	0.08	6	96
	Subgraph	0.17	0.16	32	96
	One-point	0.18	0.16	6	128
	Arithmetic	0.05	0.10	16	128
Nguyen-7	$(1 + \lambda)$	0.13	–	64	32
	None	0.11	0.18	12	96
	Block	0.10	0.10	6	48
	Subgraph	0.22	0.10	6	192
	One-point	0.09	0.28	64	256
	Arithmetic	0.16	0.12	4	48
Pagie-1	$(1 + \lambda)$	0.05	–	2	384
	None	0.10	0.10	64	768
	Block	0.10	0.20	4	1536
	Subgraph	0.09	0.06	4	256
	One-point	0.05	0.08	8	1024
	Arithmetic	0.09	0.22	32	512

The ideal elitism rate was similar across all problems and types of crossover. For the sweep, it has been set to the overall average of 15%. Combined with the fixed tournament size of four, this means that during the sweep, there would be 52.2% chance none of the individuals in a tourney would be elites from the previous generation. The ideal genotype length was highly variable and largely depended on the problem, rather than the type of crossover used. For the sweep, the genotype length was set up individually for each problem.

4.3   Boolean Functions

We have chosen to evolve both single and multiple output Boolean functions. 2-bit digital adder and multiplier were used as our multiple output problems. Former work by White et al. [15] proposed these, as suitable alternatives to the overused parity problems. Their fitness was defined as a hamming distance between the resulting truth table, and the ideal solution. To increase the speed of the evaluation, we have used compressed truth tables.

For single output problems, we used 8-bit bent and 1-resilient Boolean functions. These functions find their use in cryptography, where they can provide an LFSR based key-stream generator of a stream cipher with resistance to linear and correlation attacks [16].

Bent Boolean functions possess the maximum possible degree of nonlinearity, defined as the Hamming distance between the truth table of a given function, and truth tables of all linear function and their negations. For an 8-bit function, maximum degree of nonlinearity is 120 [17]. We defined their fitness, as the difference between its actual degree of nonlinearity and the optimal value.

1-resilient functions are highly nonlinear functions that are balanced and have correlation immunity of the first degree. Balancedness means that the function's truth table contains the same number of ones and zeros. Correlation immunity, means that if the truth table was split in half based on the value of a specific input, the two halves of the truth table would each remain balanced. To the best of our knowledge, the maximum possible nonlinearity of an 8-bit 1-resilient

Table 3. Configuration of the Boolean function parameter sweep.

Property	Adder	Multiplier	Bent	Resileint
Input bits	5	4	8	8
Output bits	3	4	1	1
Genotype length	512	96	256	192
Mutation rate	$0.002-0.02$	$0.005-0.05$	$0.01-0.1$	$0.01-0.1$
Population size	$2-48^{\star}$	$2-48^{\star}$	$2-48^{\star}$	$2-48^{\star}$
Fitness evaluations	10000	5000	2000	5000
Tournament size	2	2	2	2
Percentage of elites	0.15	0.15	0.15	0.15

$^{\star}\{2, 3, 4, 6, 8, 12, 16, 24, 32, 48\}$.

Table 4. Results of the parameter sweep for Boolean functions.

Problem	Crossover type	Mutation rate	Pop. size	Mean fitness	SD	Q1	Median	Q3
Adder	$(1 + \lambda)$	0.010	2	**4.26**	3.3923	1	4	6
	None	0.008	3	6.61^b	3.4638	4	6	9
	Block	0.010	3	6.88^b	3.2358	5	7	8
	Subgraph	0.010	3	6.60^b	3.9029	4	6	9
	One-point	0.014	2	6.99^b	3.5604	4.75	6.5	8.25
	Arithmetic	0.010	3	6.96^b	3.0975	5	7	9
Multiplier	$(1 + \lambda)$	0.035	2	**1.13**	1.0016	0	1	2
	None	0.020	4	2.09^b	1.4777	1	2	3
	Block	0.035	3	2.14^b	1.5441	1	2	3
	Subgraph	0.015	3	1.85^b	1.4240	1	2	3
	One-point	0.020	4	2.03^b	1.4997	1	2	3
	Arithmetic	0.025	2	2.23^b	1.5166	1	2	3
Bent	$(1 + \lambda)$	0.05	2	**2.92**	3.8604	0	0	8
	None	0.06	24	4.10	4.3705	0	4	8
	Block	0.04	8	3.89	4.0098	0	4	8
	Subgraph	0.05	32	4.28	4.1974	0	4	8
	One-point	0.05	16	4.04	3.9182	0	4	8
	Arithmetic	0.05	3	4.88	4.0931	0	8	8
Resilient	$(1 + \lambda)$	0.07	2	16.89	19.6612	4	4	20
	None	0.07	4	5.84^b	5.0667	4	4	4
	Block	0.08	6	6.64^b	5.6916	4	4	4
	Subgraph	0.04	6	6.24^b	5.4627	4	4	4
	One-point	0.09	4	6.12^b	5.2863	4	4	4
	Arithmetic	0.04	3	8.48^a	6.9464	4	4	14

^{a} p-value is less than 0.05. ^{b} p-value is less than 0.01.

function is not known, but it can not be higher than 116 [18]. We defined the fitness, as the difference between the actual degree of nonlinearity and the optimal value, and if the evolved function was not resilient, its fitness was further penalized by 58, half the known limit.

Table 3 shows the setting used for the parameter sweep of Boolean functions. Each setting was run one hundred times, for every problem and type of crossover. All problems used the following function set {AND, OR, XOR, AND with one input inverted}. Because the best performing population size was usually very small, we have reduced the tournament size to two, to avoid repeating the issue from the first round of meta-evolution. For problems where the optimized setting was routinely able to find the ideal solution, we have also reduced the number of fitness function evaluations to get more telling results.

Table 4 shows the results of the parameter sweep. For each problem and crossover operator, we have selected combination of mutation rate and population size which provided the best mean fitness over the hundred runs. Operators that performed significantly different from $(1+\lambda)$ have their mean values marked. The table also shows standard deviation (SD) and three quantiles.

Figure 2 provides visual comparison using box plots. The Arithmetic crossover, originally intended for use in symbolic regression, performs the worst when used for Boolean function design. For adder and multiplier problems, the $(1 + \lambda)$ strategy has significantly surpassed all other approaches. However, for the bent function, there was no statistically significant difference, and for the 1-resilient function, the $(1 + \lambda)$ has performed significantly worse than the other options. Here, even with an optimal setting, some of the runs failed to produce a resilient function, resulting in significant deterioration of the average fitness.

4.4   Symbolic Regression

For symbolic regression, we have chosen four problems from the work of Clegg et al. [8] and McDermott et al. [19] for better GP benchmarks, and the Pagie-1 one problem which has been proposed by White et al. [15] as an alternative to the heavily overused Koza-1 ("quartic") problem. The analytic functions of the problems are shown in Table 5. The training data set $U[a, b, c]$ refers to c uniform random samples drawn from a to b inclusive and $E[a, b, c]$ refers to a grid of points evenly spaced with an interval of c, from a to b inclusive.

The fitness of the individuals was represented by a cost function value, defined as the sum of the absolute differences between the correct function values and

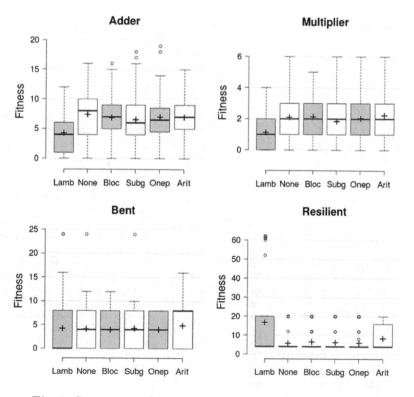

Fig. 2. Comparison of crossover operators for Boolean functions.

Table 5. Symbolic regression problems used in the experiment.

Problem	Objective function	Vars	Training set
Koza-3	$x^6 - 2x^4 + x^2$	1	U[−1, 1, 20]
Nguyen-4	$x^6 + x^5 + x^4 + x^3 + x^2 + x$	1	U[−1, 1, 20]
Nguyen-7	$ln(x + 1) + ln(x^2 + 1)$	1	U[0, 2, 20]
Pagie-1	$1/(1 + x^{-4}) + 1/(1 + y^{-4})$	2	E[−5, 5, 0.4]

Table 6. Configuration of the symbolic regression parameter sweep.

Property	Koza-3	Nguyen-4	Nguyen-7	Pagie-1
Genotype length	48	128	128	512
Mutation rate	0.02 − 0.2	0.02 − 0.2	0.02 − 0.2	0.02 − 0.2
Population size	4 − 96[★]	4 − 96[★]	4 − 96[★]	4 − 96[★]
Fitness evaluations	10000	10000	10000	10000
Tournament size	4	4	4	4
Percentage of elites	0.15	0.15	0.15	0.15

★ {4, 6, 8, 12, 16, 24, 32, 48, 64, 96}.

Table 7. Results of the parameter sweep for symbolic regression.

Problem	Crossover type	Mutation rate	Pop. size	Mean fitness	SD	Q1	Median	Q3
Koza-3	$(1 + \lambda)$	0.16	24	0.0664	0.0774	0.0119	0.0504	0.0839
	None	0.12	16	0.0642	0.0815	0.0092	0.0376	0.0849
	Block	0.06	12	0.0636	0.0755	0.0200	0.0455	0.0784
	Subgraph	0.16	64	0.0692	0.0819	0.0168	0.0483	0.0852
	One-point	0.14	96	0.0617	0.0640	0.0154	0.0333	0.0878
	Arithmetic	0.12	12	**0.0435**	0.0405	0.0146	0.0311	0.0764
Nguyen-4	$(1 + \lambda)$	0.12	6	**0.3120**	0.2658	0.1574	0.2478	0.3745
	None	0.10	8	0.3307	0.2326	0.1672	0.2865	0.4130
	Block	0.08	16	0.3485	0.2800	0.1800	0.2850	0.4125
	Subgraph	0.10	6	0.3709[a]	0.2692	0.1940	0.3393	0.4652
	One-point	0.10	12	0.3282	0.2351	0.1579	0.2864	0.4219
	Arithmetic	0.08	8	0.3231	0.2305	0.1560	0.2558	0.4318
Nguyen-7	$(1 + \lambda)$	0.18	64	**0.6722**	0.4215	0.4364	0.5935	0.7682
	None	0.10	6	0.6871	0.3736	0.4464	0.6055	0.8073
	Block	0.12	24	0.7601[a]	0.3352	0.5522	0.7101	0.9163
	Subgraph	0.12	32	0.7724[a]	0.4461	0.5090	0.7155	0.9613
	One-point	0.16	16	0.7136	0.3741	0.4405	0.6984	0.8439
	Arithmetic	0.14	6	0.8132[a]	0.4978	0.5502	0.7027	0.8288
Pagie-1	$(1 + \lambda)$	0.08	8	130.8812	48.2214	93.7397	122.7972	160.8945
	None	0.06	96	134.5053	46.6960	96.3143	140.6268	170.5598
	Block	0.04	96	126.1124	45.7809	87.4737	122.3703	161.1563
	Subgraph	0.08	64	150.4739[b]	46.9169	115.0119	161.9589	181.6550
	One-point	0.06	8	130.6106	48.9600	96.4414	122.4861	169.5678
	Arithmetic	0.04	8	**120.1536**	45.7169	84.6019	114.4325	152.5632

[a] p-value is less than 0.05. [b] p-value is less than 0.01.

the values of an evaluated individual. The configuration of the experiment is shown in Table 6. All problems used the following set of mathematical functions $\{+, -, *, /, \sin, \cos, \ln(|n|), e^n\}$.

Table 7 shows the parameter sweep results. Same as before, the primary selected criterion was the best average fitness over one hundred runs. Crossover operators that performed significantly different from $(1 + \lambda)$ have their mean values marked. As can be seen in Fig. 3, the arithmetic crossover performs very well, when used for symbolic regression, as originally designed.

5 Discussion

In our meta-evolutionary experiments, we dealt with significant problems in order to make a fair comparison. We were able to determine optimal parameter settings for the $(1 + \lambda)$-CGP as the tuning consists of only three parameters: population size, mutation rate, and genotypic length. However, determining optimal parameter settings for the canonical crossover algorithms is more complex. There are three additional parameters to contend with: crossover rate, elitism rate, and tournament size, which makes obtaining an optimal parameter setting for the respective problems significantly more difficult.

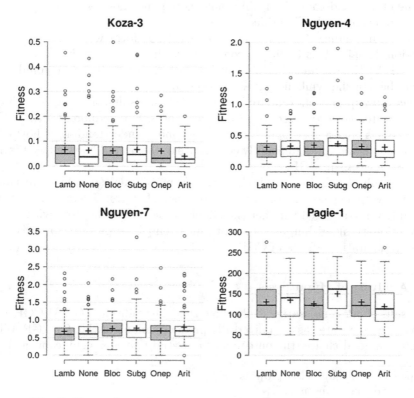

Fig. 3. Comparison of crossover operators for symbolic regression.

Furthermore, former studies on the traditional $(1 + \lambda)$-CGP algorithm have shown that large genotypes are very effective for the performance of CGP for certain problems. Consequently, we have to deal with a big parameter space in CGP in order to determine the optimal parameters and to make a fair comparison.

For this paper, we only used the meta-evolution framework of the Java Evolutionary Computation Toolkit (ECJ)[1]. However, we think that including other *state-of-the-art* methods for parameter tuning of evolutionary algorithms, like Iterated Race for Automatic Algorithm Configuration (IRace)[2] or Sequential-parameter-optimization (SPOT)[3], can provide more insight into well-performing algorithm settings in CGP, and help to provide fair and profound comparisons.

Another point which should be discussed is the observation that each type of crossover works best with different settings. Our findings indicate that there exists no general parameterization pattern for CGP when the crossover is in use. We think it should be investigated if there are similar behaviors like exploration abilities which could be obtained by fitness and search space analyses.

The results of our study show that the parameter settings vary for different problems in the respective problem domain, and indicate that there is no general pattern to parametrize the $(1+\lambda)$-CGP in a well-performing way. These findings also open up a new question, which conditions or types of problems have the need for bigger or smaller population sizes. A preliminary assumption could be that the fitness landscape of certain problems requires more exploration abilities in order to overcome local optima.

Our results indicate that bigger populations perform well in the symbolic regression domain. This finding is consistent with a recent study on mutation-only CGP by Kaufmann et al. [20] which also indicates that bigger populations perform best in the symbolic regression domain.

Since our experiments validate Kaufmann et al. results, this behavior should be investigated through more detailed experiments. Furthermore, we think that these findings offer a good opportunity to get more understanding of how CGP works in detail and can significantly contribute to the overall knowledge of fitness landscape analysis in CGP.

Specifically, Kaufmann et al. show that a mutational $(\mu + \lambda)$ evolutionary algorithm with big population size can be very effective. Therefore, we think it should be investigated whether the Block crossover can be used with a $(\mu + \lambda)$ evolutionary algorithm, as a part of our attempts to proceed towards more precise comparative studies in CGP.

5.1   Analysis of Hypothesis

The results of our comparative study show that the traditional $(1+\lambda)$-CGP algorithm can not be stated as the universally predominant algorithm for CGP. While it is often a good choice, the outcome of our study gives a significant evidence

---

[1] https://cs.gmu.edu/~eclab/projects/ecj/.

[2] http://iridia.ulb.ac.be/irace/.

[3] http://www.spotseven.de/category/sequential-parameter-optimization/.

that the $(1+\lambda)$-CGP can not be considered as the most efficient CGP algorithm in the boolean function domain. The experiments on 1-resilient Boolean function proves that the $(1+\lambda)$-CGP may indeed be significantly inferior to the other CGP algorithms.

6   Conclusion and Future Work

The first comprehensive comparative study on crossover in CGP has been proposed. We also proposed a new Block crossover technique, inspired by embedded CGP, for use in standard CGP. We have performed a comparative study using our new crossover technique, two evolutionary methods that only use mutation, and three other crossover operators that have been suggested in the literature. Simple One-point Crossover, Arithmetic crossover, used in the field of real-valued Genetic Algorithms, and Subgraph Crossover that recombines parts of the parent chromosome phenotypes.

We have formulated a hypothesis that the traditional $(1 + \lambda)$-CGP algorithm would not perform significantly worse than the crossover operators. We performed a comparison on eight selected tasks from the areas of Boolean function design and symbolic regression. We have used meta-evolution to determine the most important evolutionary parameters and find common values for the parameters of lower importance.

Next, we have performed a series of parameter sweeps, to determine the settings most suitable for every type of crossover and every task, and performed a comparison. Finally, we have performed a non-parametric statistical test to prove our hypothesis false, and shown that the $(1+\lambda)$-CGP is significantly outperformed by all other approaches, when designing 1-resilient Boolean functions.

Our results show, that it is possible for crossover operators to outperform the standard $(1 + \lambda)$ strategy. However, if both methods have their parameters fine-tuned, the $(1 + \lambda)$ strategy often remains as the overall best strategy. The question of finding a universal crossover operator is CGP therefore remains open.

Our study opens a new perspective on comparative studies on the use of crossover in CGP and its challenges. The experiments with meta-evolution in CGP have shown that it is difficult to obtain well-performing parameter settings for crossover algorithms in CGP.

These results are the first step toward a fair comparison and a more clear understanding of the function of crossover in CGP. Our future work will focus on exploring ways to make comparisons between crossover techniques and algorithms in CGP more fair, including the investigation of suitable parameter optimization techniques for CGP, widening the spectrum of problem domains on which comparison is made, and using crossover operators from other areas. We will especially focus on investigating the possibility of combining the Block crossover with the $(\mu + \lambda)$ evolutionary algorithm, and on exploring the domain of cryptographically significant boolean functions, where the $(1 + \lambda)$ algorithm faces great difficulty.

Acknowledgments. This work was supported by the Czech science foundation project 16-17538S.

References

1. Koza, J.: Genetic programming: a paradigm for genetically breeding populations of computer programs to solve problems. Technical Report STAN-CS-90-1314, Department of Computer Science. Stanford University, June 1990
2. Koza, J.R.: Genetic programming: on the programming of computers by means of natural selection. MIT Press, Cambridge (1992)
3. Koza, J.R.: Genetic Programming II: Automatic Discovery of Reusable Programs. MIT Press, Cambridge (1994)
4. Miller, J.F., Thomson, P.: Cartesian genetic programming. In: Poli, R., Banzhaf, W., Langdon, W.B., Miller, J., Nordin, P., Fogarty, T.C. (eds.) EuroGP 2000. LNCS, vol. 1802, pp. 121–132. Springer, Heidelberg (2000). https://doi.org/10. 1007/978-3-540-46239-2_9
5. Luke, S., Spector, L.: A Comparison of Crossover and Mutation in Genetic Programming. In: Proceedings of the Second Annual Conference on Genetic Programming 1997, pp. 240–248. Morgan Kaufmann, Stanford University, CA, USA, 13–16 July (1997)
6. Luke, S., Spector, L.: A revised comparison of crossover and mutation in genetic programming. In: Proceedings of the Third Annual Conference on Genetic Programming 1998, pp. 208–213. Morgan Kaufmann, University of Wisconsin, Madison, Wisconsin, USA, 22–25 July (1998)
7. White, D.R., Poulding, S.: A rigorous evaluation of crossover and mutation in genetic programming. In: Vanneschi, L., Gustafson, S., Moraglio, A., De Falco, I., Ebner, M. (eds.) EuroGP 2009. LNCS, vol. 5481, pp. 220–231. Springer, Heidelberg (2009). https://doi.org/10.1007/978-3-642-01181-8_19
8. Clegg, J., Walker, J.A., Miller, J.F.: A new crossover technique for cartesian genetic programming. In: GECCO 2007: Proceedings of the 9th annual Conference on Genetic and Evolutionary Computation, vol. 2, pp. 1580–1587. ACM Press, London, 7–11 July (2007)
9. Slaný, K., Sekanina, L.: Fitness landscape analysis and image filter evolution using functional-level CGP. In: Ebner, M., O'Neill, M., Ekárt, A., Vanneschi, L., Esparcia-Alcázar, A.I. (eds.) EuroGP 2007. LNCS, vol. 4445, pp. 311–320. Springer, Heidelberg (2007). https://doi.org/10.1007/978-3-540-71605-1_29
10. Walker, J.A., Miller, J.F., Cavill, R.: A multi-chromosome approach to standard and embedded cartesian genetic programming. In: GECCO 2006: Proceedings of the 8th Annual Conference on Genetic and Evolutionary Computation, vol. 1, pp. 903–910. ACM Press, Seattle, 8–12 July (2006)
11. Cai, X., Smith, S.L., Tyrrell, A.M.: Positional independence and recombination in cartesian genetic programming. In: Collet, P., Tomassini, M., Ebner, M., Gustafson, S., Ekárt, A. (eds.) EuroGP 2006. LNCS, vol. 3905, pp. 351–360. Springer, Heidelberg (2006). https://doi.org/10.1007/11729976_32
12. Walker, J.A., Miller, J.F.: Evolution and acquisition of modules in cartesian genetic programming. In: Keijzer, M., O'Reilly, U.-M., Lucas, S., Costa, E., Soule, T. (eds.) EuroGP 2004. LNCS, vol. 3003, pp. 187–197. Springer, Heidelberg (2004). https:// doi.org/10.1007/978-3-540-24650-3_17

13. Kaufmann, P., Platzner, M.: Advanced techniques for the creation and propagation of modules in cartesian genetic programming. In: GECCO 2008: Proceedings of the 10th Annual Conference on Genetic and Evolutionary Computation, pp. 1219–1226. ACM, Atlanta, 12–16 July (2008)

14. Kalkreuth, R., Rudolph, G., Droschinsky, A.: A new subgraph crossover for cartesian genetic programming. In: McDermott, J., Castelli, M., Sekanina, L., Haasdijk, E., García-Sánchez, P. (eds.) EuroGP 2017. LNCS, vol. 10196, pp. 294–310. Springer, Cham (2017). https://doi.org/10.1007/978-3-319-55696-3_19

15. White, D.R., McDermott, J., Castelli, M., Manzoni, L., Goldman, B.W., Kronberger, G., Jaskowski, W., O'Reilly, U.M., Luke, S.: Better GP benchmarks: community survey results and proposals. Genet. Program Evolvable Mach. 14(1), 3–29 (2013)

16. Carlet, C.: Boolean functions for cryptography and error correcting codes. Boolean Models Methods Math. Comput. Sci. Eng. 2, 257–397 (2010)

17. Picek, S., Jakobovic, D., Miller, J.F., Batina, L., Cupic, M.: Cryptographic boolean functions: One output, many design criteria. Appl. Soft Comput. 40, 635–653 (2016)

18. Sarkar, P., Maitra, S.: Nonlinearity bounds and constructions of resilient boolean functions. In: Bellare, M. (ed.) CRYPTO 2000. LNCS, vol. 1880, pp. 515–532. Springer, Heidelberg (2000). https://doi.org/10.1007/3-540-44598-6_32

19. McDermott, J., White, D.R., Luke, S., Manzoni, L., Castelli, M., Vanneschi, L., Jaśkowski, W., Krawiec, K., Harper, R., Jong, K.D., O'Reilly, U.M.: Genetic programming needs better benchmarks. In: Proceedings of the 14th International Conference on Genetic and Evolutionary Computation Conference, GECCO 2008, pp. 791–798. ACM, Philadelphia (2012)

20. Kaufmann, P., Kalkreuth, R.: Parametrizing cartesian genetic programming: an empirical study. In: Kern-Isberner, G., Fürnkranz, J., Thimm, M. (eds.) KI 2017. LNCS (LNAI), vol. 10505, pp. 316–322. Springer, Cham (2017). https://doi.org/10.1007/978-3-319-67190-1_26

Evolving Better RNAfold Structure Prediction

William B. Langdon[1(✉)], Justyna Petke[1], and Ronny Lorenz[2]

[1] CREST, Computer Science, UCL, London WC1E 6BT, UK
w.langdon@cs.ucl.ac.uk
[2] Institute for Theoretical Chemistry, University of Vienna, 1090 Vienna, Austria

Abstract. Grow and graft genetic programming (GGGP) evolves more than 50000 parameters in a state-of-the-art C program to make functional source code changes which give more accurate predictions of how RNA molecules fold up. Genetic improvement updates 29% of the dynamic programming free energy model parameters. In most cases (50.3%) GI gives better results on 4655 known secondary structures from RNA_STRAND (29.0% are worse and 20.7% are unchanged). Indeed it also does better than parameters recommended by Andronescu, M., et al.: Bioinformatics **23**(13) (2007) i19–i28.

Keywords: Genetic improvement · Genetic algorithms
Genetic programming · Software engineering · SBSE
Software maintenance of empirical constants · Bioinformatics
Local search · Genomic and phenotypic tabu restrictions
Genetic repair

1 Background: RNA, Genetic Improvement, RNAfold

The central dogma of biology [2] is essentially about the flow of information in all forms of life. In its simple form it says that this fundamental information is transcribed from DNA into messenger RNA, which in turn is translated into protein. Like DNA, RNA is a long chain biomolecule composed of 4 bases (A, C, G and U). An RNA molecule's sequence of bases is known as its primary structure. Much of the interesting biology occurs when RNA is a single strand (unlike the more stable double stranded DNA helix). Like DNA the four bases can form relatively weak temporary bonds with their complementary base. (E.g., C pairs with G, and A with U.) How an RNA chain folds up on itself to form these complementary pairing is known as its secondary structure (e.g. diagrams (1), (2) and (3) in Fig. 1). The tertiary, three dimensional structure, in turn relies on the secondary structure. (See solid colour in diagram (4) in Fig. 1 for two examples.) In people about $\frac{3}{4}$ of the DNA is transcribed into RNA but less than 3% is translated into protein. Other than conveying information for protein manufacture, there are some well known biological uses of RNA. E.g., enzymes which catalyse reactions between biomolecules. Also some transcribed RNA regulates

© Springer International Publishing AG, part of Springer Nature 2018
M. Castelli et al. (Eds.): EuroGP 2018, LNCS 10781, pp. 220–236, 2018.
https://doi.org/10.1007/978-3-319-77553-1_14

MCC = -0.008222   MCC = 0.856324

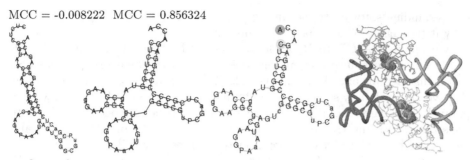

1 Original   2 genetic programming   3 true RNA_STRAND   4 three dimensional

Fig. 1. (1), (2) and (3) are secondary structures (i.e. folding patterns) for RNA molecule PDB_01001. (1) Prediction made original RNAfold does not match well true structure 3. For example the highlighted hairpin loop (red) is not in 3. (2) Prediction made with parameter changes given in Sect. 3.3. (3) True structure. (4) Three dimensional structure. Two (blue, orange) RNA molecules in a Yeast protein complex [1, Fig. 2. A]. (MCC explained on page 7.)

gene expression. Much of the chemistry of biomolecules is governed by their three dimensional shape. These areas are relatively new, and this, and other uses of RNA, have sparked renewed interest in RNA and its structure.

While tertiary structure prediction for RNA is still in its infancy and is limited to very small molecules, the hierarchical nature of RNA folding allows one to infer most of an RNA molecule's function from its secondary structure. Computer programs have had some success at predicting RNA secondary structure, i.e., the folding patterns of real RNA molecules (see Fig. 1). Mostly these are based on estimating the free energy associated with each possible secondary structure using dynamic programming and assuming the molecule will adapt the structure with the lowest energy. In principle, considering all possible RNA folding patterns is not feasible, but many patterns can be discarded as not being biologically plausible. For example, the structure of many RNA molecules is known and very few known structures have knots. Indeed, in RNA molecules of known structure, on average 95% of the structure is also free of pseudo-knots [3, Table 1]. It is common for structure prediction software to assume that RNA contains no knots [4]. Such dynamic programming based approaches scale approximately as $O(n^3)$, where n is the number of bases in the RNA molecule. (In [5] we showed great savings can be made by running such algorithms on low cost parallel GPUs.) RNAfold [6] is the widely used de facto state-of-the-art in RNA secondary structure prediction. It is a key component of the popular internet based medical research game EteRNA [7]. However RNAfold is only as good as its underlying model allows. For example it assumes only standard RNA base-to-base binding are possible. (In panel 3 of Fig. 1 the red line (g↔A) indicates a non-standard RNA base-to-base binding.)

Grow and graft genetic programming (GGGP) [5,8–13] builds on genetic improvement (GI) [14,15]. GI has been used to improve the performance of existing software, e.g. by reducing runtime [16], energy [17] and memory footprint [18], but (excluding software transplanting [19] and automatic bug repair [20]) typically it tries not to change programs' outputs.

We applied GGGP to RNAfold's C code [21]. Using traditional methods to identify performance critical components, recoding them using Intel's SSE vector instructions and then using GP [22–24] to further improve the new code. However, evolutionary search found only small increments on the human written parallel code. Nevertheless, the manually written code has been included into the standard ViennaRNA package since version 2.3.5 (14 April 2017)[1]. It is also being used by the EteRNA development team internally [25].

After speeding up RNAfold by 30% [21], the next stage was to apply GGGP to improve the accuracy of RNAfold's predictions. In technical report [26] we applied GP to the C source code and obtained a small improvement, whereas here we apply it directly to the (internal) parameters of RNAfold's dynamic programming RNA energy model. Notice here and (in the tech report) we allow (nay encourage, require) evolution to change the output of the program. I.e. to make functional changes.

See the references for introductions to GP and GGGP in particular. The next section describes our variable length linear GP system. We train it on a subset of known RNA structures from RNA_STRAND [3]. Whilst Sect. 3 describes the results of applying GP to RNAfold and show the improvements generalise to unseen RNA molecules. We conclude (Sect. 4) that evolution can improve prediction of RNA secondary structure and potentially Genetic Improvement could be widely applied to legacy chemical, physical and Bioinformatics [27] software containing empirically generated constants since maintaining such constants often lags behind knowledge in the model's target domain.

2   Genetic Improvement System

In earlier analysis [21, 26] we had established that RNAfold uses dynamic programming to both calculate the minimum free energy of each RNA molecule's secondary structure and the structure itself. To do this it uses numeric constants which specify different aspects of the energy calculation. E.g. the binding energy between C and G bases and how tightly RNA can fold up on itself to form hairpin loops (an example hairpin loop is shown in red in the first RNA structure in Fig. 1). These parameters are held in C strings (4), `float` (1) and `int` (51 521) variables. For simplicity we only allow evolution to change the `int` values. (Our approach is summarised in Table 1.) The `int` values are stored in 31 named variables and arrays, see Table 2. Also profiling with GNU gcov [26] we had showed that all 31 variables were read at some point when RNAfold is run on the training data. Although these variables are derived from others, e.g. to compensate for changes in temperature, they are the ones directly used by dynamic programming to predict secondary structures.

---

[1] The ViennaRNA package must first be configured with `./configure --enable-sse`.
https://www.tbi.univie.ac.at/RNA/documentation.html.

Table 1. GGGP to improve RNAfold's secondary structure predictions by mutating its 51 521 `int` dynamic programming parameters.

Representation:	Variable length list of 3 types (>, <,+=) of mutations (Sect. 2.3)
Fitness:	Apply mutations in order to the parameters (Table 2) before running RNAfold on training data from RNA_STRAND with less than 155 bases (681 molecules). Compare its answers with the real structure and with the default parameters' answers. Calculate the MCC between the mutated parameters' predictions and the real answers. See Sect. 2.4
Population:	Panmictic, non-elitist, generational. 2000 members
Parameters:	Initial population of random single mutants. 50% truncation selection. 50% two point crossover, 50% mutation. In generations 1–100 half mutations simply append an additional >, < or += gene whilst the others apply creep mutation (±1 to ±5, or ±10 to ±50) to on average at least 20% of replacement values. No size limit

2.1   Representation

Each member of the population is a variable length list of mutations (Sect. 2.3). These are applied one at a time in left to right order. Each mutation applies to one of the 31 variables and arrays in Table 2 but can potentially change many values in it. Once the whole individual has been processed, the final parameter values are loaded into RNAfold, which is then run on a training set of 681 RNA molecules (<155 base pairs long) and its predictions of their structure is compared with their known structure to give the individual's fitness.

2.2   Initial Population

2000 individuals each containing one randomly chosen mutant were created. In later generations, mutations can be changed, one more mutation can be added, and individuals can be recombined using linear two point crossover. The mutations are split approximately equally between the three primary mutation operators. Our new mutation operators are designed to respect the existing characteristics of the energy model's parameters (see following sections and Table 2).

Several of the arrays store parameters that are dependent on each other due to symmetry [28, page 6170]. As far as the code is concerned this is behind the scenes but it reduces the number of independent variables. In particular, interior loop contributions should be symmetric since evaluation should yield the same result no matter from which side you are looking at it. Currently mutation does not enforce this. Effectively we rely on the fitness function. In future perhaps each mutation could enforce symmetry. Alternatively we can envision additional mutation operators which do respect symmetry or indeed mutation operators which remove asymmetry. E.g., by replacing asymmetric pairs by their mean value. Adding more mutation operators, rather than more careful design

Table 2. 31 (10 scalars + 21 arrays) RNAfold parameters which can be optimised. Data structures marked[E] hold energy values which are always multiples of 10. (Mutation ensures they remain multiples of ten.) The original values of Tetraloop_E[E] and Triloop_E[E] are mostly zero [a] and so mutation of Tetraloop_E[E] is limited to the first 15 elements and in Triloop_E[E] to just the first element. NBPAIRS=7 and MAXLOOP=30.

noLP		mismatchM^E	[NBPAIRS+1][5][5]
uniq_ML		mismatchExt^E	[NBPAIRS+1][5][5]
dangles		dangle5^E	[NBPAIRS+1][5]
min_loop_size		dangle3^E	[NBPAIRS+1][5]
rtype	[8]	mismatchH^E	[NBPAIRS+1][5][5]
gquad		stack^E	[NBPAIRS+1][NBPAIRS+1]
special_hp		bulge^E	[MAXLOOP+1]
pair	[21][21]	int11^E	[NBPAIRS+1][NBPAIRS+1][5][5]
noGUclosure		int21^E	[NBPAIRS+1][NBPAIRS+1][5][5][5]
TerminalAU^E		internal_loop^E	[MAXLOOP+1]
MLintern^E	[NBPAIRS+1]	ninio[2]^E	
MLclosing^E		mismatch1nI^E	[NBPAIRS+1][5][5]
MLbase		int22^E	[NBPAIRS+1][NBPAIRS+1][5][5][5][5]
hairpin^E	[31]	mismatch23I^E	[NBPAIRS+1][5][5]
Tetraloop_E^E	[200] (15)	mismatchI^E	[NBPAIRS+1][5][5]
Triloop_E^E	[40] (1)		

total 51521 `int`

[a] The energy contributions for Tetraloop and Triloop are only used under special circumstances. They represent tabulated exceptions of small hairpin loops that do not follow the values provided in hairpin. They are only used when the sequences in question match the corresponding patterns stored in the character arrays Tetraloop and Triloop.

of the existing ones, might perhaps be beneficial [29]. Another alternative, which would be more like traditional optimisation, would be to adjust the independent variables directly outside of RNAfold.

2.3 Genetic Search Operators: Mutation and Crossover

To create a new mutation, one of the 31 data structures (Table 2) is chosen uniformly at random. If one of the ten scalars is chosen, it is assigned a new value. Scalars with value 0 or 1 are inverted, those with values of 2 or 3 are give a new value chosen uniformly between 0 and 1 or between 0 and 2 and otherwise it is incremented by a multiple of 10 between −50 and +50 (not zero).

If one of the 21 arrays is chosen, one of the three array mutations (>, <, +=) is chosen uniformly at random.

Replace Values Mutation. > The *array name* value1>value2 mutation operator is interpreted to mean every element of *array name* whose value is currently
value1 is overwritten by value2.

Notice we can build individuals composed of multiple mutations. These are
applied strictly in left-right order.

An array element is chosen uniformly at random and its default value is
noted. Then another element is similarly chosen. If the second value is small
(i.e. 0, 1, ..., or 8) then the second value is used. If it is not small or it is
negative, then 50% of the time it is used and 50% of the time a random energy
value which is a multiple of 10 between -50 and +50 (not zero) is added to it
before it is used. In all cases the second value must be different from the first.

Overwrite Mutation. < The *array name* index<value2 mutation operator is
interpreted to mean every element of *array name* which matches index is overwritten by value2.

Having chosen an array, < next chooses one or more elements in the array.
When the array has multiple indexes (Table 2) each is processed independently.
Half the time every element in that particular index is selected (denoted by *)
and the other half one of the legal indexes is chosen uniformly at random. (It
appears some of the arrays are coded with index 0, i.e. the standard C convention,
but element 0 is never used. Our mutation operator does not take notice of this
and so mutating [0] may be a silent mutation.)

value2 is chosen as in > mutation (see previous section). However, if multiple
parts of the array are to be updated (i.e. there is one or more * in the array
index) then there is no check that value2 if different from the existing value.

Increment Mutation. += The *array name* index+=value2 mutation operator
is interpreted to mean every element of *array name* which matches index is to
be replaced by its default value incremented by value2.

The array index is chosen in the same way as with < mutation. As before
value2 is given by the default of a uniformly random chosen element of the array.
If is small, a value between −5 and +5 (not zero) is chosen, otherwise a multiple
of ten between −50 and +50 (not zero) is chosen as value2.

Creep Mutation. Creep mutation changes the value2 in existing mutations.
Therefore it is not used in the initial generation. In subsequent generations half
the children are created by mutation and half by crossover. Half the mutants are
created by appending an additional mutation of one of the three primary types
(>, <, or +=) to the parent whilst creep mutation is applied to the existing genes
in the parent in the other 50% of the time. Creep mutation is applied uniformly
at random to the existing mutations. As an anti-bloat mechanism, it is applied
at least once and then on average to 20% of existing genes. Note, as individuals
increase in size, they will tend to be modified to a greater extent. (However, this
proved to be insufficient to prevent bloat, see Fig. 3 page 9.)

If the existing value2 is INF (i.e. 10000000) then no change is made. If creep
mutation is applied to a scalar with a current value2 which is small (i.e. ≤3)

then, if its value is 0 it is changed to 1. Otherwise the value is either increased by 1 or by -1.

If value2 is not small or we are dealing with an array then the size of the change to value2 is given by a tangent distribution [30], here $\pm \lfloor \tan(\frac{\pi}{4}(1 + \frac{3}{4}r) \rfloor$ (where r is chosen uniformly at random between 0.0 and 1.0.) This gives a non-uniform chance of ± 1 (54.6%), ± 2 (24.1%), ± 3 (13.0%) ± 4 (8.1%) and very little chance of ± 5 (0.178%).

Tabu: Preventing Genotypic Convergence. As we did in [11,31], we insist that each chromosome in the whole run must be unique. I.e., we impose a genotypic Tabu restriction that the same individual is never created twice. In an effort to prevent bloated individuals side stepping this by adding genes which simply redo previous changes, each individual is reduced to a canonical form. For example, if a scalar is mutated more than once, the newest value2 is used and the earlier genes are removed from the individual. However, (as noted above) this failed to prevent bloat and it turns out that in our implementation, as programs get bigger, reducing in particular crossover to canonical form, gets increasingly time consuming.

2.4 Fitness Function

Each member of the population is interpreted as a series of mutations (previous sections) to give the final values for the parameters to be modified. As mentioned above it is impossible to use `+=` to change INF and so such mutations are ignored. If all mutations are ignored, then the individual is invalid and its fitness is not evaluated and it cannot be a parent of the next generation. Value2 can be increased only up to INF. During evolution, the INF restriction effected 0.2% of individuals.

The released code `RNAfold.c` was tweaked so that before running the dynamic programming code the original parameters of the energy model are overwritten with the mutated values.

The tweaked exe is run on all the training data. I.e., one $1/3^{\rm rd}$ of RNA_STRAND which are less than 155 bases long. This means running the mutant's exe up to 681 times. That is, once for each of the short training RNA molecules. These are the same sequences as we used in [26]. In retrospect this is perhaps too many. For example, in [16] we used just five but these were randomly changed every generation.

RNAfold was run with option `--noPS` to suppress the production of nice pictures of the predicted structure. (The defaults were used for all other options.)

RNAfold produces its prediction as a text string made of nested brackets (to indicate pairs of bases which bind together) and "." (for unbound bases). As we did in [26] this is piped into the standard ViennaRNA (2.3.0) utility `b2ct` which converts the bracket string into .ct file format. The output from `b2ct` is piped into a comparison gawk script which calculates the Matthew's correlation coefficient $MCC = \frac{(TP \times TN - FP \times FN)}{\sqrt{(TP+FP)(TP+FN)(TN+FP)(TN+FN)}}$. Where:

- TP = true positives, number of predicted pairs which are in RNA_STRAND's .ct file.
- TN = true negatives, total number of possible pairings not in TP, FP or FN.
  I.e. $TN = n(n-1)/2 - TP - FP - FN$ (where n is the length of the RNA molecule).
- FP = false positives, number of predicted pairs which are not in RNA_STRAND's .ct file.
- FN = false negatives, number of pairs in RNA_STRAND's .ct file but not in the mutant's prediction.

Naturally, TN tends to be large, hence we follow Lorenz et al. [6] and use Matthew's correlation coefficient as it deals well with large class imbalances [6]. The gawk script also counts the number of cases where the predicted base pair binding is different between the mutated parameters and the default (unmutated) parameters. A mutant must make at least one change to stand a chance of being selected to be a parent.

The average MCC is computed. If it is more than 0.1 worse than the mean MCC calculated for the unmutated parameters, the individual cannot be a parent. The eligible individuals in the current generation are sorted by their average MCC. And the top half are selected to be parents of the next generation.

Tabu: Preventing over Searching the Same Fitness. In order to try and encourage diversity in the evolutionary search, we apply a phenotypic Tabu limit: Each fitness value, i.e. average MCC value, can only be used as a parent $0.01\times$ the population size ($0.01 \times 2000 = 20$) in the whole run. Once this limit has been reached, individuals of exactly this MCC are passed over and individuals with a lower fitness are selected to be parents.

No Sandbox Protection Against Rogue Mutants. Since evolution is not permitted to change any of the code, no particular precautions were taken against badly behaved mutants.

About 2.2% of mutants caused RNAfold to fail, 90% of them with a segmentation error. For example, in the initial generation, all six mutations which change rtype (excluding rtype[0]) to a value outside the range 0..10 cause a segmentation error. Mutants which fail at runtime are not permitted to be parents of the next generation.

3 Results

The variable length representation evolutionary computation GI system was run with a population of 2000 for 100 generations (see Table 1). The training improvement in average MCC is shown in Fig. 2 and together with the evolution of size (bloat) in Fig. 3. The best individual from the last generation had an average MCC on the training set of 0.737044 (RNAfold release 2.3.0 scores 0.663946) and had bloated to size 2849.

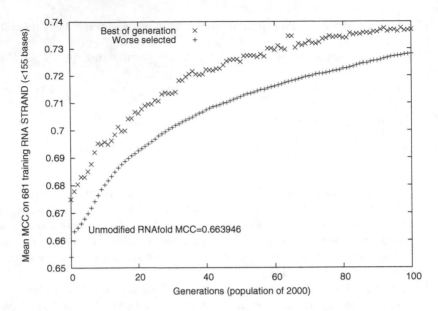

Fig. 2. Evolution of fitness (mean MCC). 1000 children whose fitness lies between best and worst are chosen to be parents of the next generation. (See also Fig. 3.)

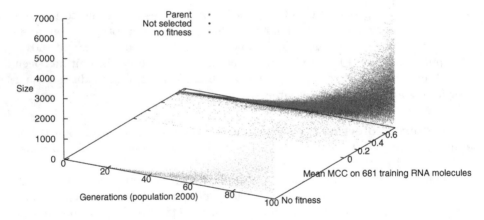

Fig. 3. Evolution of fitness and size (vertical). Children selected to be parents of the next generation shown in red (1000 per gen). Purple 4566 children which failed during fitness evaluation. Blue others also not selected. (Color figure online)

3.1 Post Evolution Tidy

As bloat is common, there is often a post evolution phase where each part of the best individual is tested one at a time to see if that component can be removed without loosing the overall benefit [16]. Weimer et al. [32] use delta debugging to trim their bug fixing patches but we use a simple hill climber. Starting at the front of the best evolved individual, we progressively remove each mutation and test the new individual on the whole training set. If it fails or it performs worse than the evolved individual, the deleted mutation is restored, otherwise it is deleted permanently. Then we test the next gene and so on, until we reach the end of the evolved individual. By which point each part of it has been checked. This reduced the evolved individual from 2849 mutations to 49 and the mean MCC had increased very marginally (by 0.000533) to 0.737577.

Here we ran the hill climbing a second time which further reduced it from 49 mutations to 42, with a final fitness of 0.737752. Again a very slight increase (0.000175) in performance on the training molecules. (No more changes were made when a third pass was tried.)

3.2   Generalisation Performance

The cleaned up individual (i.e. with 42 mutations) retains its performance when tried on similar length RNA molecules not used during training (average MCC 0.737752 training, 0.730137 on 682 holdout examples containing less than 155 bases). Indeed it extrapolates well to the whole of the holdout set (1553 RNA molecules from RNA_STRAND of any length). When including the larger RNA molecules, RNAfold's performance falls (release 2.3.0's mean MCC is 0.541106) but our mutant is still better, mean MCC = 0.568323. Figure 4 (page 12) compares the performance of the new RNAfold against the released code across all 1553 RNA molecules of the holdout set.

RNAfold has the ability (via its -P option) of loading other parameter settings. Andronescu et al. [33] optimised the setting and their "better optimized" values have been included in the ViennaRNA package (v2.3.0) in the file misc/rna_andronescu2007.par. In Fig. 4 we show our 42 mutant GGGP parameters also do better than Andronescu et al. [33] (solid v. dotted line).

3.3   Changes to the Energy Model Parameters

The 42 changes cover 19 of the 31 data structures. All but 2 (ninio [2] and TerminalAU) are arrays. Together they change 14732 int parameters (29% of them all). Table 3 summarises these by data structure. The data structures are sorted by their individual impact in Table 3, but, of course, the changes are interlinked and cannot readily be treated in isolation. Next, we describe a few of the changes which seem to have most impact and try and explain how they work.

mismatchH. Array mismatchH has three indexes (see Table 2). The first, type, is calculated via a look up from the other two. The second, si1, is given by the base after the current active i position along the RNA molecule, the third, sj1, is similarly given by the base before the current active j position. Thus mismatchH *,*,*+=−90 mismatchH *,*,3<−130 mismatchH *,1,2<−80 corresponds to mismatchH[*,A,C] set to −80, mismatchH[*,*,G] set to −130, and all others being reduced by −90.

Table 3. Impact of the 42 components of the cleaned up evolved patches to 51 521 **int** paramters of RNAfold's dynamic programming model of RNA secondary structure. First column: components grouped by data structure (order in group is still significant). 2nd number of **int** changed. 3rd responsibility for fitness change (mutations build on each other, so isolated changes only give an indiaction of their importance). 4th again impact, this time on number of bonds changes across the whole training set. Last column describes changes with impact >2%. See also Sect. 3.3.

internal_loop *+=−40	29	−6.91%	667	Add 40 to internal_loop[2..30] ([0] and [1] are INF and so cannot be incremented)
MLintern *+=10 MLintern 3<−150	8	−3.25%	437	MLintern[0..7] were all −90, now -80 except [3] is −150
ninio[2] 80		−2.50%	501	Was 60 now 80
mismatch23I 70>10000000	108	−1.40%	131	
dangle5 *,*+=60	40	−1.27%	101	
int22 260>80 int22 180>280 int22 *,*,2,*,*,*+=10 int22 280>200 int22 200>10000000				
	10454	0.05%	37	
mismatchI *,*,0<100 mismatchI *,*,1+=−10 mismatchI 2,3,1+=−100 *,4,*+=−40				
	96	0.05%	617	
int11 *,*,*,*<200 int11 6,*,*,2+=−70	1600	1.22%	1306	
dangle3 5,*+=−80	5	1.28%	13	
mismatch1nI 70>110	125	1.89%	173	
TerminalAU 80		3.04%	759	Was 50 now 80
rtype 6<6 rtype 2+=1	2	3.05%	1257	[2] 1←2 and [6] was 5 becomes 6, page 14
mismatchExt *,*,*+=80 mismatchExt *,*,1<−40	200	3.90%	320	+80 is added to all elements, except 1 in 5 is set to −40
stack −100>60 stack −140>0 stack 2,2+=−20 stack *,4<−50				
	14	6.08%	2135	[0,4] 10000000←−50 [1,4] −140←−50 [1,7] −140←0 [2,2] −340←−360 [2,4] −150←−50 [3,5] −140←0 [4,1] −140←0 [4,4] 30←−50 [4,6] −100←60 [5,3] −140←0 [5,4] −60←−50 [6,4] −100←−50 [7,1] −140←0 [7,4] 30←−50
int21 230>260 int21 *,*,*,*,3+=−70 int21 220>10000000	1669	6.51%	287	283 values that were 230 replaced by 260. 161 values of 220 replaced by INF. And 1225 cases (of a possible 1600) where int21[*,*,*,3] is reduced by 70
bulge *+=40	30	7.53%	635	All bulge[1..30] increased by 40. ([0]is INF and so cannot be incremented)
mismatchM −70>−130 mismatchM *,3,*+=20 mismatchM *,1,*+=−40 mismatchM -110>−130 mismatchM *,0,*+=−170 mismatchM -60>−40	142	10.70%	1227	15 cases where −70 is replaced by −130. 2 cases where −110 is replaced by −130. 20 cases where -60 is replaced by −40. 40 cases where [*,0,*] is reduced by −170, 35 [*,1,*] by −40, and 30 [*,3,*] by −40
hairpin *<560	30	14.75%	1217	All hairpin[*] are set to 560 (Fig. 5)
mismatchH *,*,*+=−90 mismatchH *,*,3< −130 mismatchH *,1,2<−80	180	16.30%	1610	39 cases where mismatchH [*,*,3] is set to -130. 8 cases mismatchH [*,1,2] becomes −80 and 133 where other values in mismatchH are reduced by −90
Total:	14732			

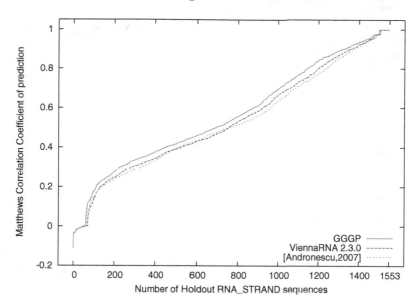

Fig. 4. Performance of best GGGP run (after two hill climbing passes) on 1553 RNA_STRAND molecules not used in training (all lengths). Dashed line performance of unmodified RNAfold 2.3.0 on same molecules. GGGP gives better predictions on 769 RNA molecules, worse 471 and same 313, $p < 10^{--16}$. GGGP also does better than parameters from the RNA_STRAND team [33], RNAfold -P ViennaRNA-2.3.0/misc/rna_andronescu2007.par, dotted line, $p < 10^{--15}$.

mismatchH[*,A,C]← −80 means: where the base after i is A and the base before j is a C, the energy predicted for a hairpin loop is mutated to −80. In both the cases which matter, [6,A,C] and [7,A,C], the hairpin energy was −30. I.e., the hairpin energy has been reduced by −50. Thus GI has made pairs 6,A,C and 7,A,C appear more beneficial by −50.

mismatchH[*,*,G]← −130: in the fifteen cases which matter, mismatchH [(1 4 and 7),*,G], by default holds values from −240 to −10. All 15 are over written with −130.

mismatchH *,*,*+=−90 reduces by −90 elements of mismatchH which were −250 to +20. That is, these 152 elements of mismatchH now have values of -340 to -70.

Since the dynamic programming calculation works on relative changes, it is the differences in changes between all the components of the energy calculation which determine which of the possible RNA folds are taken to be RNAfold's final prediction. In summary, GI changed 180 values in the mismatchH array, which has changed the attractiveness of 169 types of i,j pairings (those adjacent to 6,A,C (1), 7,A,C (1), (1 4 and 7),*,G (15) and 152 others, i.e. *,*,*).

hairpin hairpin (see Table 2) holds the penalty (i.e. a positive value) for forming a loop of a given size. Loops longer than 30 lie outside hairpin's valid index

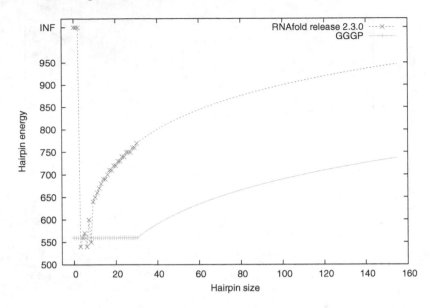

Fig. 5. Reduced energy penalty in RNAfold ($\times$ original, + evolved) for forming hairpin loops of varying sizes (x-axis). The tightest loop allowed in RNA limits x to be more than 3. (Remember training data is <155 bases long.)

range and are given by a log term. The GI change hairpin *<560 sets all values of hairpin[0..30] to 560. Figure 5 shows, in all but 3 cases, the evolved version has a lower penalty, thus encouraging the formation of hairpin folds.

To try and asses the importance of the hairpin mutations by themselves, we tried restoring the 31 default values of hairpin. As expected this performs less well on the 681 training RNA molecules (now average MCC is 0.730457 v. 0.737752 for 42 mutations). On the 1553 molecules of the holdout set it also does less well: better 344, worse 455, same 754, $p < 10^{--4}$. In summary, evolution has found a way of simplifying the contents of the hairpin array (i.e. setting the whole array to one value, 560) which is significantly better when included but in only 7% of cases is the change in MCC more than 0.1.

mismatchM. The C `int` array mismatchM has the same three index as mismatchH (above). It stores energy values associated with the stabilising effect of a base pairing being adjacent to a free end or a multiloop (called a dangling end). Like mismatchH it supplies an energy value according to the bases adjacent to the two active positions i and j and their types. By default (like mismatchH) all values are negative (actually between −160 and −30), except in mismatchH the first 25 values (i.e. mismatchH[0,*,*]) are INF, whereas with mismatchM they are zero. Again mismatchM does not use array elements with index 0 also type 7 also does not seem to be used, meaning the mutations (given in Table 3) affect 59 index positions. Five increase (i.e. penalise) bond pairs by +40 (all −80 to −40, and G↔G or G↔A). Twelve increase (i.e. penalise) bond pairs by +20 (−140 to −120, −120 to −100 and −60 to −40, C↔C, C↔U, G↔*, U↔U).

The other 22 cases reduce the penalty by −40. Changes to C↔C or U↔C encourage bonds by reducing the energy by −70 to −130. Whilst changes for A↔* dangling end bonds are also treated more favourably by −40 but cover a large range of initial values (−160 to −30, including −70).

Manual Removal of rtype. Array type does not hold energy values but (together with array pair) correspond to the internal coding of base pairs. For example, a C-G pair encoding is 1, and its reverse type (rtype array at position 1) is 2, which is the same encoding as that of a G-C pair. We were therefore surprised that evolution had changed rtype.

Of the 42 mutations, two affect rtype: The first, rtype[2]←2, (value 2 refers to G-C pairings) so rtype[G-C] now contains the code for itself (rather than for C-G). The second, rtype[6]←6, (value 6 refers to U-C pairings) so rtype[U-C] now also contains the code for itself (rather than for C-U). Notice rtype is no longer a permutation.

Table 3 shows the two rtype mutations by themselves gave a small improvement (MCC 0.666190 v. 0.663946) averaged over the 681 training molecules compared to no mutations. Suggesting during the run mutating rtype had an evolutionary advantage. However, this does not sustain to the end of the GGGP process.

As a post-hoc experiment, we manually removed both changes to rtype. On the training set of 681 short molecules it has an average MCC of 0.724700, (i.e. slightly worse than the end of the run best mutant, 0.737044, and worse than the cleaned up 42 mutant). However, on the 1553 RNA molecules in the holdout set the average MCC is now 0.569085 (remember the 42 mutant's mean MCC is slightly lower at 0.568323) but a non-parametric two sided sign test does not show a significant difference. We should perhaps remove the 2 evolved rtype changes, since removing them does not make the prediction worse and it certainly makes the mutants simpler, however, the statistics do not allow us to claim it is better.

4   Conclusions

Our previous work [26] suggested that the parameters of the dynamic programming model of the energy changes used by folding RNA were a suitable route for making non-function preserving changes to RNAfold. These parameters are derived from detailed scientific measurement of RNA. However, they are not set in stone and have been manually updated in the past to incorporate new scientific knowledge of how RNA behaves. Andronescu et al. [33] fitted the RNAfold free energy parameters by formulating a constraint optimization problem, which is quite complicated, time consuming and tedious and our GI does better (see Sect. 3.2 and Fig. 4).

It is typical for RNA molecules with more challenging non-standard bindings to be excluded when testing RNA prediction software [6]. However we have attempted to evolve the state-of-the-art program to match all of the known RNA structures. The new version does better overall, in some cases its predictions are much better, but there are some (albeit a smaller number) where it does worse.

Genetic programming is routinely used to generate from scratch small models of physical systems (e.g. Eureqa [34]) but here we have shown it can potentially be widely used to automatically update constants within sizeable programs which have taken years to develop and are in daily use but where the task of keeping up with the latest empirical data is highly skilled, labour intensive and liable to drag months behind current scientific knowledge.

Tuned RNAfold parameters
The complete changes to RNAfold's default parameters are given in Table 3. However, for ease of use we manually converted them into a free energy parameter file rna_gi.par compatible with all programs of the ViennaRNA Package. Thus, they can be optionally loaded at runtime, e.g. RNAfold -P rna_gi.par. This required removing rtype (page 14) and ensuring matrices stack, int11 and int22, are symmetric. In the case of stack, Table 3, it can be made symmetric by adding another mutation: stack 4,*<−50. Similarly int11 is made symmetric by adding int11 *,6,2,*+=−70 and int22 by adding int22 *,*,*,*,2,*+=10 immediately after int22 *,*,2,*,*,*+=10. The parameters evolved will be shipped with the next ViennaRNA package.

Acknowledgements. I am grateful for the assistance of Rhiju Das and Fernando Portela, and our anonymous reviewers.

References

1. Tsunoda, M., et al.: Structural basis for recognition of cognate tRNA by tyrosyl-tRNA synthetase from 3 kingdoms. Nucleic Acids Res. **35**(13), 4289–4300 (2007). https://doi.org/10.1093/nar/gkm417
2. Crick, F.: Central dogma of molecular biology. Nature **227**, 561–563 (1970). https://doi.org/10.1038/227561a0
3. Andronescu, M., et al.: RNA STRAND: The RNA secondary structure and statistical analysis database. BMC Bioinformatics **9**(1), 340 (2008). https://doi.org/10.1186/1471-2105-9-340
4. Reeder, J., et al.: pknotsRG: RNA pseudoknot folding including near-optimal structures and sliding windows. Nucleic Acids Res. **35(Suppl 2)**, W320–W324 (2007). https://doi.org/10.1093/nar/gkm258
5. Langdon, W.B., Harman, M.: Grow and graft a better CUDA pknotsRG for RNA Pseudoknot free energy calculation. In: GI 2015 Workshop, pp. 805–810 (2015). http://www.cs.bham.ac.uk/~wbl/biblio/gp-html/langdon_2015_gi_pknots.html
6. Lorenz, R., et al.: ViennaRNA package 2.0. Algorithms Mol. Biol. **6**(1), 26 (2011). https://doi.org/10.1186/1748-7188-6-26
7. Lee, J., et al.: RNA design rules from a massive open laboratory. PNAS **111**(6), 2122–2127 (2013). https://doi.org/10.1073/pnas.1313039111

8. Harman, M., Jia, Y., Langdon, W.B.: Babel Pidgin: SBSE can grow and graft entirely new functionality into a real world system. In: Le Goues, C., Yoo, S. (eds.) SSBSE 2014. LNCS, vol. 8636, pp. 247–252. Springer, Cham (2014). https://doi.org/10.1007/978-3-319-09940-8_20

9. Jia, Y., Harman, M., Langdon, W.B., Marginean, A.: Grow and serve: growing Django citation services using SBSE. In: Barros, M., Labiche, Y. (eds.) SSBSE 2015. LNCS, vol. 9275, pp. 269–275. Springer, Cham (2015). https://doi.org/10.1007/978-3-319-22183-0_22

10. Kocsis, Z.A., Swan, J.: Genetic programming + proof search = automatic improvement. J. Autom. Reasoning **60**(2), 157–176 (2018). http://www.cs.bham.ac.uk/ wbl/biblio/gp-html/PolyfinicJAR.html

11. Langdon, W.B., Lam, B.Y.H., Petke, J., Harman, M.: Improving CUDA DNA analysis software with genetic programming. In: GECCO, pp. 1063–1070 (2015). http://www.cs.bham.ac.uk/~wbl/biblio/gp-html/Langdon_2015_GECCO.html

12. Langdon, W.B.: Genetic improvement of software for multiple objectives. In: Barros, M., Labiche, Y. (eds.) SSBSE 2015. LNCS, vol. 9275, pp. 12–28. Springer, Cham (2015). https://doi.org/10.1007/978-3-319-22183-0_2

13. Langdon, W.B., Lam, B.Y.H., Modat, M., Petke, J., Harman, M.: Genetic improvement of GPU software. GP & EM **18**(1), 5–44 (2017). http://www.cs.bham.ac.uk/ wbl/biblio/gp-html/Langdon_2016_GPEM.html

14. Langdon, W.B: Genetically improved software. In: Gandomi, A.H., et al. (Eds.): Handbook of Genetic Programming Applications, pp. 181–220. Springer, Cham (2015). http://www.cs.bham.ac.uk/~wbl/biblio/gp-html/langdon_2015_hbgpa.html

15. Petke, J., Haraldsson, S.O., Harman, M., Langdon, W.B., White, D.R., Woodward, J.R.: Genetic improvement of software: a comprehensive survey. IEEE Transactions on Evolutionary Computation (In press). http://www.cs.bham.ac.uk/~wbl/biblio/gp-html/Petke_gisurvey.html

16. Langdon, W.B., Harman, M.: Optimising existing software with genetic programming. IEEE Trans. Evol. Comput. **19**(1), 118–135 (2015). http://www.cs.bham.ac.uk/ wbl/biblio/gp-html/Langdon_2013_ieeeTEC.html

17. Bruce, B.R., Petke, J., Harman, M.: Reducing energy consumption using genetic improvement. In: GECCO, pp. 1327–1334. ACM (2015)

18. Wu, F., Weimer, W., Harman, M., Jia, Y., Krinke, J.: Deep parameter optimisation. In: Silva, S., et al., (Eds.) GECCO, pp. 1375–1382 (2015). http://www.cs.bham.ac.uk/~wbl/biblio/gp-html/Wu_2015_GECCO.html

19. Marginean, A., Barr, E.T., Harman, M., Jia, Y.: Automated transplantation of call graph and layout features into kate. In: Barros, M., Labiche, Y. (eds.) SSBSE 2015. LNCS, vol. 9275, pp. 262–268. Springer, Cham (2015). https://doi.org/10.1007/978-3-319-22183-0_21

20. Le Goues, C., Nguyen, T., Forrest, S., Weimer, W.: GenProg: a generic method for automatic software repair. IEEE Trans. Softw. Eng. **38**(1), 54–72 (2012). http://www.cs.bham.ac.uk/ wbl/biblio/gp-html/DBLP_journals_tse_GouesNFW12.html

21. Langdon, W.B., Lorenz, R.: Improving SSE parallel code with grow and graft genetic programming. In: Petke, J., et al. (Eds.) GI-2017, pp. 1537–1538. ACM (2017). http://www.cs.bham.ac.uk/~wbl/biblio/gp-html/Langdon_2017_GI.html

22. Koza, J.R.: Genetic Programming. MIT press, Cambridge, MA (1992). http://www.cs.bham.ac.uk/ wbl/biblio/gp-html/koza_book.html

23. Banzhaf, W., Nordin, P., Keller, R.E., Francone, F.D.: Genetic Programming - An Introduction. Morgan Kaufmann, San Francisco (1998). http://www.cs.bham.ac.uk/ wbl/biblio/gp-html/banzhaf_1997_book.html

24. Poli, R., Langdon, W.B., McPhee, N.F.: A field guide to genetic programming. Lulu Enterprises, UK (2008). http://www.gp-field-guide.org.uk
25. Das, R.: Personal Communication (2017)
26. Langdon, W.B.: Evolving better RNAfold C source code. Technical Report RN/17/08, University College, London, (2017). http://www.cs.bham.ac.uk/~wbl/biblio/gp-html/langdon_RN1708.html
27. MacKerell Jr., A.D., Banavali, N., Foloppe, N.: Development and current status of the CHARMM force field for nucleic acids. Biopolymers **56**(4), 257–265 (2000). https://doi.org/10.1002/1097-0282(2000)56:4⟨257::AID-BIP10029⟩3.0.CO;2-W
28. Zuber, J., et al.: A sensitivity analysis of RNA folding nearest neighbor parameters identifies a subset of free energy parameters with the greatest impact on RNA secondary structure prediction. Nucleic Acids Res. **45**(10), 6168–6176 (2017). https://doi.org/10.1093/nar/gkx170
29. Angeline, P.J.: Multiple interacting programs: a representation for evolving complex behaviors. Cybern. Syst. **29**(8), 779–803 (1998). http://www.cs.bham.ac.uk/ wbl/biblio/gp-html/angeline_1998_mips3.html
30. Langdon, W.B.: Genetic Programming and Data Structures. Kluwer, Norwell (1998). http://www.cs.bham.ac.uk/ wbl/biblio/gp-html/langdon_book.html
31. Langdon, W.B., Lam, B.Y.H.: Genetically improved BarraCUDA. BioData Min., 20(28) (2017). http://www.cs.bham.ac.uk/~wbl/biblio/gp-html/Langdon_2017_BDM.html
32. Weimer, W., Nguyen, T., Le Goues, C., Forrest, S.: Automatically finding patches using genetic programming. In: ICSE, pp. 364–374 (2009). http://www.cs.bham.ac.uk/~wbl/biblio/gp-html/Weimer_2009_ICES.html
33. Andronescu, M., et al.: Efficient parameter estimation for RNA secondary structure prediction. Bioinformatics **23**(13), i19–i28 (2007). https://doi.org/10.1093/bioinformatics/btm223
34. Schmidt, M., Lipson, H.: Distilling free-form natural laws from experimental data. Science **324**(5923), 81–85 (2009). http://www.cs.bham.ac.uk/ wbl/biblio/gp-html/Science09_Schmidt.html

Geometric Crossover in Syntactic Space

João Macedo[1,2(✉)] ⓘ, Carlos M. Fonseca[2] ⓘ, and Ernesto Costa[2] ⓘ

[1] ISR, Department of Electrical and Computer Engineering,
University of Coimbra, 3030 290 Coimbra, Portugal
[2] CISUC, Department of Informatics Engineering,
University of Coimbra, 3030 290 Coimbra, Portugal
{jmacedo,cmfonsec,ernesto}@dei.uc.pt

Abstract. This paper presents a geometric crossover operator for Tree-Based Genetic Programming that acts on the syntactic space, where each expression tree is represented in prefix notation. The proposed operator is compared to the standard subtree crossover on a symbolic regression problem, on the Santa Fe Ant Trail and on a classification problem. Statistically validated results show that the individuals produced using this method are significantly smaller than those produced by the subtree crossover, and have similar or better performance in the target tasks.

Keywords: Genetic Programming · Geometric operators · Crossover

1 Introduction

Geometric variation operators have been available for Genetic Algorithms (GA) for some time now [1]. They are representation-independent operators based on a distance on the search space interpreted as a metric space. The geometric description of the variation operators uses the notions of line segment (crossover) and ball (mutation). In the case of the crossover operator, the resulting offspring is on a shortest path, i.e., line segment, linking its parents. In the case of mutation, the resulting individual is in the neighbourhood of the original individual, i.e., within a ball centred on the individual and with a given radius, which defines the magnitude of the mutation. More formally, considering a given distance d defined over the search space, and provided that the parent individuals A and B are different, a geometric crossover operator will produce an offspring O such that $d(A, B) = d(A, O) + d(O, B)$.

Uniform crossover is an instance of a geometric operator devised for GAs. As an example, consider the binary strings: $A = 00000$ and $B = 11111$. One possible offspring generated by uniform crossover could be: $O = 10101$. Considering the Hamming distance as the metric we have, $d(A, O) = 3, d(O, B) = 2, d(A, B) = 5$. This example can be easily modified to accommodate other types of search spaces, the only requirement being the definition of an appropriate distance between individuals.

© Springer International Publishing AG, part of Springer Nature 2018
M. Castelli et al. (Eds.): EuroGP 2018, LNCS 10781, pp. 237–252, 2018.
https://doi.org/10.1007/978-3-319-77553-1_15

Moraglio [1] introduced abstract definitions of geometric crossover and mutation operators that are independent of the individuals' representation. One such operator is Geometric Uniform Crossover, *UX*. In this operator, all individuals between the parents have equal probability of being an offspring. Using this and other definitions, Moraglio showed that it is possible to devise geometric variation operators for different representations, e.g., binary strings, real value vectors and permutations.

Constructing a geometric variation operator for expression-trees, a structure commonly used in Genetic Programming (GP), is not straightforward, as it is not clear what a suitable distance would be. Moreover, making a small modification to the genotype of a GP individual may lead to a big change in its behaviour. This is known as the low-locality problem inherent to the representation of GP individuals [2].

Moraglio and Poli [3] have provided a theoretical study on how homologous crossover is geometric. When using homologous crossover, the topologies of both trees are compared, and the common rooted structures are found. Then, genetic material from the common regions is exchanged. One point crossover [4] is a special case of homologous crossover. With this operator, only one node is selected from the common region of each parent and the subtrees rooted by them are exchanged. Despite having been proposed some time ago, these operators never achieved much popularity. The fact that the alignment of the parent trees is a time-consuming task may be one of the reasons for their lack of usage. Another reason may be that, with extremely different trees, the common region may be very small, leading to very big syntactic changes which, together with the low-locality inherent to this tree representation, may have a great impact on the semantics of the offspring, possibly leading to their death.

Krawiec and Lichocki [5] proposed the Approximately Geometric Semantic Crossover (SX). They consider the semantics of a GP individual as the set of input-output mappings created by it. Based on this idea, they propose a binary variation operator that tries to approximate a geometric crossover in semantic space by producing an offspring that has a behaviour as close as possible to the linear combination of the semantics of its parents. This is done by applying multiple times a usual crossover operator to the parents, creating a set of candidate offspring. The semantics of the created offspring is assessed and compared to those of the parents. The resulting offspring is one with the closest semantics to both parents. This approach has several drawbacks. Firstly, it does not guarantee that the individuals created are on a shortest path between their parents. As the authors point out, there is a low probability of producing semantically geometric offspring, depending on the parents, the crossover operator, the terminal and function sets, and the size of the set of candidate offspring. Secondly, the evaluation of the many GP individuals produced usually becomes a time consuming process. Thus, the execution time of the algorithm is expected to increase proportionally to the size of the generated sets of candidate offspring.

Moraglio et al. [6] further developed the idea of geometric variation operators in semantic space, and later proposed Geometric Semantic Genetic Programming (GSGP), where the crossover operation consists of making a weighted average of both parents. More formally, the crossover operation is defined by:

$$o = \alpha \cdot p_1 + (1 - \alpha) \cdot p_2,$$

where o denotes the offspring, p_1 and p_2 are the parent individuals, and α is a random value sampled uniformly from the interval $[0, 1]$. This operation guarantees that the semantics of the generated individual, i.e., its input-output mapping, is a blend of the semantics of both its parents. GSGP offers the advantage of, under certain conditions, inducing a unimodal fitness function over the semantic space, which makes the search process much easier. Also, by acting on the semantic space, geometric semantic crossover controls how much the behaviour of the individual is changed. Conversely due to the low locality of the tree-based representation, making modifications at the syntactic level often leads to large modifications to the semantics of the individuals, which may deem them unfit and prevent them from surviving into the next generations. However, GSGP is not without its faults, the most significant being the exponential growth of the individuals, which makes it unfeasible to employ this method for a large number of generations. Vanneschi et al. [7] propose a different, more efficient implementation that remedies this problem, but does not solve it.

A different approach to the development of geometric operators for GP would be to adopt a simpler individual representation and, like the traditional variation operators, act on the syntactic space. There are already some works on Evolutionary Algorithms (EAs) that use an alternative representation to encode the individuals. Brameier and Banzhaf [8] describe Linear Genetic Programming, which uses a linear representation of the individuals to evolve computer programs. Another work that represents the individuals linearly is Gene Expression Programming (GEP) [9]. The genotype of GEP individuals is represented by a string that can be decoded into more complex structures, such as expression trees, graphs and neural networks. The notation used for the genotypes is the sequence of nodes visited during a breadth-first traversal of the phenotype. The length of the genotypes is fixed and chosen a priori for all individuals. GEP is able to evolve different individuals using various types of functions, which may have different arities. In order to guarantee their validity, each genotype is divided into head and tail. While the head may contain functions and terminal symbols, the tail must contain only terminal symbols. The sizes of the head and tail are pre-computed to ensure that even in the presence of a head composed only of functions with maximum arity, the tail is long enough to generate a valid individual. A consequence of this is that there will often exist redundant genes at the end of the tail. As an example, consider that the genotype of an individual is Q*+−abcd, where the head contains Q*+−, and the tail contains abcd. Figure 1 depicts the phenotype of this individual. However, the genotype Q*+−abcd**defgh** would lead to the same phenotype, with the genes in bold not being decoded. The individuals are evolved using a set of variation operators that are not known to be geometric. The GEP author claims that the advantage of this method lies in using a simple representation that is easy to manipulate, and yet originates complex structures.

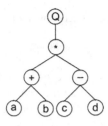

Fig. 1. Phenotype of an individual evolved by GEP. Figure extracted from [9].

In this paper, we propose a geometric crossover operator for GP that acts on the syntactic space. We use a linear, prefix representation of the GP individuals with no redundant genes. This way, we can avoid tree alignment algorithms and, instead, rely on string edit distances between the individuals. We compute the longest common subsequence of the parents, and use the result to make a controlled modification in one of the parents producing an offspring that lies in between those parents in the syntactic, i.e., genotype, space. The proposed operator is compared to the standard subtree crossover on a symbolic regression problem, on the Santa Fe Ant Trail and on a classification problem. Statistically validated results show that the individuals produced using this method are significantly smaller than those produced by subtree crossover, and have equivalent or better performance on the target tasks.

The rest of this paper is organised as follows: Sect. 2 describes the proposed recombination operator, Sect. 3 presents the experimental setup, Sect. 4 presents and discusses the experimental results obtained and Sect. 5 presents the conclusions and provides some insight into the future work.

2   Geometric Crossover on the Syntactic Space

We propose to perform a geometric crossover operation between two GP individuals in the syntactic space. The genotype of each individual is a string that encodes an expression tree in prefix notation. There are no redundant genes. The crossover operation aligns the genomes of the two parents and performs the necessary operations so that one becomes more similar to the other. In our implementation, the alignment is performed by computing the Longest Common Subsequence. The modifications are made by inserting or deleting pairs of symbols of different types, i.e., a terminal and a non-terminal symbol, or by deleting a symbol and inserting another one of the same type. The distance between the two parent individuals is the number of operations that convert one individual into the other.

A flow chart of the crossover operator is depicted on Fig. 2. We start by using a dynamic programming algorithm to compute the Longest Common Subsequence between two parent individuals, A and B, obtaining a matrix C. This matrix contains the information about the common and non-common genetic

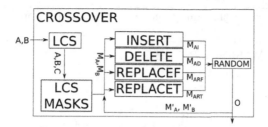

Fig. 2. Geometric crossover operator.

material to A and B. Using that information, Algorithm 1 (LCS_MASKS), constructs two modification masks M_A, M_B. These masks contain the aligned symbols from the longest common subsequence, along with blank spaces denoting the locations where insertions and deletions must be made, and the non-common symbols marked for deletion or insertion. Using these masks, it is possible to make a copy of the parent A more similar to parent B by inserting and/or deleting genetic material. The resulting offspring O is chosen randomly from the set of valid candidate individuals resulting from each of those operations. This can be repeated for a number of steps, in order to generate individuals farther from the first parent (and closer to the second one). In that case, the previously chosen individual takes the place of A and M_A and M_B are replaced by M'_A and M'_B, respectively, which are updated versions that reflect the operations performed.

2.1   String Edit Distance

The string edit distance is a family of metrics that reflect the number of operations needed to transform a string A into a string B. The Longest Common Subsequence (LCS) [10] is the longest possibly non consecutive subsequence that is common to both strings. From it, it is possible to compute the distance between the two strings, i.e., the number of operations required to modify the non-common symbols from the two strings. We refer the interested reader to [11], where a dynamic programming algorithm to compute the LCS is presented. This algorithm outputs a matrix C that holds in each position (i, j) the length of the longest common subsequence contained up to the character i of string A and character j of string B. After the algorithm terminates, a longest common subsequence can be obtained by going through the matrix C from the bottom right cell to the upper left cell. A symbol in position (i, j) is part of a longest common subsequence if $C[i][j] = C[i-1][j-1] + 1$ and $C[i][j] \neq C[i-1][j]$, $C[i][j] \neq C[i][j-1]$.

2.2   Crossover Operator

The proposed geometric crossover operator (GSynGP) works by modifying a copy of the first parent, A, to make it more similar to the second parent, B,

Algorithm 1. Modification masks generated from the Longest Common Subsequence.

```
 1: function LCS_MASKS(A, B, C)
 2:     M_A, M_B ← []
 3:     i ← len(C) − 1
 4:     j ← len(C[0]) − 1
 5:     while i >= 1 or j >= 1 do
 6:         if i > 0 and j > 0 and C[i − 1][j − 1] = C[i][j] then
 7:             M_A ← get_symbol(A[i − 1], function_set)
 8:             M_B ← get_symbol(B[j − 1], function_set)
 9:             i ← i − 1
10:             j ← j − 1
11:         else if i > 0 and C[i − 1][j] = C[i][j] then
12:             M_A ← get_symbol(A[i − 1], function_set)
13:             M_B ← ' '
14:             i ← i − 1
15:         else if j > 0 and C[i][j − 1] = C[i][j] then
16:             M_A ← ' '
17:             M_B ← get_symbol(B[j − 1], function_set)
18:             j ← j − 1
19:         else if i > 0 and j > 0 and C[i − 1][j − 1] = C[i][j] − 1 then
20:             M_A ← A[i − 1]
21:             M_B ← B[j − 1]
22:             i ← i − 1
23:             j ← j − 1
24:     return reverse(M_A), reverse(M_B)
```

by inserting and/or removing nodes while preserving the syntactic validity of the intermediate individuals. It starts by computing the LCS to determine the similarity between the two parent individuals (A, B). That information is contained in matrix C. Then, using matrix C, Algorithm 1 (LCS_MASKS) computes two modification masks, M_A and M_B, that contain the symbols in the longest common subsequence, as well as those that must be modified and blank spaces denoting locations that must be filled. In other words, Algorithm 1 computes an alignment of the two parent strings. The masks are created by going through matrix C, aligning the common genetic material and inserting blank spaces or markers in the positions where symbols must be added or removed from A. The markers are provided by function get_symbol, which returns the symbol passed as a parameter along with a prefix to denote whether it belongs to the terminal $(T\_)$ or function set $(F\_)$. As an example, consider two parent individuals A: $/ - */bcaa*aa$ and B: $*b/cb$. A possible LCS between them is $*bc$. The masks produced by Algorithm 1 are:

M_A: $F\_/, F\_-, *, F\_/, b, \ '\ ', \ c, T\_a, T\_a, F\_*, T\_a, T\_a$
M_B: $'\ ', \ '\ ', \ *, \ '\ ', \ b, F\_/, c, \ '\ ', \ '\ ', \ '\ ', \ '\ ', \ T\_b$

where, the symbols in M_A that have the prefixes T_- and F_- are not present in the individual B, and must be deleted. The symbols with those prefixes present in M_B represent genetic material that must be inserted into A. The locations for insertions are marked by blank spaces in M_A, and the locations for deletions are marked by blank spaces in M_B.

Algorithm 2. Geometric Syntactic Crossover Operator

1: **function** CROSSOVER(M_A, M_B)
2:      $candidates \leftarrow []$
3:      **if** 'F_' in M_A and 'T_' in M_A **then**
4:          $candidates \leftarrow candidates \cup delete(M_A, M_B, function\_set)$
5:      **if** not' F_' in M_A and 'T_' in M_A **then**
6:          $candidates \leftarrow candidates \cup replaceT(M_A, M_B, function\_set)$
7:      **if** 'F_' in M_B and 'T_' in M_B **then**
8:          $candidates \leftarrow candidates \cup insert(M_A, M_B, function\_set)$
9:      **if** 'F_' in M_B and 'F_' in M_A **then**
10:         $candidates \leftarrow candidates \cup replaceF(M_A, M_B, function\_set)$
11:      **return** $random(candidates)$

The remaining steps of the crossover operator, presented in Algorithm 2, consist simply in checking four conditions and performing the corresponding operations, where appropriate. The four possible operations are: inserting a function and a terminal (Algorithm 3), removing a function and a terminal (Algorithm 4), removing a terminal and inserting another one and removing a function and inserting another one (Algorithm 5). The operation of deleting a terminal and inserting another one is identical to what is presented in Algorithm 5, with the difference that F_- should read T_- and function_set should read terminal_set. The function $all\_combinations()$ returns all pairs of symbols to be tested in each case, that is, in Algorithm 3, the combinations of all function and terminal symbols

---

Algorithm 3. Insertion of a function and a terminal symbol

1: **function** INSERT(M_A, M_B)
2:      $combs \leftarrow all\_combinations()$
3:      **while** $len(combs) > 0$ **do**
4:          $(f, t) \leftarrow random(combs)$
5:          $M_A[t] \leftarrow M_B[t]$
6:          $M_A[f] \leftarrow M_B[f]$
7:          **if** $check\_indiv(M_A, function\_set)$ **then return** M_A
8:          **else**
9:             $M_A[t] \leftarrow '\ '$
10:            $M_A[f] \leftarrow '\ '$
11:      **return** M_A

---

that are present in B and absent in A; in Algorithm 4, all pairs of terminal and function symbols that are present in A and absent in B and; for the operations of deleting and inserting symbols of the same type, all pairs of symbols of the desired type that are present in one parent and absent in the other one.

Algorithm 4. Deletion of a function and a terminal symbol

```
 1: function DELETE(M_A, M_B)
 2:     combs ← all_combinations()
 3:     while len(combs) > 0 do
 4:         (f, t) ← random(combs)
 5:         v ← [M_A[f], M_A[t]]
 6:         M_A[t] ←' '
 7:         M_A[f] ←' '
 8:         if check_indiv(M_A, function_set) then return M_A
 9:         else
10:             M_A[t] ← v[1]
11:             M_A[f] ← v[0]
12:     return M_A
```

An individual is then selected from the set of valid generated individuals. These operations create an individual that is one step away from the first parent. In order to create offspring at different distances from each parent, Algorithm 2 may be iterated over a number of times, with the resulting individual of one iteration taking the place of A in the following iteration. For example, if the operator is applied twice, on the first iteration it will be applied to parents A and B, outputting an offspring O_1. In the second iteration, the offspring O_1 will take the place of A in the crossover, creating the individual O_2. In general, in an iteration where all operations are possible, the offspring created only has a 25% chance of becoming larger than its parent. That growth will only by 2 nodes and, at most, by one depth level.

An individual is valid if it can be converted into a valid expression tree, without any exceeding genes. It is also possible to test this validity without converting the string into the corresponding tree. Consider a counter c, that holds the number of necessary terminal symbols for an individual to be valid. For an empty string, $c = 1$, i.e., a terminal symbol is required in order to create a valid individual. To test an individual, start with $c = 1$ and go through the string. Increment c for each function symbol read, and decrement it for each terminal read. If c reaches 0 and there are still unread symbols, the individual has redundant genes and, thus, is invalid. On the other hand, if the string ends and $c > 0$, the individual can not be converted into a valid tree. In the given example, two valid candidates for offspring are:

1. $/-*/bcaaa$, obtained by removing the last $*$ and one of the marked a from A
2. $/-*/b/caaaa$, obtained by removing the last $*$ from A and inserting the $/$ from B

Algorithm 5. Deletion of a function and insertion of another function symbol

```
 1: function REPLACEF(M_A, M_B)
 2:     combs ← all_combinations()
 3:     while len(combs) > 0 do
 4:         (f_b, f_a) ← random(combs)
 5:         if M_A[f_b] =' ' then
 6:             M_A[f_b] ← M_B[f_b]
 7:             v ← M_A[f_a]
 8:             M_A[f_a] ← ' '
 9:             if check_indiv(M_A, function_set) then return M_A
10:             else
11:                 M_A[f_b] ← ' '
12:                 M_A[f_a] ← v
13:         else if 'F_' in M_A[f_b] then
14:             M_A[f_b] ← M_B[f_b]
15:         else
16:             v ← M_A[f_a]
17:             M_A[f_a] ← M_B[f_b]
18:             if check_indiv(M_A, function_set) then return M_A
19:             else
20:                 M_A[f_a] ← v
21:     return M_A
```

Table 1. Parameters of the SGP.

Parameter	Value
Population size	400
Elite size	1
Tournament size	10
Maximum tree depth	10
Generations	1000
Crossover rate	0.7
Mutation rate	0.3
Number of immigrants	120

3   Experimental Setup

In order to assess the usefulness of the proposed approach, we performed 30 independent runs on a Symbolic Regression problem, on the Santa Fe Ant Trail and on a Classification problem, comparing the performance of the proposed approach to that of Standard Genetic Programming (SGP). In the following, we shall refer to the geometric syntactic approach as GSynGP.

3.1 Standard Genetic Programming Algorithm

We implemented a version of SGP using subtree crossover and two types of mutation: point mutation and subtree crossover with a randomly generated individual. Point mutation selects one symbol from the genotype and replaces it by another of the same type. The initial population is created using the method known as ramped-half-and-half [12]. The parents are selected using tournaments and survivor selection is generational, with an elite individual. In order to maintain a diverse population throughout the run, at each generation a set of immigrant individuals are introduced into the population. These immigrants have a 50% chance of being elitist or randomly generated. Elitist immigrants are mutated copies of a good quality individual that is selected from the population by tournament. The random immigrants are generated using the ramped-half-and-half method. The parameters used for this algorithm are presented on Table 1.

3.2 Geometric Syntactic Approach

This algorithm uses the same parameters as those used for the SGP, differing only in the crossover operator employed, which has been described in Algorithm 2. Due to the low-locality problem inherent to expression trees in GP, we are not interested in making modifications that are too disruptive. However, we are still interested in generating individuals that have different distances to each parent. For these reasons, the crossover operator is performed for a random number of steps, which is uniformly sampled from $\{1, 2, 3\}$.

3.3 Symbolic Regression

Dataset. The performance of the two algorithms is assessed on a dataset generated using Eq. 1, as proposed by Keijzer [13]. The dataset contains 50 randomly generated points, with x_1, x_2 sampled uniformly from the interval $[-10, 10]$.

$$y = x_1^4 - x_1^3 + \frac{x_2^2}{2} - x_2 \tag{1}$$

Terminal and Function Sets. The SGP and the GSynGP use the same terminal and function sets. For this problem, the terminal set is composed only of the variables from the dataset (i.e., x_1, x_2) and the function set is composed of the basic arithmetic functions (i.e., $+, -, *, /$). / stands for protected division, where $x/0 = 1$.

Fitness Evaluation. The fitness of each individual was assessed using the Mean Squared Error (MSE), as defined in Eq. 2, thus making this a minimisation problem.

$$MSE = \frac{\sum_{i=1}^{N}(\hat{Y}_i - Y_i)^2}{N} \tag{2}$$

where N is the number of samples, $\hat{Y}_i$ is the ith predicted value and Y_i is the corresponding target value.

3.4   Santa Fe Ant Trail

The Santa Fe Ant Trail is a path planning benchmark problem where an artificial ant must follow a deceptive trail collecting food pellets. We used the traditional version of this problem, as used by Koza in [12]. The map is a 32×32 toroidal grid with 89 food pellets. The ant starts facing east, at the upper left cell. The simulation ends when the ant completes 400 moves.

Terminal and Function Sets. The SGP and the geometric approach use the same terminal and function sets. For this problem, the terminal set is composed of the basic actions of the ant $\{left, right, move\}$, where $left$ and $right$ rotate the ant 90^0 in each direction and $move$ makes it move forward one cell. The function set is composed of the functions $ifFoodAhead$ and $Progn2$. Both functions have an arity of 2. Function $ifFoodAhead$ checks if there is a food pellet directly in front of the ant and, if there is, executes its first argument, otherwise executes the second argument. $Progn2$ is a progression function that executes both arguments in sequence.

Fitness Evaluation. The fitness of each individual is measured as the number of food pellets eaten by the ant, within the 400 steps limit. Thus, this is a maximisation problem.

3.5   Classification

In order to understand how our approach performs in real world problems, we decided to test it in a classification problem with real data. The chosen dataset is the Wisconsin Breast Cancer [14], available at the UCI repository. It is a binary classification problem that aims at determining whether a sample, described by 30 features, represents a benign or malign tumour. This dataset has 569 samples, from which, on the beginning of each run, the algorithm chose 70% to be used to evolve the individuals. As the dataset was unbalanced, prior to this split a balancing of the data was made, by randomly discarding samples of the larger class until both had an equal amount of examples. Thus, the balanced dataset was left with 424 samples, from which 296 were used in the evolutionary process.

Terminal and Function Sets. As before, both the SGP and the GSynGP use the same terminal and function sets. The terminal set is composed only of the variables from the dataset (i.e., $x_1, x_2, ..., x_{30}$) and the function set is composed of the basic arithmetic functions (i.e., $+, -, *, /$).

Fitness Evaluation. The performance of each individual was measured with the F1-Score, a commonly used criterion in classification problems. As the Santa Fe Ant Trail, this is a maximisation problem.

4 Experimental Results

For each problem, we characterise each algorithm according to three features: the fitness of the best individual at the end of each run, and the average depth and average number of nodes of the individuals in the population, also at the end of each run. The results were validated using statistical tests, at a significance level $\alpha = 0.05$. The results of those tests are presented on Tables 2 and 3.

4.1 Symbolic Regression

We start by applying the Kolmogorov-Smirnov test to the data, in order to be able to decide whether we should apply parametric or non-parametric tests. The results of this test show that the data for the average depth and average number of nodes of the individuals of both algorithms follow normal distributions, with $p > 0.05$, whereas the fitness data follows non-normal distributions ($p < 0.05$). As we have two sets of paired samples per feature, we apply the Wilcoxon Rank-Sum test to the data that follow non-normal distributions, i.e., to the fitness data, and the Paired Samples T-Test to the data of the average number of nodes and average depth. Both tests were applied as SGP-GSynGP. For the fitness data, the Wilcoxon test shows that there are no statistically significant differences between the performances of the individuals evolved by each

Table 2. Kolmogorov-Simrnov test applied to the data of the three problems.

		Symbolic regression			Santa Fe Ant Trail			Classification		
		Depth	Size	Fitness	Depth	Size	Fitness	Depth	Size	Fitness
SGP	Z	0.88	0.74	1.64	0.65	0.47	1.51	0.79	0.93	0.75
	p	0.416	0.648	0.006	0.794	0.979	0.014	0.556	0.358	0.628
GSynGP	Z	0.83	0.93	1.37	2.00	2.26	2.41	1.40	1.74	0.75
	p	0.498	0.358	0.035	0.0	0.0	0.0	0.028	0.003	0.628

Table 3. Wilcoxon applied to the fitness of the Symbolic Regression problem, to all data of the Santa Fe Ant Trail and to the average size and depth data of the Classification. Paired Samples T-Test applied to the data of the average depth and size of the Symbolic Regression problem and to the fitness data of the Classification problem.

Wilcox. (Dep. T)	Symbolic regression			Santa Fe Ant Trail			Classification		
	Depth	Size	Fitness	Depth	Size	Fitness	Depth	Size	Fitness
Z (t)	14.11	3.40	−1.22	−4.78	−4.78	−3.15	−4.62	−4.78	−1.16
p	0.0	0.0	0.221	0.0	0.0	0.002	0.0	0.0	0.257
P Ranks ($\overline{SGP}$)	119.34	4149.04	15	30	30	3	28	30	0.96
N Ranks ($\overline{GSynGP}$)	13.12	108.02	15	0	0	18	2	0	0.97
Effect size	0.934	0.931	−0.223	−0.873	−0.873	−0.575	−0.843	−0.873	0.211

algorithm. However, the Paired Samples T-Test revealed that the geometric app-roach evolved significantly smaller individuals, both in depth and in number of nodes (both $p = 0.0$), with large effect sizes of respectively 0.934 and 0.931. Analysing the average values, SGP evolved individuals with an average depth of 119.34 levels and an average size of 4149.04 nodes. The geometric syntactic approach, on the other hand, evolved much smaller individuals, with an average depth of 13.12 levels (only 3.12 levels higher than the maximum depth of the initial individuals) and an average size of 108.02 nodes.

4.2   Santa Fe Ant Trail

This section describes the analysis of the data collected on the Santa Fe Ant Trail. The Kolmogorov-Smirnov test yielded $p < 0.05$ for the data of all three features of the geometric approach. For that reason, we applied the Wilcoxon test to the data of the three features, in a SGP-GSynGP manner. It yielded $p = 0.0$ for the two features related to the size of the evolved individuals, leading us to conclude that the geometric approach evolved significantly smaller indi-viduals, with a large effect size of -0.873. In fact, the individuals evolved by the SGP had an average depth of 107,60 levels and an average size of 3450.31 nodes, while those evolved by the GSynGP had an average depth of 10.94 and an average size of 61.05 nodes. Regarding the fitness of the best individuals, the Wilcoxon test output a p-value of 0.002, which lets us know that there are statistically significant differences between the performance of the individuals of each algorithm. Moreover, the 18 negative ranks found by this test lead us to conclude that the geometric approach evolved individuals that perform signif-icantly better in this task than those evolved by the SGP, with a large effect size of -0.575. On average, the individuals evolved by the SGP collected 77.87 food pellets, while those evolved by the GSynGP collected 86.77, out of the 89 available in the world.

4.3   Classification

In the classification problem, the results of the Kolmogorov-Smirnov test do not allow us to reject the null hypothesis that the fitness data follow normal distri-butions. For that reason, we applied the Paired Samples T-Test which revealed no statistically significant differences, with the individuals evolved by the SGP achieving an average F1-Score of 0.96 and those evolved by the GSynGP achieving 0.97. Carrying on to assess the average size of the individuals in the last popula-tions of each algorithm, the results of the Kolmogorov-Smirnov, allow us to assume that the data of the SGP follows a normal distribution, while the data for the GSynGP does not. For that reason, we applied the Wilcoxon test, which yielded significant differences ($p=0.00$) and, with all ranks being positive, it shows that the individuals evolved by the SGP are significantly larger, with an average number of nodes of 5278.39, while those evolved by the GSynGP have an average size of 283.98 nodes. Finally, the Kolmogorov-Smirnov test yielded the same results for the average depth as it did for the average size. Applying a Wilcoxon test to the depth data, we conclude that the individuals evolved by the SGP are significantly deeper (p=0.0), with a large effect size of −0.843. The individuals evolved by the SGP have an average depth of 122.54 levels, while those evolved by the GSynGP have an average depth of 26.86 levels.

4.4   Population Diversity

We now focus on studying the diversity of the individuals from the population in each generation. We repeated the experiments without using immigrants, as they are an artificial method of increasing the population diversity. Due to space and time constraints, we focused on a single benchmark problem, the Santa Fe Ant Trail, and reduced the population size to 50 individuals. The tournament size was reduced to 3, as this is expected to increase the population diversity. The other parameters remained unchanged from the previous experiments.

Figure 3 presents the distances between each pair of parent individuals evolved by the SGP (left) and the GSynGP (right), sampled with a 50 generations period. Each distance is represented by a blue circle with high transparency. Thus, darker circles represent many pairs of parents with equal distances. At the beginning of the experiments, both algorithms present similar diversity (note the different scales). However, over time, the SGP seems to achieve a much greater diversity than the GSynGP. This is due to bloat. Bloated populations contain individuals with more diverse genotypes, but that diversity does not necessarily transfer into different behaviours. Moreover, over the entire run there are very large individuals in the population that result in big differences between the parents. However, these individuals are usually unable to survive many generations, leading the population to converge on a set of more similar individuals. The GSynGP behaves differently, gradually converging its population into a good quality area of the search space, with the most different individuals being phased out over the generations. However, this does not necessarily mean that the population has completely lost its diversity, as the parents of the last population still have an average distance of approximately 6.8 operations and a

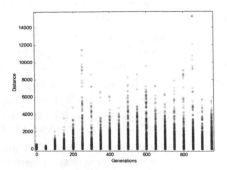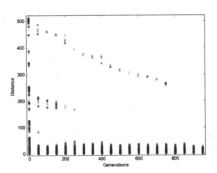

Fig. 3. Distances between each pair of parent individuals of the SGP (left) and GSynGP (right), sampled with a period of 50 generations.

standard deviation of 5.8. The bloat present in the populations of the SGP leads to more diverse populations in the genotype space, with the pairs of parents of the last population having an average distance of 1151 operations and a standard deviation of roughly 1095.3.

5 Conclusions

This work presented a novel geometric crossover operator that acts on the syntactic space of expression trees. The method was implemented and compared to SGP on problems from the domains of symbolic regression, path planning and classification. Our approach was able to consistently evolve smaller individuals than the SGP, both in size and depth. This reduction in size of the individuals does not imply a loss in quality, as our approach outperformed the SGP in the only test problem where there were statistically significant differences. The diversity experiments showed that the geometric operator led the population into a good quality region of the search space, without completely loosing its diversity. SGP seemed to have a much more diverse population, but that diversity was due to bloat. Future work includes adapting this approach to function sets containing symbols with different arities. Moreover, the crossover implementation should be improved to avoid having to test individuals after each insertion and/or deletion pair before a more thorough experimental study is carried out.

Acknowledgement. This article is based upon work from COST Action CA15140: Improving Applicability of Nature-Inspired Optimisation by Joining Theory and Practice (ImAppNIO), supported by COST (European Cooperation in Science and Technology), www.cost.eu. Support by national funds through the Portuguese Foundation for Science and Technology (FCT) and by the European Regional Development Fund (FEDER) through COMPETE 2020 – Operational Program for Competitiveness and Internationalization (POCI) is also acknowledged. J. Macedo acknowledges the Portuguese Foundation for Science and Technology for Ph.D. studentship SFRH/BD/129673/2017.

References

1. Moraglio, A.: Towards a Geometric Unification of Evolutionary Algorithms. Ph.D. thesis, Department of Computer Science, University of Essex (2007)
2. Galván-López, E., McDermott, J., O'Neill, M., Brabazon, A.: Defining locality as a problem difficulty measure in genetic programming. In: Genetic Programming and Evolvable Machines (2011)
3. Moraglio, A., Poli, R.: Geometric landscape of homologous crossover for syntactic trees. In: 2005 IEEE Congress on Evolutionary Computation (2005)
4. Poli, R., Langdon, W.B.: Genetic programming with one-point crossover. In: Chawdhry, P.K., Roy, R., Pant, R.K. (eds.) Soft Computing in Engineering Design and Manufacturing. Springer, London (1998). https://doi.org/10.1007/978-1-4471-0427-8_20

5. Krawiec, K., Lichocki, P.: Approximating geometric crossover in semantic space. In: Proceedings of the 11th Annual Conference on Genetic and Evolutionary Computation (2009)
6. Moraglio, A., Krawiec, K., Johnson, C.G.: Geometric semantic genetic programming. In: Coello, C.A.C., Cutello, V., Deb, K., Forrest, S., Nicosia, G., Pavone, M. (eds.) PPSN 2012. LNCS, vol. 7491, pp. 21–31. Springer, Heidelberg (2012). https://doi.org/10.1007/978-3-642-32937-1_3
7. Vanneschi, L., Castelli, M., Manzoni, L., Silva, S.: A new implementation of geometric semantic GP and its application to problems in pharmacokinetics. In: Krawiec, K., Moraglio, A., Hu, T., Etaner-Uyar, A.Ş., Hu, B. (eds.) EuroGP 2013. LNCS, vol. 7831, pp. 205–216. Springer, Heidelberg (2013). https://doi.org/10.1007/978-3-642-37207-0_18
8. Brameier, M., Banzhaf, W.: Linear Genetic Programming. Springer, Boston (2007). https://doi.org/10.1007/978-0-387-31030-5
9. Ferreira, C.: Gene expression programming in problem solving. In: Roy, R., Köppen, M., Ovaska, S., Furuhashi, T., Hoffmann, F. (eds.) Soft Computing and Industry: Recent Applications. Springer, London (2002). https://doi.org/10.1007/978-1-4471-0123-9_54
10. Paterson, M., Dančík, V.: Longest common subsequences. In: Prívara, I., Rovan, B., Ruzička, P. (eds.) MFCS 1994. LNCS, vol. 841, pp. 127–142. Springer, Heidelberg (1994). https://doi.org/10.1007/3-540-58338-6_63
11. Cormen, T.H., Leiserson, C.E., Rivest, R.L., Stein, C.: Introduction to Algorithms. The MIT Press, Cambridge (2001)
12. Koza, J.R.: Genetic programming: On the Programming of Computers by Means of Natural Selection. MIT press, Cambridge (1992)
13. Keijzer, M.: Improving Symbolic regression with interval arithmetic and linear scaling. In: Ryan, C., Soule, T., Keijzer, M., Tsang, E., Poli, R., Costa, E. (eds.) EuroGP 2003. LNCS, vol. 2610, pp. 70–82. Springer, Heidelberg (2003). https://doi.org/10.1007/3-540-36599-0_7
14. Lichman, M.: UCI Machine Learning Repository (2013). http://archive.ics.uci.edu/ml

Investigating a Machine Breakdown Genetic Programming Approach for Dynamic Job Shop Scheduling

John Park[1](✉), Yi Mei[1], Su Nguyen[1,2], Gang Chen[1], and Mengjie Zhang[1](✉)

[1] Evolutionary Computation Research Group,
Victoria University of Wellington, PO Box 600, Wellington, New Zealand
{John.Park,Yi.Mei,Su.Nguyen,Aaron.Chen,Mengjie.Zhang}@ecs.vuw.ac.nz
[2] La Trobe University, Melbourne, Australia

Abstract. Dynamic job shop scheduling (JSS) problems with dynamic job arrivals have been studied extensively in the literature due to their applicability to real-world manufacturing systems, such as semiconductor manufacturing. In a dynamic JSS problem with dynamic job arrivals, jobs arrive on the shop floor unannounced that need to be processed by the machines on the shop floor. A job has a sequence of operations that can only processed on specific machines, and machines can only process one job at a time. Many effective genetic programming based hyper-heuristic (GP-HH) approaches have been proposed for dynamic JSS problems with dynamic job arrivals, where high quality dispatching rules are automatically evolved by GP to handle the dynamic JSS problem instances. However, research that focus on handling multiple dynamic events simultaneously are limited, such as both dynamic job arrivals and machine breakdowns. A machine breakdown event results in the affected machine being unable to process any jobs during the repair time. It is likely that machine breakdowns can significantly affect the effectiveness of the scheduling procedure unless they are explicitly accounted for. Therefore, this paper develops new machine breakdown terminals for a GP approach and evaluates their effectiveness for a dynamic JSS problem with both dynamic job arrivals and machine breakdowns. The results show that the GP approaches with the machine breakdown terminals do show improvements. The analysis shows that the machine breakdown terminals may indirectly contribute in the evolution of high quality rules, but occur infrequently in the output rules evolved by the machine breakdown GP approaches.

1 Introduction

Job shop scheduling (JSS) problems are combinatorial optimisation problems that have been studied over the past 60 years [1]. Due to their direct application in important real-world manufacturing systems, extensive research has been carried out for JSS problems to find effective and practical techniques which may be incorporated to a real-world scenario for the manufacturers so that they gain a

© Springer International Publishing AG, part of Springer Nature 2018
M. Castelli et al. (Eds.): EuroGP 2018, LNCS 10781, pp. 253–270, 2018.
https://doi.org/10.1007/978-3-319-77553-1_16

competitive edge in the respective markets [1]. In a JSS problem instance, there are *machines* on the *shop floor* that are used to process arriving *jobs*, and the manufacturer needs to make intelligent decisions to process the jobs as effectively as possible. In other words, machine resources need to be optimally allocated (given a specific criterion) by determining the *sequence* in which the jobs are processed. However, optimal allocation of machines can be a difficult task. Most job shop scheduling (JSS) problems are NP-hard [1], and mathematical opti-misation techniques that return optimal solutions for problem instances do not scale effectively with the problem size. In addition, in a dynamic JSS problem instance there are unforeseen events that affect the properties of the shop floor, e.g., dynamic job arrivals and machine breakdowns [2]. To handle dynamic JSS problems, various heuristic approaches have been proposed to generate good solutions to problem instances while coping with the unforeseen events. For this paper, we handle dynamic JSS problems with dynamic job arrivals, where the jobs' properties and their arrival times are unknown until the job arrival times are reached during processing [3]. Dispatching rule approaches are the most prominent method of handling dynamic JSS problems with dynamic job arrivals due to their short reaction times and their ability to cope with the dynamic environment [4].

In addition to manually designing effective dispatching rules for dynamic JSS problems with dynamic job arrivals, researchers have proposed various genetic programming based hyper-heuristic (GP-HH) approaches to automat-ically evolving dispatching rule from heuristic subcomponents [5]. GP evolved dispatching rules generally perform better than man-made dispatching rules for JSS problems [6]. However, GP approaches that have been proposed for dynamic JSS problems have mainly focused on dynamic job arrivals [3–7]. In a real-world scenario, it is likely that there are different types of dynamic events that occur during processing. An example is machine breakdown, where the machines need to be serviced and repaired [2]. It is likely that disruptions caused by machine breakdowns can likely impact the performance of the scheduling algorithm if they are not specifically accounted for. The only GP approach that explicitly accounts for machine breakdowns in the literature deals with a single machine JSS prob-lem with no dynamic job arrivals [8]. Developing machine breakdown specific GP approaches may allow us to improve the overall quality of rules evolved by GP for dynamic JSS problems with dynamic job arrivals and machine breakdowns (DJSS-MB).

1.1 Goal

The goal of this paper is to develop new machine breakdown terminals for a GP terminal set commonly used in the literature [4,7] to handle a DJSS-MB. By incorporating machine breakdown terminals into the GP terminal set, it may be possible to evolve rules that can account for machine breakdown information. This may result in the evolved rules being able to make better decisions for both machine breakdown and non-machine breakdown JSS problem instances than rules evolved without machine breakdown information, and generate better solutions overall. In other words, developing new machine breakdown terminals may allow GP to consistently evolve high quality rules for DJSS-MB. Afterwards, by analysing specific machine breakdown GP evolved rules, it may be possible to develop an insight into how the rules behave in DJSS-MB, allowing us to potentially develop more effective machine breakdown GP approaches in the future. Overall, this paper carries out the following objectives:

(a) Develop and evaluate new machine breakdown terminals for an existing GP approach [4, 7].
(b) Carry out a structural analysis of the machine breakdown GP rules to gain an understanding of the useful features and properties of GP rules that are evolved under machine breakdown.

1.2   Organisation

First, we cover the background to dynamic JSS in Sect. 2, which includes the problem definition and outlines existing GP approaches for dynamic JSS problems. Afterwards, Sect. 3 describes the existing GP approach used in the literature [4, 7], the benchmark GP terminals, and the machine breakdown GP terminals investigated in this paper. Section 4 describes the dynamic JSS simulation model used in this paper, and the GP parameters. Finally, Sect. 5 gives the results and an analysis of the findings, and Sect. 6 gives the concluding remarks and the future works.

2   Background and Related Work

This section covers the problem definition for the DJSS-MB, and the GP approaches for dynamic JSS problems in the literature.

2.1   Problem Definition

In a dynamic JSS problem instance, an arriving job j has a sequence of operations $\sigma_{1j}, \ldots, \sigma_{N_j j}$. The operations must be processed sequentially (e.g. σ_{1j} must be processed before σ_{2j}) and need to be processed on specific machines $m(\sigma_{1j}), \ldots, m(\sigma_{N_j j})$. An operation σ_{ij} needs to be processed on machine $m(\sigma_{ij})$ for time p_{ij} (which is called the *processing time*), and a machine can only process one job at a time. The time that a job arrives at the machine for an operation σ_{ij} is denoted as r_{ij}, and the time that the job arrives on the shop floor (r_{1j}) is called the *job arrival time* r_j. For this paper, the objective is to minimise total weighted tardiness (MWT). MWT objective has been studied extensively in the literature [1], and tardiness related objectives have been shown to be quite sensitive to machine breakdown events [9, 10]. In a MWT objective, an arriving job has a due date d_j and a weight w_j. If job j's completion time C_j (the time when the last operation of job j completes) is below due date d_j, i.e., $C_j \leq d_j$, then no penalty is incurred. Otherwise, job j is considered *tardy*, and has tardiness

$T_j = d_j - C_j$ [11]. After all N arriving jobs are completed, the MWT of the schedule is given by $\frac{1}{N} \sum_{j=1}^{N} w_j T_j$.

The two types of dynamic events that occur during processing are dynamic job arrivals and machine breakdowns. With a dynamic job arrival, the information about a job j, including its properties, are not known in advance until the job arrives at time r_j. With a machine breakdown event, a machine m breaks down at time b^m and the machine needs to be repaired for time r^m. During the repair time, the machine is unavailable to process any operations. If a job's operation is currently being processed at the machine at the time of breakdown, then the job's operation is resumed from the time it was interrupted after the machine is repaired. This means that if a job j's operation σ_{ij} is started at time s_t and is interrupted by machine m's breakdown, then the operation completes at time $s_t + p_{ij} + r^m$. A job operation resuming from the point of interruption is consistent with the machine breakdown definition in the literature [8,9] to handle dynamic JSS problems with machine breakdown. In addition, a common assumption in the literature is that machine breakdowns events are unforeseeable [2]. However, for this paper we simplify the problem by allowing the shop floor to know when the machine breakdown occurs in advance. This is because GP approaches that handle both dynamic job arrivals and machine breakdowns have not yet been proposed in the literature. By doing this, we can carry out a preliminary investigation of machine breakdown terminals that incorporate informations about future machine breakdowns (described in Sect. 3.2). After analysing these terminals that take full information about future machine breakdowns into account, it may be possible to develop machine breakdown terminals in the future that can cope effectively even when the machine breakdown events are unforeseen events.

2.2   GP for Dynamic JSS Problems

GP approaches have been extensively applied to dynamic JSS problems to evolve dispatching rules [5,6]. Many GP approaches use single arithmetic function trees as priority dispatching rules [5,6]. Geiger et al. [12] showed that GP can evolve optimal priority dispatching rules for static JSS problems that are not NP-hard. They also showed that priority dispatching rules evolved for NP-hard static JSS problems and dynamic JSS problems perform better than benchmark man-made dispatching rules. Hildebrandt et al. [3] provided a GP approach for a dynamic JSS problem with the flowtime minimisation objective, and showed that the GP evolved rules outperform state-of-the-art man-made dispatching rules. Branke et al. [5] and Nguyen et al. [6] both carry out extensive survey of GP-HH approaches to scheduling problems in the literature. Nguyen et al. [13] also provide a unified framework for GP-HH to scheduling problems and categorises existing GP approaches using the framework.

The following are GP approaches that evolve scheduling rules for dynamic JSS problems with machine breakdowns. Yin et al. [8] proposed a two-tree GP approach for a single machine JSS problem. The first tree acts as a dispatching rule, and the second tree is used to calculate the idle time to add in between processing different jobs on a machine. They showed that the evolved rules outperform the benchmark man-made heuristics designed for JSS problems with machine breakdowns. Park et al. [10] carried out an investigation on the generality of GP over a JSS problem with different frequencies and durations of machine breakdowns, and found that the proportion of time the machines are being repaired is a significant factor in the qualities of the evolved rules. In addition, they showed that GP is not general enough to cover for all different scenarios effectively, and that it is likely that machine breakdown specific information is required to improve the generality of GP.

3 Machine Breakdown GP for the Dynamic JSS Problem

As machine breakdown GP approach for DJSS-MB have not been proposed, this paper proposes simple but novel machine breakdown terminals which are incorporated to a GP approach. This allows GP to evolve dispatching rules that may make better decisions during decision situations, potentially leading to better performance than GP evolved rules which do not incorporate machine breakdowns. First, we describe the GP representation, the benchmark terminal and function sets. This is followed by the descriptions and justifications for the machine breakdown GP terminals. The first approach replaces existing terminals related to operation processing times and add repair time of machines if necessary. This approach is denoted as "augmented" approach, as it attempts to *improve* certain benchmark terminals by incorporating machine breakdown information. The second approach adds *new* machine breakdown terminals, which "react" to the machine breakdowns happening on the shop floor, to the existing set of GP terminals.

3.1 GP Representation, Terminal Set and Function Set

For this paper, we use a tree-based GP representation [14]. The GP individuals represent arithmetic function trees that calculate the priorities of jobs during decision situations. Arithmetic GP representation has been used prominently in the literature to evolve effective priority dispatching rules for JSS problems [5]. A GP terminal corresponds to a job, machine and shop floor attribute value at a decision situation, or combines multiple base level shop floor attributes as a part of the terminal. For example, the RT terminal returns the sum remaining total processing times of job j waiting at a machine to process operation σ_{ij}, i.e., $\mathrm{RT}(j) = \sum_{k=i}^{N_j} p_{kj}$. The GP terminals and the arithmetic operators used by the benchmark GP approach are listed in Table 1, which is based off existing terminal and function sets used by GP approaches to dynamic JSS problems in the literature [7,10]. The function set consists of the arithmetic operators $+$, $-$, $\times$, protected $/$, binary operators max, min and a ternary operator if. The protected $/$ returns one if the denominator is zero, and carries out the standard division operator otherwise. max and min returns the maximum and the minimum value of the two arguments respectively. if returns the value of the second argument if the value of the first argument is greater than or equal to zero, or returns the value of the third argument otherwise.

Table 1. Terminal set for GP, where a job j is waiting at the available machine m at a decision situation.

Terminal	Description
RJ	Operation ready time of job j
PT	Operation processing time of job j
RO	Remaining number of operations of job j
RT	Remaining total processing times of job j
RM	Machine m's ready time
WINQ	Work in next queue for job j
DD	Job's due date d_j
SL	Slack of job j
W	Job's weight w_j
NPT	Next operation processing time of job j
NNQ	Number of idle jobs waiting at the next machine
NQW	Average waiting time of last 5 jobs at the next machine
AQW	Average waiting time of last 5 jobs at all machines

A GP individual ω is evaluated over a set of dynamic JSS training instances to calculate its fitness as follows. GP individual ω is applied to a JSS problem instance γ as a *non-delay* dispatching rule [11]. During a simulation when a machine m becomes available, the jobs that are currently waiting at machine m are assigned priority values by the dispatching rule. The job that has the highest priority value is selected to be processed at machine m. This continues until the termination criteria for the simulation has been reached (e.g. after a certain number of jobs have been completed), at which point the objective function value $Obj(\omega, \gamma)$ is calculated from the solution generated by individual ω. The individual ω is applied to all problem instances in the training sets, obtaining objective values $Obj(\omega, \gamma_1), \ldots, Obj(\omega, \gamma_{T_{train}})$.

After the GP individual ω is applied to the problem instances in the training set, the objective values obtained are normalised to reduce the likelihood that the GP individuals are biased towards specific instances [3,15]. In other words, an objective value $Obj(\omega, \gamma)$ for a solution generated by individual ω for instance γ is normalised using *reference* objective value $Obj(R, \gamma)$ as shown in Eq. (1). The reference objective value $Obj(R, \gamma)$ is calculated from the solution generated by applying a *reference rule* R to the problem instance γ. The reference rule used is the weighted apparent tardiness cost (wATC) rule [16], a man-made dispatching rule effective for weighted tardiness related objectives. This was also used by Park et al. [10] in the fitness function to evolve GP rules for the DJSS-MB.

$$Obj'(\omega, \gamma) = \frac{Obj(\omega, \gamma)}{Obj(R, \gamma)} \tag{1}$$

3.2   Augmented GP Terminals

The following terminals in the original GP terminal set (as described in Table 1) are replaced by terminals that add repair times of the machines: job's *operation processing time* (PT), job's *next operation processing time* (NPT) and *work in next queue* (WINQ). The replaced GP terminals return the original value if the job's operation is not interrupted by a machine breakdown, and adds the repair time of the machine otherwise. The terminals that incorporate the machine breakdown information is denoted with the prefix 'MB-' (e.g. MBPT for machine breakdown adjusted processing time). The GP approach that incorporates the MBPT, MBNPT and MBWINQ terminals is denoted as GP-Aug.

Machine breakdown adjusted processing time (MBPT): The machine breakdown adjusted processing time terminal (MBPT) replaces the processing time terminal (PT) in Table 1. Given that the current time during the decision situation is t and the processing time of job's current operation j is p_{ij}, MBPT terminal returns the actual duration of time required to process the job's operation by factoring the machine breakdown interruption into account. In other words, if the job is *not* interrupted by a machine breakdown, i.e., if the operation completes earlier than the breakdown time b_t^m of the current machine m, then the job's actual processing time p'_{ij} is equal to the expected processing time p_{ij}. Otherwise, the actual processing time is the sum of the processing time and the machine repair time r_t^m required to get the machine back up and running before the operation is resumed. The value returned by MBPT$(j) = p'_{ij}$, where the calculation for p'_{ij} is shown in Eq. (2).

$$p'_{ij} = \begin{cases} p_{ij} & \text{if } t + p_{ij} < b_t^m \\ p_{ij} + r_t^m & \text{otherwise} \end{cases} \tag{2}$$

Machine breakdown adjusted next processing time (MBNPT): The machine breakdown adjusted next processing time terminal (MBNPT) replaces the next processing time terminal (NPT) in Table 1. MBNPT terminal returns zero if the job j's current operation σ_{ij} is the last operation before job j's completion. Otherwise, given that the next operation $\sigma_{(i+1)j}$ is processed on machine m', the repair time of m' is added to the next processing time $p_{(i+1)j}$ if it is expected to be interrupted by a breakdown at machine m' at operation $\sigma_{(i+1)j}$ earliest possible completion at machine m'. The earliest possible time that job j can be completed is if operation σ_{ij} is selected immediately by machine m, and then the successive operation $\sigma_{(i+1)j}$ is then processed by machine m' as soon as operation σ_{ij} is completed. The time when operation σ_{ij} completes is given by the current time t and the actual processing time p'_{ij}, which depends on whether the operation is interrupted by machine breakdown (Eq. (2)). In other words, the earliest time operation $\sigma_{(i+1)j}$ can be processed at machine m'

is at $t + p'_{ij}$ after operation σ_{ij} is expected to complete. Therefore, if machine m' breaks down before time $t + p'_{ij} + p_{(i+1)j}$, then repair time $r_t^{m'}$ of machine m' is added to the operation $\sigma_{(i+1)j}$'s processing time $p_{(i+1)j}$ as shown in Eq. (3).

$$\text{MBNPT}(j) = \begin{cases} p_{(i+1)j} & \text{if } t + p'_{ij} + p_{(i+1)j} < b_t^{m'} \\ p_{(i+1)j} + r_t^{m'} & \text{otherwise} \end{cases} \tag{3}$$

Machine breakdown adjusted work in next queue (MBWINQ): The machine breakdown adjusted work in next queue terminal (MBWINQ) replaces the work in next queue terminal (WINQ) in Table 1. Both WINQ and MBWINQ terminals return zero if the job j's current operation σ_{ij} is the last operation before job j's completion. Otherwise, given that machine m' is required by operation $\sigma_{(i+1)j}$, the standard WINQ terminal returns the sum processing times of the jobs that are currently waiting at machine m' plus the remaining time required to process the operation currently being processed by machine m', i.e., the work remaining. MBWINQ modifies the work remaining time calculated by WINQ, and adds the machine m''s repair time if the work is interrupted by machine breakdown at time $b_t^{m'}$. In other words, $\text{MBWINQ}(j) = wr'_{m',j'} + \sum_{i=1}^{N_{m'}} p_{j_i}$, where $p_{j_1}, \ldots, p_{j_{N_{m'}}}$ are the processing times of jobs waiting at machine m', $wr'_{m'j'}$ is the actual work remaining required on j' being processed on machine m' before it becomes available. The calculation for actual work remaining $wr'_{m'j'}$ is given in Eq. (4), where $s_{j'}$ denotes the time when j' started, $p_{j'}$ is the processing time required by j' at machine m' and t is the current time.

$$wr'_{m'j'} = \begin{cases} s_{j'} + p_{j'} - t & \text{if } s_{j'} + p_{j'} < b_t^{m'} \\ s_{j'} + p_{j'} - t + r_t^{m'} & \text{otherwise} \end{cases} \tag{4}$$

In summary, the augmented GP approach replaces three existing terminals (PT, NPT and WINQ) with equivalent terminals that incorporate information about future machine breakdowns. The existing terminals are related to the processing times of the jobs waiting on the shop floor, where repair times need to be added onto the processing times if we expect the jobs to be interrupted by machine breakdowns. By doing this, we expect the GP rule to be able to use the "actual" processing times of the jobs to make better decisions on what job should be processed next by a machines during decision situations.

3.3 Reactive GP Terminals

Reactive machine breakdown terminals are added to the GP terminal set described in Table 1 and incorporate information about current machine status. As the two terminals incorporate informations about the potential wait time of a job waiting at a machine for the next machine it visits, they are investigated separately. The two terminals being investigated are the *repair time remaining next machine terminal* (RTR) and the *minimum wait time next machine terminal* (WT). The two reactive GP terminals may allow rules to make better decisions by prioritising jobs with low expected wait time compared to jobs with high expected wait time. This may lead to jobs spending less time waiting at busy machines, and the evolved rules may generate higher quality schedules. The GP approach that incorporates the RTR terminal is denoted as GP-RTR, and the GP approach that incorporates the WT terminal is denoted as GP-WT.

Repair Time Remaining Next Machine (RTR): The repair time remaining next machine RTR returns zero if a job j waiting at a machine at time t is currently on its last operation or the next machine m' visited by j is currently not broken down. Otherwise, given that machine m' broke down at time $b_t^{m'}$ and the repair time is $r_t^{m'}$, the value given by RTR $= b_t^{m'} + r_t^{m'} - t$.

Minimum Wait Time Next Machine (WT): The minimum wait time next machine WT returns the earliest time that the machine to be visited by job j next becomes available. If the current operation of j is the last operation before completion, then WT returns zero. In addition, if the next machine m' that job j visits is currently not busy and is not broken down, i.e., is completely available, then WT returns zero. Otherwise, the WT returns the duration of time required for machine m' to be available. If machine m' is currently processing a job j' or is broken down with an interrupted job, then it returns the actual work remaining $wr'_{m'j'}$ which is given in Eq. (4). Otherwise, if the m' is broken down and a job was not interrupted by the machine breakdown, WT returns the remaining repair time of machine m' as given by the terminal RTR.

4   Experimental Design

This section describes the setup used to evaluate the different GP approaches to tackle the DJSS-MB. To evaluate the machine breakdown GP approaches, a simulation model that is slightly modified from existing simulation models in the literature [9,10] is used to both evolve and evaluate the evolved rules. Afterwards, we provide the parameter used by the GP approaches.

4.1   Dynamic Simulation Model with Machine Breakdown

Discrete-event simulations are the most prominent method of generating dynamic JSS problem instances [5,6]. In a discrete-event simulation, the dynamic events such as the job arrivals and the machine breakdowns are stochastically generated from a set of parameters. A simulation configuration is the set of parameters required, along with a seed value, to generate a dynamic JSS problem instance. In a dynamic JSS problem instance, there are $M = 10$ machines on the shop floor. The problem instance has a "warm-up" period of 500 jobs, where the first 500 jobs completed do not contribute towards the objective. The simulation is terminated after 2500 jobs are completed, and the objective function value is calculated from the 2000 jobs completed after the warm-up phase. The job arrivals times follow a Poisson process with arrival rate $\lambda = \rho \times \mu \times p_M$. In the equation, ρ is the utilisation rate, μ the mean processing time, and p_M

the mean number of job operations to machine ratio. Utilisation rate (ρ) is the expected proportion of time the machines are spent processing job operations, and $\rho = 80\%$ for all problem instances. If the utilisation rate plus the machine breakdown level is too high, it is very likely that the shop will be *unstable* [11], i.e., job arrival rate is greater than the rate at which the shop floor can process them. Therefore, 80% utilisation rate is used instead of higher utilisation rates used in the literature (e.g. 90% or 95% [4,9]) to accommodate for the high level of machine breakdown used by the simulation model (described below). The mean processing time (μ) is used in a uniform distribution with the interval $[1, 2\mu - 1]$ that the jobs' processing times follow, and $\mu = 25$ for all problem instances. The mean number of job operations to machine ratio (p_M) is the expected number of machines that a job will visit divided by the total number of machines. The number of operations a job has follows a uniform distribution in the interval $[2, 10]$, i.e., the minimum and the maximum number of operations that a job can have is 2 and 10 respectively. Therefore, the expected number of operations is 6 for a job arriving on the shop floor and $p_M = 0.6$ for all problem instances. In addition, a job's weight has the value of 1, 2 or 4 with 20%, 60% or 20% probabilities respectively. Given a job j's arrival time r_j, total processing time $\sum_{i=1}^{N_j} p_{ij}$ and the due date tightness simulation parameter h, the due date of job j is $d_j = r_j + h \times \sum_{i=1}^{N_j} p_{ij}$.

For generating machine breakdown events, the inter-breakdown times of the machines follow an exponential distribution, and the expected breakdown rate is given by $\eta = r_t / \pi - r_t$. In the equation, π is the breakdown level and r_t is the machine repair time. The breakdown level is the expected proportion of time for the machine to be broken down over the course of the simulation, and varies between the different simulation configurations used to generate the problem instances. For a problem instance the machine repair time is the same across all machines for a problem instance.

The dataset parameters for generating job arrivals and machine breakdowns are shown in Table 2. First, the simulation configurations have the possible breakdown levels $\pi = 0\%$, 2.5%, 5%, 10%, or 15% for a simulation configuration. Second, fixed repair times for the machine breakdowns are either $r_t = 37.5$, 137.5, or 262.5 for a simulation configuration. These parameters were selected after running the benchmark GP approach on different breakdown levels and durations of repair times as part of a preliminary experiment. The simulation configuration consists of a combination of the dataset parameters, which means that there are $3 \times 5 \times 2 = 30$ different simulation configurations available in the dataset. We use the simulation configuration with $\pi = 15\%$, $r_t = 262.5$ and $h = 3$ to generate the training problem instances. In addition, a different seed is used with the training simulation configuration every generation during the GP process, resulting in different dynamic JSS problem instances being used every generation.

Table 2. Dynamic JSS parameter settings

Simulation model parameter	Value
Number of machines (M)	10
Utilisation rate (ρ)	80%
Mean processing time (μ)	25
Weight/probability ((w, p))	$\{(1, 20\%), (2, 60\%), (4, 20\%)\}$
Due date tightness (h)	3 or 5
Machine breakdown level (π)	0%, 2.5%, 5%, 10% or 15%
Repair time (r_t)	37.5, 137.5 or 262.5

4.2 GP Parameters

The GP parameters are used by the GP approaches are shown in Table 3. The GP parameters are the same as the parameters that are same as the ones used by Park et al. [10] in their investigation into GP approaches for a DJSS-MB, which allows our benchmark GP approach to be consistent with the GP approach that was used during their investigation.

Table 3. GP parameter settings

GP Parameter	Value
Population size	1024
Number of generations	51
Crossover rate	80%
Mutation rate	10%
Reproduction rate	10%
Max initial depth	2
Max depth	8
Initialisation method	Ramped half-and-half
Selection method	Tournament selection
Selection size	2

5 Experimental Results

In this investigation, we first carry out a performance evaluation of the GP approaches. The performance evaluation first compares the GP approaches and how consistently they can evolve high quality dispatching rules for the dynamic JSS problem, i.e., measures the effectiveness of the GP approaches. This is done by evolving a set of rules for each approach and applying them over the dynamic JSS simulation model. Afterwards, the best rules are extracted from the sets of evolved rules for the GP approaches and compared individually to determine whether an individual machine breakdown rule can outperform an evolved rule generated by the benchmark GP approach. Finally, we carry out a structural analysis of the best rules evolved by the machine breakdown GP approaches to find out the useful properties from the evolve rules.

5.1   Performance Evaluation

For the performance evaluation, each GP approach is applied to a training set (described in Sect. 4.1) thirty times to evolve thirty independent rules. Afterwards, each of the rule is applied to the dynamic JSS simulation model as follows.

Performance Measure: First, an evolved rule ω is run multiple times over each simulation configuration in the simulation model. A single run consists of a seed value and a simulation configuration, which are used to generate a dynamic JSS problem instance. The rule is then applied to the problem instance and generates a schedule, which has a MWT objective value. Afterwards, the subsequent runs over the simulation configuration use unique seeds so that new problem instances are generated from the same simulation configuration. In other words, given a simulation configuration sim and rule ω, schedules with MWT values $Obj(\omega, \gamma_{(sim)1}), \ldots, Obj(\omega, \gamma_{(sim)30})$ are generated by the rule for the given simulation configuration. These are used slightly differently for the rule set evaluation and best rule evaluation, which are described below.

Rule Set Results: In the rule set evaluation, the MWT values $Obj(\omega, \gamma_{(sim)1})$, $\ldots, Obj(\omega, \gamma_{(sim)30})$ generated by a rule ω for a simulation configuration sim is averaged out to obtain the "performance" $Perf$ of the rule over the simulation configuration, i.e., $Perf(\omega) = \frac{1}{30} \sum_{i=1}^{30} Obj(\omega, \gamma_{(sim)i})$. The rule performances are then used to compare between the different sets of rules evolved by the GP approaches.

The results of the performance evaluation is shown in Table 4. In the table, $\langle \pi, r_t, h \rangle$ denotes that the simulation configuration has the respective breakdown level π, repair time r_t, and due date tightness h. In addition, each entry $\mu \pm \sigma$ is the mean (μ) and standard deviation (σ) of the performance $Perf$ of the rules for the simulation configuration respectively. If a set of GP evolved rules that use the machine breakdown terminals is *significantly* better than the set of benchmark GP rules for a simulation configuration by satisfying the two sided Student's t-test at $p = 0.05$, then the particular entry is highlighted.

Although the differences are not significant, the results show that the three machine breakdown approaches (GP-Aug, GP-WT and GP-RTR) have slightly better performances than the benchmark GP for some simulation configurations. In particular, the GP-WT rules have slightly better performances for all simulation configurations than the benchmark GP rules. In addition, the GP-RTR rules

Table 4. Comparison of the rule sets evolved by the GP approaches over the simulation configurations. Rules are evolved from $\langle 0.15, 262.5, 3\rangle$.

Model Subset		MB GP-Aug	MB GP-WT	MB GP-RTR	GP
MWT ($\times 10^2$)	$\langle 0, 37.5, 5\rangle$	0.74 ± 0.17	0.66 ± 0.07	0.67 ± 0.13	0.67 ± 0.16
	$\langle 0, 37.5, 3\rangle$	1.12 ± 0.16	1.05 ± 0.07	1.06 ± 0.12	1.07 ± 0.15
	$\langle 0, 137.5, 5\rangle$	0.60 ± 0.16	0.53 ± 0.06	0.53 ± 0.11	0.53 ± 0.14
	$\langle 0, 137.5, 3\rangle$	1.31 ± 0.18	1.24 ± 0.07	1.24 ± 0.12	1.26 ± 0.16
	$\langle 0, 262.5, 5\rangle$	0.74 ± 0.17	0.66 ± 0.07	0.66 ± 0.13	0.66 ± 0.16
	$\langle 0, 262.5, 3\rangle$	1.36 ± 0.19	1.28 ± 0.08	1.29 ± 0.13	1.30 ± 0.16
	$\langle 0.025, 37.5, 5\rangle$	1.60 ± 0.89	1.54 ± 0.52	1.55 ± 0.75	1.59 ± 0.69
	$\langle 0.025, 37.5, 3\rangle$	2.92 ± 0.98	2.78 ± 0.52	2.86 ± 0.76	2.95 ± 0.88
	$\langle 0.025, 137.5, 5\rangle$	38.39 ± 11.90	36.07 ± 11.10	38.91 ± 10.38	39.20 ± 12.65
	$\langle 0.025, 137.5, 3\rangle$	42.53 ± 12.23	40.49 ± 11.10	42.96 ± 10.87	43.23 ± 12.89
	$\langle 0.025, 262.5, 5\rangle$	88.51 ± 25.15	89.94 ± 23.00	92.91 ± 23.63	94.43 ± 27.18
	$\langle 0.025, 262.5, 3\rangle$	92.11 ± 24.34	93.81 ± 21.81	96.36 ± 22.91	98.17 ± 26.54
	$\langle 0.05, 37.5, 5\rangle$	1.64 ± 0.44	1.53 ± 0.20	1.57 ± 0.35	1.58 ± 0.36
	$\langle 0.05, 37.5, 3\rangle$	2.76 ± 0.57	2.68 ± 0.30	2.74 ± 0.52	2.78 ± 0.50
	$\langle 0.05, 137.5, 5\rangle$	9.04 ± 2.17	8.29 ± 2.03	8.91 ± 1.89	9.05 ± 2.42
	$\langle 0.05, 137.5, 3\rangle$	11.14 ± 2.18	10.65 ± 1.87	11.04 ± 1.97	11.28 ± 2.29
	$\langle 0.05, 262.5, 5\rangle$	36.43 ± 11.35	34.32 ± 10.75	37.00 ± 10.13	37.38 ± 12.40
	$\langle 0.05, 262.5, 3\rangle$	36.56 ± 11.29	34.33 ± 10.20	36.74 ± 9.61	37.27 ± 12.09
	$\langle 0.1, 37.5, 5\rangle$	3.69 ± 0.61	3.53 ± 0.27	3.60 ± 0.50	3.63 ± 0.60
	$\langle 0.1, 37.5, 3\rangle$	4.60 ± 0.54	4.51 ± 0.26	4.57 ± 0.44	4.63 ± 0.51
	$\langle 0.1, 137.5, 5\rangle$	6.11 ± 1.21	5.79 ± 0.35	5.89 ± 0.81	6.07 ± 1.26
	$\langle 0.1, 137.5, 3\rangle$	8.15 ± 1.28	7.91 ± 0.44	8.02 ± 0.86	8.29 ± 1.39
	$\langle 0.1, 262.5, 5\rangle$	11.72 ± 1.95	11.07 ± 1.56	11.43 ± 1.56	11.74 ± 2.22
	$\langle 0.1, 262.5, 3\rangle$	13.14 ± 1.50	12.61 ± 1.52	13.08 ± 1.21	13.29 ± 1.76
	$\langle 0.15, 37.5, 5\rangle$	6.20 ± 0.67	5.89 ± 0.21	6.07 ± 0.50	6.06 ± 0.80
	$\langle 0.15, 37.5, 3\rangle$	7.73 ± 0.63	7.52 ± 0.18	7.66 ± 0.50	7.74 ± 0.82
	$\langle 0.15, 137.5, 5\rangle$	9.02 ± 0.94	8.53 ± 0.59	8.69 ± 0.64	8.81 ± 1.17
	$\langle 0.15, 137.5, 3\rangle$	11.73 ± 0.81	11.38 ± 0.42	11.55 ± 0.58	11.71 ± 1.10
	$\langle 0.15, 262.5, 5\rangle$	12.55 ± 1.23	12.01 ± 1.29	12.31 ± 0.95	12.48 ± 1.44
	$\langle 0.15, 262.5, 3\rangle$	16.60 ± 1.27	16.08 ± 1.33	16.37 ± 0.97	16.59 ± 1.50

have slightly better performance than the benchmark GP rules for most simulation configurations except configurations $\langle 0, 37.5, 5\rangle$ and $\langle 0.15, 37.5, 5\rangle$. Finally, the results of the comparison between the GP-Aug rules and the benchmark GP rules is most mixed, where GP-Aug rules are slightly better or worse than the benchmark rules on roughly the equal number of simulation configurations. However, due to the lack of statistical significance, no significant conclusions can be drawn on whether the machine breakdown GP approaches is more consistent in evolving higher quality dispatching rules than the benchmark GP approach. However, by analysing the rules further, it may be possible to gain a better understanding of how GP can be applied effectively to the machine breakdown problem.

Best Rule Results: The best rule from each GP approach are compared against each other after the performances of the rules are compared. The best rule is defined to be the rule that has the lowest average performance values over all the simulation configurations out of the evolved rules. The best rules are then compared on each simulation configuration by the MWT values from the generated schedules. In other words, best rule comparison uses the results

$Obj(\omega, \gamma_{(sim)1}), \ldots, Obj(\omega, \gamma_{(sim)30})$ from the rules being applied to the 30 problem instances generated by each simulation configuration sim. The results of the best rules being applied to each simulation configuration is shown in Table 5, where each entry $\mu \pm \sigma$ is the mean (μ) and standard deviation (σ) of the MWT values generated by the best rule after being applied to 30 independent problem instances generated from the simulation configuration.

Table 5. Comparison of the best rules over the simulation configurations.

Data Subset		MB GP-Aug	MB GP-WT	MB GP-RTR	GP
MWT ($\times 10^2$)	$\langle 0, 37.5, 5 \rangle$	0.69 ± 0.31	0.63 ± 0.36	0.63 ± 0.31	0.63 ± 0.29
	$\langle 0, 37.5, 3 \rangle$	1.06 ± 0.27	1.01 ± 0.28	1.01 ± 0.27	1.01 ± 0.25
	$\langle 0, 137.5, 5 \rangle$	0.55 ± 0.17	0.47 ± 0.16	0.49 ± 0.14	0.50 ± 0.15
	$\langle 0, 137.5, 3 \rangle$	1.24 ± 0.34	1.21 ± 0.34	1.23 ± 0.37	1.18 ± 0.33
	$\langle 0, 262.5, 5 \rangle$	0.67 ± 0.32	0.60 ± 0.30	0.61 ± 0.27	0.63 ± 0.29
	$\langle 0, 262.5, 3 \rangle$	1.30 ± 0.46	1.25 ± 0.46	1.27 ± 0.43	1.24 ± 0.41
	$\langle 0.025, 37.5, 5 \rangle$	1.42 ± 0.96	1.72 ± 1.49	1.13 ± 0.70	1.19 ± 0.56
	$\langle 0.025, 37.5, 3 \rangle$	2.47 ± 2.17	2.54 ± 2.14	2.15 ± 2.04	2.33 ± 2.48
	$\langle 0.025, 137.5, 5 \rangle$	15.84 ± 9.08	16.73 ± 10.11	15.46 ± 8.12	17.85 ± 7.83
	$\langle 0.025, 137.5, 3 \rangle$	19.36 ± 11.74	21.39 ± 12.86	18.30 ± 12.22	21.28 ± 11.67
	$\langle 0.025, 262.5, 5 \rangle$	44.44 ± 24.05	51.07 ± 23.93	48.60 ± 26.00	46.87 ± 24.62
	$\langle 0.025, 262.5, 3 \rangle$	50.49 ± 35.66	56.92 ± 37.17	54.08 ± 37.06	52.61 ± 35.43
	$\langle 0.05, 37.5, 5 \rangle$	1.59 ± 0.63	1.63 ± 0.79	1.41 ± 0.53	1.42 ± 0.49
	$\langle 0.05, 37.5, 3 \rangle$	2.92 ± 1.16	2.89 ± 1.25	2.54 ± 0.94	2.55 ± 0.88
	$\langle 0.05, 137.5, 5 \rangle$	4.59 ± 2.45	4.44 ± 2.91	4.61 ± 2.41	5.62 ± 2.84
	$\langle 0.05, 137.5, 3 \rangle$	7.00 ± 3.67	7.06 ± 4.24	7.01 ± 3.47	8.03 ± 3.90
	$\langle 0.05, 262.5, 5 \rangle$	14.29 ± 7.32	15.76 ± 8.02	14.78 ± 7.12	16.50 ± 6.88
	$\langle 0.05, 262.5, 3 \rangle$	14.30 ± 5.66	14.56 ± 6.10	14.97 ± 4.69	16.87 ± 4.78
	$\langle 0.1, 37.5, 5 \rangle$	3.83 ± 1.34	3.83 ± 1.26	3.29 ± 1.10	3.40 ± 1.24
	$\langle 0.1, 37.5, 3 \rangle$	4.96 ± 1.64	4.84 ± 1.42	4.45 ± 1.23	4.38 ± 1.18
	$\langle 0.1, 137.5, 5 \rangle$	5.65 ± 1.13	5.41 ± 1.45	5.23 ± 1.12	5.58 ± 1.28
	$\langle 0.1, 137.5, 3 \rangle$	7.71 ± 1.41	7.62 ± 1.60	7.16 ± 1.50	7.64 ± 1.69
	$\langle 0.1, 262.5, 5 \rangle$	8.39 ± 2.58	7.56 ± 3.37	8.18 ± 2.77	9.28 ± 2.95
	$\langle 0.1, 262.5, 3 \rangle$	10.12 ± 1.38	9.27 ± 1.59	10.44 ± 1.52	11.12 ± 1.48
	$\langle 0.15, 37.5, 5 \rangle$	5.89 ± 1.48	5.52 ± 1.59	5.60 ± 1.31	5.56 ± 1.16
	$\langle 0.15, 37.5, 3 \rangle$	7.70 ± 1.33	7.52 ± 1.27	7.37 ± 1.26	7.07 ± 0.99
	$\langle 0.15, 137.5, 5 \rangle$	7.66 ± 1.27	6.79 ± 1.30	7.55 ± 1.15	7.90 ± 1.16
	$\langle 0.15, 137.5, 3 \rangle$	10.97 ± 1.56	10.62 ± 1.88	10.51 ± 1.34	10.74 ± 1.11
	$\langle 0.15, 262.5, 5 \rangle$	10.27 ± 1.25	8.85 ± 1.04	9.89 ± 1.11	10.91 ± 1.03
	$\langle 0.15, 262.5, 3 \rangle$	13.76 ± 1.99	13.10 ± 2.15	13.90 ± 1.75	14.81 ± 1.67

The best rules from the machine breakdown GP approaches show greater difference in the performance to the best rule from the benchmark GP approach. The best machine breakdown GP rules are significantly better than the best benchmark GP rule for certain simulation configurations, e.g., all three machine breakdown GP rules perform better than the GP rule for the $\langle 0.15, 262.5, 3 \rangle$ simulation configuration. Therefore, it is likely for GP approaches with the machine breakdown terminals to evolve high quality individual rules than the benchmark GP approach.

5.2   Rule Analysis

The evolved machine breakdown GP rules are analysed further by carrying out a qualitative analysis based on the structures of the evolved rules. First, the best rules are simplified to remove any redundant branches (e.g. if an `if` will only return the "if" sub branch, then the `if` operator is replaced with the "if" branch) before analysing the structures of the rules. The simplified best rules for GP-Aug, GP-WT, and GP-RTR are shown in Fig. 1a, b and c respectively.

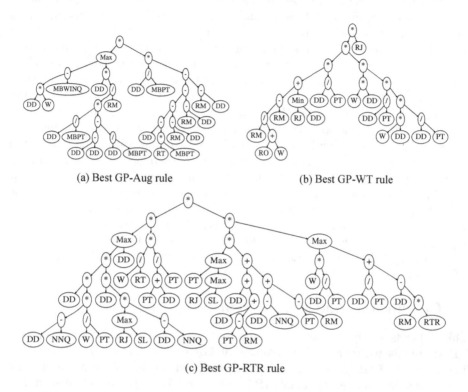

(a) Best GP-Aug rule   (b) Best GP-WT rule

(c) Best GP-RTR rule

Fig. 1. The structures of the best rules found by the GP approaches.

An important observation from the best rules evolved by GP-WT and GP-RTR is the lack of machine breakdown terminals that make up the best rules. The best rule from GP-WT has *no* occurrence of the WT terminal that is incorporated into the terminal set, and the best rule from GP-RTR has *one* occurrence of the RTR terminal. Therefore, it may be the case that the machine breakdown terminals do not directly contribute towards the qualities of the final evolved rules. Instead, the machine breakdown terminals may facilitate the evolution of good GP rules, and are discarded from the best GP individuals near the end of the GP process. This may explain the lack of machine breakdown terminals in best GP-WT and GP-RTR rules, but why the best rules generally perform better than the best benchmark rule In addition, it may also explain why GP-WT and GP-RTR rules also perform slightly better than the benchmark GP rules.

For the best rules from the GP approaches, the method in which the non-machine breakdown related terminals are combined may also be a factor in the effectiveness of the rules. These include the frequent occurrence of important terminals such as the job's weights and processing time in the best evolved rules. Intuitively, important jobs with short processing time should be prioritised out of the jobs waiting at the available machine. However, in all three machine breakdown GP rules (and the best benchmark GP rule), there are many segments of the tree that form DD/PT, which indicates that the best rules prioritise jobs with high due date and low processing time. This is contrary to the expectation that jobs with low due date (i.e. jobs that are more urgent) should be prioritised first. A possible explanation is that the due date terminal is *time variant*, i.e., expected due dates of jobs steady increases with the duration of the simulation. On the other hand, the processing time terminal is *time invariant*, i.e., the expected processing times of jobs remains relatively the same over the whole duration of the simulation. Therefore, the relative differences in the due date between an urgent job and a non-urgent job waiting on a machine late in the simulation may not be as big as the differences in their processing time, due to the large due date values of both the urgent and non-urgent jobs. This may result in the due date of a job for long simulations being used as an arbitrary large value that can be combined with the processing time terminal using the protected / operator to form a composite that prioritises short processing times. Further experiments can be carried out to determine whether the same phenomenon occurs by replacing the processing time terminal with 1/PT terminal in future GP approaches.

6   Conclusions and Future Work

This paper is a very first piece of work that develops new machine breakdown GP terminals to improve the qualities of GP evolved rules for a DJSS-MB. The first set of GP terminals (called "augmented terminals") replace existing processing time related terminals (PT, NPT and WINQ) with equivalent terminals that take potential machine breakdown into account. The second set of GP approaches (called "reactive terminals") add new terminals (RTR and WT) that gives information on current state of the shop floor. The machine breakdown GP approach does not evolve significantly better rules overall, but the best rules evolved by the machine breakdown GP significantly outperform the best rule evolved by the benchmark GP. The analysis shows very interesting results and insights, where the machine breakdown terminals appear infrequently in the best rules for GP-WT and GP-RTR. Hypotheses have been raised to explain why this is the case, and further work will be needed in this direction. We hope that this work can attract more people to start their work in this direction in the near future.

For the future work, further analysis based on the behaviours of the evolved rules will be carried out. Analysis of evolved rule behaviours in JSS problems have been carried out in the literature [17,18], and further investigation into the behaviours of rules evolved for DJSS-MB may allow better machine breakdown specific approaches to be developed. In addition, the relation between the utilisation rate of job shop scheduling problems and the machine breakdown level will be explored further by analysing rule behaviours in different shop environments. For example, a rule evolved for shop with low utilisation rate and high machine breakdown will be compared against a rule evolved for shop with high utilisation rate and low machine breakdown. This relation may help us develop further insight into machine breakdowns and how the properties of the shop changes with such disruptions.

References

1. Potts, C.N., Strusevich, V.A.: Fifty years of scheduling: a survey of milestones. J. Oper. Res. Soc. **60**(1), S41–S68 (2009)
2. Ouelhadj, D., Petrovic, S.: A survey of dynamic scheduling in manufacturing systems. J. Sched. **12**(4), 417–431 (2009)
3. Hildebrandt, T., Heger, J., Scholz-Reiter, B.: Towards improved dispatching rules for complex shop floor scenarios: a genetic programming approach. In: Proceedings of Genetic and Evolutionary Computation Conference (GECCO 2010), pp. 257–264. ACM, New York (2010)
4. Nguyen, S., Zhang, M., Johnston, M., Tan, K.C.: A computational study of representations in genetic programming to evolve dispatching rules for the job shop scheduling problem. IEEE Trans. Evol. Comput. **17**(5), 621–639 (2013)
5. Branke, J., Nguyen, S., Pickardt, C.W., Zhang, M.: Automated design of production scheduling heuristics: A review. IEEE Trans. Evol. Comput. **20**(1), 110–124 (2016)
6. Nguyen, S., Mei, Y., Ma, H., Chen, A., Zhang, M.: Evolutionary scheduling and combinatorial optimisation: applications, challenges, and future directions. In: Proceedings of IEEE Congress on Evolutionary Computation (CEC 2016), pp. 3053–3060 (2016)
7. Hunt, R., Johnston, M., Zhang, M.: Evolving "less-myopic" scheduling rules for dynamic job shop scheduling with genetic programming. In: Proceedings of Genetic and Evolutionary Computation Conference (GECCO 2014), pp. 927–934. ACM, New York (2014)
8. Yin, W.J., Liu, M., Wu, C.: Learning single-machine scheduling heuristics subject to machine breakdowns with genetic programming. In: Proceedings of IEEE Congress on Evolutionary Computation, CEC 2003, pp. 1050–1055 (2003)
9. Holthaus, O.: Scheduling in job shops with machine breakdowns: an experimental study. Comput. Ind. Eng. **36**(1), 137–162 (1999)
10. Park, J., Mei, Y., Nguyen, S., Chen, G., Zhang, M.: Investigating the generality of genetic programming based hyper-heuristic approach to dynamic job shop scheduling with machine breakdown. In: Wagner, M., Li, X., Hendtlass, T. (eds.) ACALCI 2017. LNCS (LNAI), vol. 10142, pp. 301–313. Springer, Cham (2017). https://doi.org/10.1007/978-3-319-51691-2_26
11. Pinedo, M., Hadavi, K.: Scheduling: theory, algorithms and systems development. In: Gaul, W., Bachem, A., Habenicht, W., Runge, W., Stahl, W.W. (eds.) ORP 1991. Operations Research Proceedings 1991, vol. 1991. Springer, Heidelberg (1992). https://doi.org/10.1007/978-3-642-46773-8_5
12. Geiger, C.D., Uzsoy, R., Aytuğ, H.: Rapid modeling and discovery of priority dispatching rules: an autonomous learning approach. J. Sched. **9**(1), 7–34 (2006)
13. Nguyen, S., Mei, Y., Zhang, M.: Genetic programming for production scheduling: a survey with a unified framework. Complex Intell. Syst. **3**(1), 41–66 (2017)

14. Koza, J.R.: Genetic Programming: on the Programming of Computers by Means of Natural Selection. MIT Press, Cambridge (1992)
15. Mei, Y., Zhang, M., Nguyen, S.: Feature selection in evolving job shop dispatching rules with genetic programming. In: Proceedings of the 2016 Conference on Genetic and Evolutionary Computation, pp. 365–372 (2016)
16. Vepsalainen, A.P.J., Morton, T.E.: Priority rules for job shops with weighted tardiness costs. Manag. Sci. **33**(8), 1035–1047 (1987)
17. Hildebrandt, T., Branke, J.: On using surrogates with genetic programming. Evol. Comput. **23**(3), 343–367 (2015)
18. Hart, E., Sim, K.: A hyper-heuristic ensemble method for static job-shop scheduling. Evol. Comput. **24**(4), 609–635 (2016)

Structurally Layered Representation Learning: Towards Deep Learning Through Genetic Programming

Lino Rodriguez-Coayahuitl[(✉)], Alicia Morales-Reyes, and Hugo Jair Escalante

Instituto Nacional de Astrofisica, Optica y Electronica,
72840 Tonantzintla, Puebla, Mexico
{linobi,a.morales,hugojair}@inaoep.mx

Abstract. We introduce a novel method for representation learning based on genetic programming (GP). Inspired into the way that deep neural networks learn descriptive/discriminative representations from raw data, we propose a structurally layered representation that allows GP to learn a feature space from large scale and high dimensional data sets. Previous efforts from the GP community for feature learning have focused on small data sets with a few input variables, also, most approaches rely on domain expert knowledge to produce useful representations. In this paper, we introduce the structurally layered GP formulation, together with an efficient scheme to explore the search space and show that this framework can be used to learn representations from large data sets of high dimensional raw data. As case of study we describe the implementation and experimental evaluation of an autoencoder developed under the proposed framework. Results evidence the benefits of the proposed framework and pave the way for the development of *deep genetic programming*.

Keywords: Representation learning · Deep learning
Feature extraction · Genetic programming
Evolutionary machine learning

1 Introduction

Machine learning (ML) algorithms, for tasks such as classification, prediction or clustering, require that fed training samples are described in a compact yet discriminative form, in order to deliver a correct output in a reasonable amount of time. This description is known as the data *representation*. Usually, the representation is inferred and generated by domain experts who characterize particularities of the problem under analysis. Although domain experts' knowledge is critical, in certain scenarios they may disregard certain features that could be relevant for machine learning methods. In addition, in some cases, domain experts are not even available.

© Springer International Publishing AG, part of Springer Nature 2018
M. Castelli et al. (Eds.): EuroGP 2018, LNCS 10781, pp. 271–288, 2018.
https://doi.org/10.1007/978-3-319-77553-1_17

Representation learning is a subfield of ML that aims to automatically learn models able to transform raw data (e.g., image pixels, audio signals) into a representation meaningful for ML methods approaching a particular task (e.g., image classification, speech recognition) [1,2]. Among the existent solutions, which comprise a wide variety of techniques such as matrix factorization [3], linear discriminant analysis [4], principal component analysis [5] and genetic programming (GP) [6], among others [7], deep learning based methods have proved to be among the most effective ones [1,2]. Deep learning models are those formed by stacked layers of parameters, that together, learn highly non linear functions. In essence, they are artificial neural networks (ANN) with many hidden layers. These deep neural networks (DNNs) are based on the idea that a slightly more abstract representation is generated with each additional layer. Thus, with enough of these layers, the initial representation (raw data) can be transformed into a more descriptive/discriminative one, which in most cases is also much more compact than the initial input. The learned representation is expected to be more *meaningful* for ML algorithms that take the learned representation as input.

As previously mentioned, GP has also been used for representation learning with mixed results [8–12]. In these works, a GP individual codifies a function that generates a new representation by combining the initial features. Although competitive results have been obtained with these implementations, they have only focused on small data sets with a few dozens of input variables (features). When initial representations are high dimensional (e.g., composed of hundreds or thousands of features), such as in image processing problems, GP approaches to representation learning need to leverage from human expert knowledge of the problem's domain in order to achieve competitive results, an undesirable property considering a trend towards higher degrees of automation, and unlike deep learning, which most important aspect is being domain agnostic.

In this paper, we introduce a novel methodology for representation learning based on GP and inspired into the way deep learning techniques process information. The proposed method is able to deal with both previous limitations of state of the art GP solutions for representation learning. The main novel components of our proposal are an arrangement of GP populations that processes information locally and a learning strategy inspired in online learning for feature-trees fitting. Both components can be used to learn deep genetic programs, in the sense that layers of the same model could be stacked if necessary. As a case study, we propose a GP based autoencoder. This model is evaluated on benchmark datasets. Experimental results show the feasibility of deep representation learning via GP. We foresee this paper will pave the way for the development of a new paradigm for representation learning based on GP. The main contributions of this paper are as follows:

- We introduce the idea of deep genetic programming. A methodology based on GP and inspired in deep learning.
- We propose two mechanisms that make GP suitable for learning representations in large scale and high dimensional data sets starting from raw data.

– We show the feasibility of the deep genetic programming framework by proposing an autoencoder that is based on the proposed formulation. This method is evaluated in benchmark high dimensional and large scale data showing competitive results.

These contributions and reported evidence show that a multilayered GP based representation learning is not only possible in the near future, but it might yield competitive results against those of state-of-the-art DNN.

2   Related Work

The ability of ANN to process raw data in order to generate representations useful for ML tasks such as classification has been recognized and documented in works at least as early as [13,14]. Even though researchers in the area were already aware that adding up layers to their ANN was beneficial for improving the quality of representations and/or classification performance, training ANN with more than two or three layers proved to be difficult. Several advances in the area over recent years have allowed researchers to train ANN with several layers [15–17]. *Alexnet* [15] is perhaps one of the most representative works in this regard, being a deep network for its time, and achieving unprecedented classification performance, giving birth to the field of deep learning.

Even though the ultimate goal of most DNN is to boost classification performance, these deep networks actually work by internally generating new representations for the data, where classes are easier to discriminate by classification layers of the DNN. There are other architectures of DNNs where the aim of generating new representations is clearer, e.g. *autoencoders*. Autoencoders are a type of multilayer ANN with a small neuron layer at the center of their architecture. Data is compressed (decompressed) as traverses half the network, from the input (central) layer to the central (output) layer. The purpose of autoencoders is to generate a small compact representation of the data at this small central layer. Autoencoders can be used for representation learning [18]. Hinton et al. [19] used them as a method for training DNN as well. Accordingly, in this paper we develop an autoencoder through the proposed GP based methodology and evaluated it on benchmark image datasets widely used as testbeds for DNN.

Examples of representation learning methods for image processing with GP are [11,12]; however these kind of works rely on GP individuals that process images as a whole in each primitive function; primitives are often specialized image filters. Thus the GP searches for a combination and workflow of these type of functions that render a new representation useful for classification or detection tasks. The availability of these highly specialized primitives to the GP are a form of human expert knowledge brought to the system by the designer. In contrast, our proposed method process images at individual pixel level, and only simple arithmetic and generic trigonometric functions form the set of primitives.

There are other methods for representation learning through GP that are not image processing-specific such as those presented in [8,10,20]. Recently, Limon et al. [9] developed a method for representation learning based on GP that utilized arithmetic operations along with a few statistics measures. They tested their method on a variety of datasets from different problem domains, and found that GP could learn representations that boosted classifiers performance when the input representation consisted in a few dozen features, but beyond that (one hundred or more features) other representation learning methods or just a classifier by itself performed better. Similarly, Lin et al. [21] proposed an approach to binary classification based on generating multiple layers of representations through GP. Although superficially similar to our proposed method, their approach vastly differs from ours: each new representation layer contained a single new feature compared with the previous representation layer from which was generated, the rest of the features are taken directly from such previous layer, whereas our approach aims at generating a representation composed of completely new features. Their method was in fact developed to tackle problems where initial representations consist of just a few features, whereas our method is designed to treat large-scale problems where we wish to reduce initial data dimensionality.

More recently, Tran et al. [22] also acknowledged the shortcomings of using GP on problems with high dimensional data. In [22] they presented a study on different approaches to tackle such large initial representation problems and in [23] they proposed a method based on automatically clustering similar features and picking a single representative feature from each cluster as a way for reducing representation dimensionality. Their method bears some similarity with our proposed approach in the sense that our approach also groups features in small clusters. However our method does not focuses on the way the clusters are built, neither discards features before the GP process starts.

3   Structured Layered GP for Representation Learning

In this section we introduce our proposed framework for representation learning through GP. We approach the representation learning problem, while additionally stimulating dimensionality reduction[1]. Hence, for a dataset described in representation $\mathbf{N}$, where $\mathbf{N}$ is a matrix of s rows (samples) $\times$ n columns (features), each sample z represented by feature (row) vector $\mathbf{o}_z \in \mathbb{R}^n, \forall z$, we wish to learn a new, more compact and abstract, representation $\mathbf{M}$, where $\mathbf{M}$ is a $s \times m$ matrix, such that each sample z is now represented by feature vector $\mathbf{q}_z \in \mathbb{R}^m, \forall z$ and $m \ll n$. Ideally, the learned representation should better describe the data, in terms of the goal associated to the problem at hand (e.g., classification, regression, or data reconstruction as in this paper).

In the following subsection we present a direct, straightforward approach to tackle this problem through GP, and we discuss why is not convenient to attempt such simple approach; motivating the need of the proposed structured layered GP for representation learning that is described in Sect. 3.2.

---

[1] Please note that reducing the dimensionality is not necessarily a requirement of representation learning, but herein we include this restriction so that the learned representations can be descriptive/discriminative and compact at the same time.

3.1   Straightforward GP Approach for Representation Learning

A straightforward way to learn a representation $\mathbf{M}$ via GP would be as follows. For every feature (column) $\mathbf{y}_i$ that composes $\mathbf{M}$ we set GP to find a function $f_i^M : \mathbb{R}^n \to \mathbb{R}$, such that set of input variables of $f_i^M(x_1, x_2, ..., x_n)$ is the set of features that compose the representation $\mathbf{N}$, and that $\mathbf{y}_i = f_i^M$.

Although this is a valid and direct view of the representation learning problem. It has several limitations, most notably, the fact that it poses an intractable search space for GP. Let us suppose, without loss of generality, we implement a standard genetic program with forest-based population to solve the problem. In this solution, each GP tree t_i^M is associated to a single learned feature f_i^M, which is built from 2-arity primitives, and each t_i^M is a perfect binary tree. Since t_i^M could (1) require, conceivably, access to all n original features to generate $\mathbf{y}_i$, (2) t_i^M is a perfect binary tree, and (3) input features can only be placed in leaf nodes of GP trees, then the height of tree t_i^M is, approximately at least, $\lceil log_2(n) \rceil$, and the number of internal nodes is, approximately, $2^{\lceil log_2(n) \rceil}$. For simplicity, let us assume for now on that n is a power of 2, therefore the number of internal nodes of t_i^M is n. Now let us suppose that we will use a set of K eligible primitives, then the total size of the search space the GP needs to explore is $\mathcal{O}(mK^n)$. This is an optimistic, lower bound estimate, since we are not taking into account that constants can be used as leaf nodes as well; still, this estimate shows us the complexity of the problem we are dealing with. To illustrate our point consider the following example.

Example 1. Suppose we want to process a set of images to convert them from an original feature space of 64×64 gray scale pixels into a 32-dimensional vector. Hence, $n = 4096$ and $m = 32$. We are set to search for a GP individual composed of 32 trees; each tree, potentially, of height 12. Suppose we are considering the following set of primitives $\{+, -, \times, /\}$. The GP needs to search for an optimal individual among, at least, 32×4^{4096} distinct possible solutions.

Although not precisely used for representation learning, but for classification, a similar approach was attempted in [24]. Instead of building GP trees for each feature, they built one tree per class, and then used a *max* function to discriminate classes among the outputs of such trees. They tested this method on the MNIST [25] data set and the GP yielded a notoriously low classification accuracy. This result serves as an example of our warning against attempting such a straightforward approach. In the next subsection we hypothesize about an alternative approach to representation learning through GP that attempts to overcome the issues discussed so far, and in Sect. 4 we present the details for its implementation in a particular case of study.

3.2   GP Structured Layers

We propose that considering a structural layered processing inspired in deep learning models will allow GP to significantly improve its performance in representation learning while reducing the computational burden. The idea is as follows: instead of attempting to build GP trees that convert from representation **N** to representation **M** in a single step, we generate a series of intermediate representations **L**$_l$ that allow a GP to gradually go from representation **N** to representation **M**.

Starting form initial representation **N**, and in order to generate intermediate representation **L**$_1$, features **N** are partitioned into c small indexed subsets C_i, such that $|C_i| \ll n$, $\forall i$. Representation **L**$_1$ is also partitioned into c small subsets K_i, such that $|C_i| > |K_i|$, $\forall i$. Each feature $l_{1,j} \in K_i$ is generated by a GP tree $t_j^{L_1}$, whose leaf nodes can be feature variables taken only from subset C_i as well as constant values; in other words, $t_j^{L_1}$ represents function $f_j^{L_1} : \mathbb{R}^{|C_i|} \to \mathbb{R}$, such that $f_j^{L_1}(x_1, x_2, ..., x_w)$, and $(x_1, x_2, ..., x_w) \in C_i$. Each layer **L**$_i$ is built in the same fashion as **L**$_1$, relying on the partitioned set of features in **L**$_{i-1}$, up until **L**$_z$ = **M**. In this way, the processes of representation learning and dimensionality reduction is done in a gradual manner, unlike in a straightforward single-step approach.

Intuitively, in order to generate each intermediate representation, we split the previous representation into small subsets, so each GP tree focuses into processing the features present in only one of these subsets. This allows the GP to find and exploit more efficiently the relationships existent among the features belonging to the same subset. In contrast, in the straightforward GP approach the GP evolutionary search has to find on its own the features that are useful together as parameters to the functions that generate new features, in addition to building such functions. One may question on how the subsets are supposed to be built, we argue that simple strategies that leverage directly from the nature of the problem at hand can be used, as we show in Sect. 4.

3.3   Efficient Training with Online Learning

One of the distinctive features of DNN is their support for *minibatch* based training, thanks to the use of stochastic gradient descend (SGD) and alike methods. This feature allows an ANN to train, i.e. adjust its parameters, using a small subsets of samples of the entire training dataset. This characteristic helps DNNs to achieve faster convergence speeds and deal with large scale datasets. In this work we propose that an analogy can be draw with respect to GP evolutionary form of learning: *instead of presenting the entire dataset to each individual every generation, only a very small, variable, subset of samples, a minibatch, of the dataset are presented to each individual in every generation.* The number of generations for the GP is chosen such that, *no. of generations × size of minibatches = size of complete training dataset*. This way we guarantee that the population sees each sample in the training dataset at least once. The minibatches conform a partition, in the mathematical sense, of the entire dataset. In this way, each sample of the dataset is seen only once by p individuals (where p is population size). Notice that this does not mean that all individuals ever see all samples once, because many individuals will not make it to further generations, and will never be tested against most of the samples in the dataset. Even more interesting, is the fact that the final top performer individual (the solution returned by the GP) might actually be tested against only the final minibatch, given that this individual is the result from a crossover or mutation from individuals of the penultimate generation.

Just as in the case of DNN's usage of SGD, the purpose of this form of training is to dramatically reduce the computational cost of the GP algorithm. Although for the time being we do not present a theoretical bound on the approximation ratio of minibatch form of training with respect to the regular full batch, we do present experimental evidence that supports the feasibility of this form of evolutionary machine learning.

4   GP Autoencoder

In this section we present a detailed implementation of an autoencoder, both through the straightforward GP approach and also following the guidelines of our proposed approach. To the authors best knowledge, this is the first time an autoencoder is synthesized through GP. Autoencoders were, originally, multilayered ANN that attempted to copy its input to its output while data traverses a bottleneck neuron layer formed at the middle of the network. The layers of the network that comprises from its input layer up to the bottleneck it is called "encoder", because its task is to fit the data to pass through the bottleneck layer, with the least amount of information loss possible, while the rest of the network, from the bottleneck up to the output is called "decoder", because its task is to decompress the data from its compressed representation at the bottleneck layer back to its original form. In order to perform acceptably, i.e. that reconstructions are as similar to their samples as possible, autoencoders are trained with a dataset. This means their internal parameters, or *weights*, are calibrated to perform the compression-decompression task for such a given dataset. When presented with new samples of the same nature of the dataset they were trained with, they are supposed to perform similarly as with the training dataset.

Notice how, according to the proposed GP approach, the process of representation learning is performed by the GP itself. The best performing individual that outputs the GP after the last generation serves as a feature extraction engine, that can take as input a sample in its original representation and returns as output the sample in the new representation. The GP evolves such feature extraction engine and at the same time it also discovers the features it generates as output, i.e. the actual representation learned (because it is not defined beforehand). Therefore the GP attempts to, symbiotically, learn a new representation and the mathematical functions that can generate it.

However, a problem arises when we try to define a way to evaluate the learned, more compact, representation. The answer to this problem is not trivial [1]. In this work we follow suit of ANN autoencoders through the use of a decoding mechanism that reverses the compact representation to its original form. The average discrepancy between some given dataset in its original form and its reconstructed version (returned by the decoding mechanism) provides a simple way to evaluate the learned encoded representation, in terms of abridge of information.

Therefore, we set the GP to discover three elements: an encoding mechanism that compacts the input representation; the compact representation itself, i.e. the set of features that conforms it; and a decoding mechanism, that provides a supporting role in order to evaluate the learned representation (and the mechanism that generates it). Together, encoder and decoder, form what is known as an autoencoder. It is important to remark how, in a way, the discovery of an optimal/acceptable encoding mechanism and the new representation is a byproduct of the evolution of the decoder: the output of the decoder is the only element used for fitness evaluation, both encoding and new representation are only indirectly evaluated by the performance of the decoder.

The GP autoencoder individuals design consist in two forests of GP trees connected through a bus of data. One forest is the encoding mechanism, the other forest is the decoding mechanism and the data bus is the new representation for the data. Figure 1 illustrates this concept.

Given a dataset comprised of an arbitrary number of samples, each one described by the same n features, we wish to reduce the n features to $l_1 = \alpha n$ new features, for $\alpha < 1$, losing the least possible amount of information. That is, from the learned l_1 features, it should be possible to reconstruct samples to the original n features.

The encoder will be comprised of a forest of l_1 GP trees $t_i^{L_1}$; the decoder will be comprised of a forest of n GP trees $t_i^{H_0}$. Each tree $t_i^{L_1}$ will generate feature $l_{1,i}$ of the compact representation $\mathbf{L_1}$, and each tree $t_i^{H_0}$ will generate feature h_i of reconstructed representation $\mathbf{H_0}$.

Terminals of each tree $t_i^{H_0}, \forall i$ can only be features of representation $\mathbf{L_1}$ (as well as constant values within some range), i.e. none of these trees can see any of the original features. Similarly, terminals of tree $t_i^{L_1}, \forall i$ can only be taken from the original representation, and they cannot look ahead for features from representation $\mathbf{L_1}$ they are constructing.

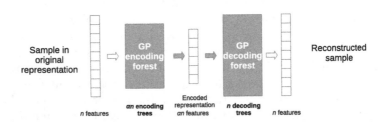

Fig. 1. Abstract depiction of a GP autoencoder individual. Each individual consist of two forests connected through a bus the size of the compact representation desired.

In a straightforward GP approach, terminals of each tree $t_i^{L_1}, \forall i$ can be any feature from the entire set of n original features. Analogously, terminals of each tree $t_i^{H_0}, \forall i$ can be any of the l_1 features from the compact representation $\mathbf{L_1}$. On the other hand, in a structured layer GP, the initial n input features are split into $c = \frac{n}{\beta}$ subsets C_i, such that $\beta > 1$. Associated to each subset C_i there is a subset K_i of features from $\mathbf{L_1}$, such that each feature $l_{1,j} \in K_i$ is generated by $t_j^{L_1}$. Tree $t_j^{L_1}$ can only *see* the small subset of features in C_i.

As stated before, our GP individual comprehends both encoder and decoder forests integrated as a single indivisible unit we call from now on autoencoder. These autoencoders never get their encoder and decoder components separated during the evolutionary process, but they exchange small bits of their structure with other autoencoders within the population through the evolutionary operators described in Subsect. 4.1.

4.1   Genetic Program

We tested three different configurations for GP autoencoders. The first setup consists of a straightforward GP approach, as described in Sect. 3.1, i.e., all GP trees that generate $\mathbf{L_1}$ can see all features from $\mathbf{N}$, as well as all GP trees in the decoder can see all features from $\mathbf{L_1}$. The compression ratio is setup to $\alpha = \frac{3}{4}$. This is our control setup, and its result will serve as a baseline for comparison purposes with the proposed approach.

The second setup consists of a structured layer GP autoencoder, as described in Sect. 4. The compression ration remains the same as in the first setup, for a valid comparison, and the n (l_1) features from representation $\mathbf{N}$ $(\mathbf{L_1})$ are split into $c = \frac{n}{4}, (k = \frac{l_1}{3} = c)$ subsets C_i (K_i), such that $|C_i| = 4, (|K_i| = 3), \forall i$. That is, four

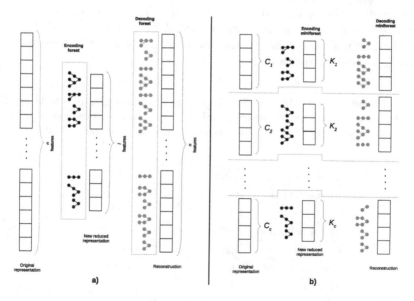

Fig. 2. Comparison between (a) straightforward GP and (b) structured layer GP.

features from input representation $\mathbf{N}$ are assigned to each subset C_i, and from this, and only this, subset of features, is that features in subset K_i can be generated.

Mirroring this configuration, the decoding forest is also partitioned in subsets of four trees, such that trees in each subset can only see the subset of three features of some subset K_i. All these subsets, (4) input features-(3) encoding trees-(3) new features-(4) decoding trees, are coupled together, to form a mini-autoencoder. Figure 2 contrasts this setup against the straightforward GP.

The third setting is exactly as the second one, except that we use a *minibatch* based form of training, as described in Sect. 3.3.

We carried all experiments on gray scale image datasets and each pixel is an input feature. We use every four neighboring pixels in the same row to be in subset C_i.

Evolutionary Parameters and Operators. We build a set of randomly generated GP autoencoders individuals, that will constitute an initial population, and through a GP evolutionary process, we search for an autoencoder that maximizes the average similarity between each sample and its reconstruction, across an entire training dataset. Table 1 shows the set of GP evolutionary parameters used across all experiments performed.

A preliminary study showed that the following setup along the specified parameters converged the fastest to acceptable quality solutions; however, a full study of different GP configurations is still required. In each generation, half of the population is chosen through binary tournament to make it to the next generation. From this half of the population, new individuals are generated with a 0.6 probability of crossover and a 0.3 probability of mutation. Thus, given that population size is fixed to 60 individuals, 30 of them are directly taken from the previous generation, 18 are generated from crossover and 9 are generated through mutation. The remaining 3 individuals are chosen directly from previous generation through elitism.

Table 1. Evolutionary parameters for the GP runs. Arithmetic operands are 2-ary and trigonometric functions are unary primitives. The division function is protected, meaning that any attempt to divide between zero returns as output 1×10^6, instead of an error.

Parameter	Value
Population size	60
Max. tree depth	4
Set of primitives	$< +, -, \times, \div, sin, cos >$
Constants range	[0,1]
Crossover Prob.	0.6
Mutation Prob.	0.3
No. generations	40/40/Variable (See Sect. 3.3)

Notice that the partitioning scheme of the structured layered GP creates in effect c independent GPs, each with its own 60 individuals. However, the computational cost is the same for both scenarios, as in the straightforward GP individuals are large single autoencoders, while in the structured layered GP individuals are c miniautoencoders.

Standard crossover and mutation operations for forest were used. In the case of crossover, when two individuals are selected to undergo crossover, a single randomly selected tree in the encoder forest from one of the individuals is exchanged with a tree from the other individual's encoder, then the same process is executed again but now for trees in the decoder forests. When an individual is selected for mutation, one randomly selected tree within encoder and decoder forests are deleted and replaced by new randomly generated ones. Notice how none of these operators operate at node level.

Objective Function. To determine similarity between an original sample and the reconstructed output from the autoencoder, we used the mean square error (MSE), defined in Eq. 1. MSE receives as input original sample x and reconstructed y vectors, and compares them feature by feature, averaging the difference across all of features. MSE output can be thought as a *distance* between a sample and its reconstruction.

$$d_{\mathbf{MSE}}(x, y) = \frac{1}{n} \sum_{i=1}^{n} (x_i - y_i)^2 \qquad (1)$$

The objective function in the first two setups described is to minimize the average MSE across all pairs sample-reconstruction from some given dataset. The evolutionary process is executed for 40 generations in both setups. In every generation, each individual of the population is tested against the entire training dataset, the resulting MSEs for every instance in the dataset are averaged, and this result is assigned as the fitness for a given individual. On the other hand, the objective function in the third setup is minimizing average MSE for all samples in the current minibatch presented to the population.

5   Experimental Results

In this section we present an experimental evaluation of the proposed autoencoder following three different proposed approaches. We compared the three approaches in ML dataset MNIST [25]. For the third setup we also performed a test varying the size of the minibatches and increasing the number of generations in order to allow the GP to see each sample more than just once. After we calibrated the size of minibatches and confirmed that giving more than one pass over training data was beneficial to the proposed method, we further tested it onto two additional ML datasets, namely LFWcrop [26] and Olivetti [27]. And finally compared the results with those obtained with conventional ANN autoencoders.

Table 2. Datasets used for experimentation. All datasets consist in grayscale images; pixel values are normalized to fall in the range [0, 1] in all cases.

Dataset	MNIST	LFWcrop	Olivetti
Images resolution (input features)	28×28 (784)	64×64 (4,096)	64×64 (4,096)
No. training samples	60,000	12,000	360
No. testing samples	10,000	1,233	40

Each dataset was split into training and testing set. The evolutionary process is carried with the training set and the top performer individual that results from the process is evaluated with the testing set, composed of images not seen during the evolutionary process. Table 2 describes the used datasets.

Figure 3 shows the results obtained by the straightforward GP, the structured layer GP and the minibatch training version of it; minibatches were composed of 100 samples. All experiments were done in a workstation with an Intel Xeon CPU with 10 physical cores at 2.9 GHz, with two virtual cores per each physical core, to amount for a total of 20 processing threads, 16 GB of RAM, running Ubuntu Linux 16.04. Algorithms implementation and setups were done by using an in-house software library developed in Python version 3.6. Accelerated NumPy library is used only in the final step of fitness evaluation (averaging the MSE of all sample-reconstruction pairs) of each individual. The straightforward GP can make use of the multiple processing cores by parallelizing the evaluation of sample-reconstructions pairs. On the other hand, both structured layer GP approaches distribute evolution of multiple miniautoencoders across most (but not all) processing threads available, and in this case the evaluations of the sample-reconstructions pairs is done sequentially for each miniautoencoder.

A visual depiction of the performance of the synthesized autoencoders is shown in Figs. 4 and 5. Figure 4 shows the gradual increase in performance obtained by the structured layer GP through the evolutionary process. Figure 5 compares the reconstruction for the first ten images in the training set, as obtained by the best autoencoders generated by the three different experimental setups. Table 3 shows results of varying the size of minibatches to 30, 60, 100, 300, and 600 samples; as well as allowing the algorithm to give one, two and five forward passes over the training data.

We picked up the best performing structured layer GP (SLGP) setup from our minibatch study (minibatches of size 60) and compared its performance against a one hidden layer, fully connected, Multilayer Perceptron (MLP) set up as an autoencoder that performs the same compression ratio. We implemented the MLP in TensorFlow Deep Learning library [28]. We also set the size of the minibatches of the MLP to 60, just as in our GP approach. Table 4 shows the results of each approach, for the different amounts of forward passes/epochs. The main idea behind these experiments is to make a 1-to-1 "layer" and "epoch-to-epoch" comparison in order the study the behavior of the new proposed method

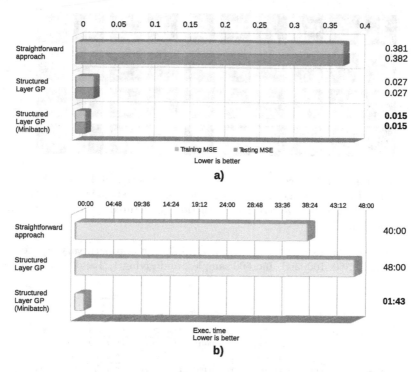

Fig. 3. Results obtained by the three different GP setups (a) MSE across all samples in training and testing datasets. (b) Execution time expressed in hh:mm.

Fig. 4. A depiction of the reconstruction generated by the top performer individuals during the evolution process of the structured layer GP. From top row to bottom:the original first ten images of the training set, best reconstruction obtained after 0, 10, 20, 30 and 40 generations.

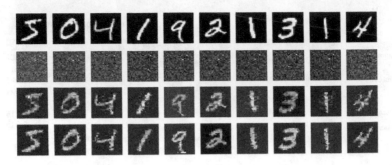

Fig. 5. Comparison of the reconstruction of the three experimental setups. From top to bottom: original first 10 images from the training set, best straightforward GP reconstruction, best structured layer GP reconstruction, structured layer GP + minibatch training reconstruction.

Table 3. Average MSEs and exec. time for a minibatch approach varying the size of the batches to 30, 60, 100, 300, and 600 samples; and giving 1, 2 and 5 forward passes over the dataset.

	Mini batch size														
	30			60			100			300			600		
Passes	Training	Testing	Time	Training	Testing	Time	Training	Testing	Time	Training	Testing	Time	Training	Testing	Time
1	0.013	0.013	04:44	0.012	0.012	03:36	0.013	0.013	03:15	0.017	0.017	02:49	0.020	0.020	02:39
2	0.010	0.010	09:31	0.011	0.011	07:24	0.011	0.011	06:32	0.014	0.014	05:30	0.017	0.017	05:31
5	0.009	0.009	23:27	**0.008**	**0.008**	18:04	0.009	0.009	16:27	0.011	0.011	14:26	0.014	0.014	13:23

and contrast it with conventional methods of deep learning. Figure 6 show a visual appreciation on the reconstructions generated by both autoencoders for LFWcrop and Olivetti samples.

5.1 Discussion

From the results presented in the previous section we can appreciate that the structured layer GP approach is one order of magnitude better than a straightforward GP approach in average MSE . In fact, the straightforward approach does not reach an acceptable solution at all given approximately the same amount of time, making further clear the advantage of proposed approach. Even though the difference, in terms of quality of solutions, is not as decisive between the structured layer GP and the its minibatch learning version, the gap in execution time between them is also one order of magnitude. Results also show that size of minibatches have to be carefully selected in order to get a balance between quality of solution and execution time. We also confirmed that the minibatch version can benefit from making several passes over the training data.

When compared with a conventional autoencoder, results show that GP behaves quite different from ANN. GP can quickly (in terms of passes over the training data) build acceptable encoding-decoding models, while an ANN that attempts to generate a representation of 3,072 (588) features from 4,096

Table 4. MSE results obtained by structured layer GP with minibatch training (SLGP) and a 1-hidden layer perceptron (MLP) autoencoders when tested with testing subsets of different ML datasets, as well as the time required for training/evolution with the training sets of each dataset.

Dataset	Passes	SLGP		MLP	
		Avg. MSE	Time	Avg. MSE	Time
MNIST	1	**0.012**	03:36:00	0.046	00:00:12
	2	**0.011**	07:24:00	0.035	00:00:24
	5	**0.008**	18:04:00	0.021	00:01:04
	50	-	-	**0.003**	00:11:03
Olivetti	1	0.032	00:08:25	**0.018**	00:00:00
	2	0.018	00:15:02	0.018	00:00:01
	5	**0.009**	00:35:23	0.018	00:00:03
	50	-	-	0.018	00:00:34
LFWcrop	1	**0.003**	04:00:00	0.027	00:00:42
	2	**0.002**	07:40:37	0.025	00:01:10
	5	**0.002**	18:30:00	0.024	00:03:30
	50	-	-	0.014	00:36:23

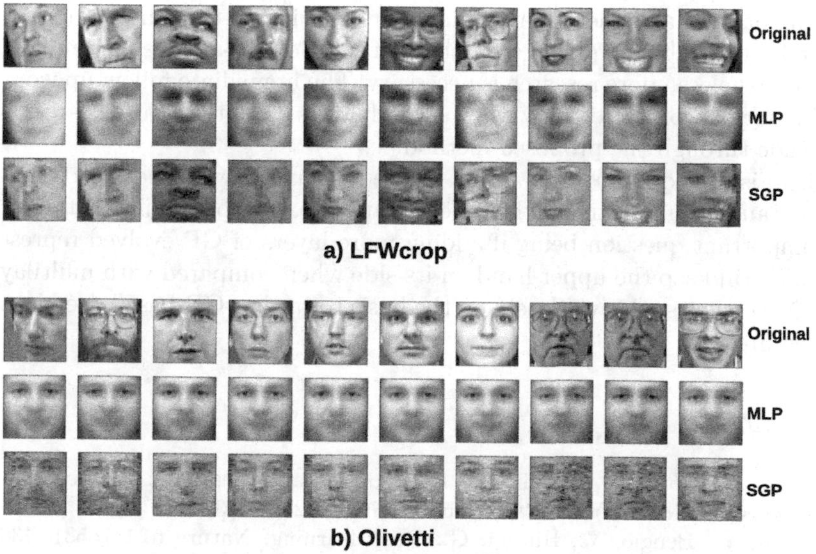

a) LFWcrop

b) Olivetti

Fig. 6. Visual results of the reconstructions generated by the SLGP and the MLP in the (a) LFWcrop (b) Olivetti faces datasets. From top to bottom (for both sets) original first ten images from the testing datasets, reconstructions generated by the MLP and reconstructions generated by the SLGP.

(785) initial features, in the case of LFWcrop and Olivetti (MNIST) datasets, has just too many parameters to adjust. This effect is further noticeable as we test datasets with fewer training samples. The GP is simply a more efficient approach in terms of data usage.

6   Concluding Remarks and Future Work

In this work we have: (1) introduced a new method that allows GP to perform representation learning on large-scale problems without the need of specialized primitives; (2) presented a detailed case of study of the proposed method, a GP based autoencoder; (3) we provided experimental results that prove the performance gains of implementing such an autoencoder with the proposed method compared against when a straightforward GP approach is used; (4) we proved experimentally that the implemented autoencoder supports minibatch training/evolution, a very important feature when considering very large (as in number of samples) datasets; (5) and finally, we compared the resulting autoencoder with a simple, yet conventional, ANN autoencoder.

We hold that the overall results show strong evidence on the possibility of a Deep Learning framework fully based on GP. We believe that our proposed method, if applied iteratively, can yield representations compact and abstract enough to be usable for other machine learning tasks such as classification or decision making, and therefore attain results previously though unreachable for GP, and comparable in quality with those of state-of-the-art deep learning methods. Nevertheless, more research is necessary. The immediate follow-up work will be to perform a complete study when several layers of representations are evolved in cascade through the proposed method.

There is still quite a road ahead before we can answer whether or not this method can be competitive with modern state-of-art deep learning methods. The most important question being if adding more layers of GP evolved representations will still keep the upper hand on its side when compared with multilayered ANN. Nevertheless this work sets up the basis towards a GP-based deep learning architecture.

References

1. Bengio, Y., Courville, A., Vincent, P.: Representation learning: a review and new perspectives. IEEE Trans. PAMI **35**(8), 1798–1828 (2013)
2. LeCun, Y., Bengio, Y., Hinton, G.: Deep learning. Nature **521**(7553), 436–444 (2015)
3. Lee, D.D., Seung, H.S.: Learning the parts of objects by non-negative matrix factorization. Nature **401**(6755), 788 (1999)
4. Mika, S., Ratsch, G., Weston, J., Scholkopf, B., Mullers, K.: Fisher discriminant analysis with kernels. In: Proceeding of Workshop on Neural Networks for Signal Processing (1999)
5. Wold, S., Esbensen, K., Geladi, P.: Principal component analysis. Chemometr. Intell. Lab. Syst. **2**(1–3), 37–52 (1987)

6. Koza, J.R.: Genetic Programming: On the Programming of Computers by means of Natural Selection, vol. 1. MIT Press, Cambridge (1992)

7. Gomez, G., Morales, E.: Automatic feature construction and a simple rule induction algorithm for skin detection. In: ICML Workshops (2004)

8. Garcia-Limon, M., Escalante, H.J., Morales, E., Morales-Reyes, A.: Simultaneous generation of prototypes and features through genetic programming. In: Proceedings of the 2014 Annual Conference on Genetic and Evolutionary Computation, pp. 517–524. ACM (2014)

9. Limón García, M., Escalante, H.J., Morales, E., Pineda, L.V.: Class-specific feature generation for 1NN through genetic programming. In: Proceeding of ROPEC (2015)

10. Bot, M.C.J.: Feature extraction for the k-nearest neighbour classifier with genetic programming. In: Miller, J., Tomassini, M., Lanzi, P.L., Ryan, C., Tettamanzi, A.G.B., Langdon, W.B. (eds.) EuroGP 2001. LNCS, vol. 2038, pp. 256–267. Springer, Heidelberg (2001). https://doi.org/10.1007/3-540-45355-5_20

11. Trujillo, L., Olague, G.: Synthesis of interest point detectors through genetic programming. In: Proceeding of GECCO, pp. 887–894. ACM (2006)

12. Shao, L., Liu, L., Li, X.: Feature learning for image classification via multiobjective genetic programming. IEEE Trans. Neural Netw. Learn. Syst. **25**(7), 1359–1371 (2014)

13. Rumelhart, D.E., Hinton, G., Williams, R.J., et al.: Learning representations by back-propagating errors. Cogn. Model. **5**(3), 1 (1988)

14. LeCun, Y., Boser, B.E., Denker, J.S., Henderson, D., Howard, R.E., Hubbard, W.E., Jackel, L.D.: Handwritten digit recognition with a back-propagation network. In: Advances in Neural Information Processing Systems, pp. 396–404 (1990)

15. Krizhevsky, A., Sutskever, I., Hinton, G.: ImageNet classification with deep convolutional neural networks. In: Advances in Neural Information Processing Systems (2012)

16. Szegedy, C., Liu, W., Jia, Y., Sermanet, P., Reed, S., Anguelov, D., Erhan, D., Vanhoucke, V., Rabinovich, A.: Going deeper with convolutions. In: Proceeding of CVPR, pp. 1–9 (2015)

17. He, K., Zhang, X., Ren, S., Sun, J.: Deep residual learning for image recognition. In: Proceedings of the IEEE Conference on Computer Vision and Pattern Recognition (2016)

18. Goodfellow, I., Bengio, Y., Courville, A.: Deep Learning. MIT Press, Cambridge (2016)

19. Hinton, G., Salakhutdinov, R.: Reducing the dimensionality of data with neural networks. science **313**(5786), 504–507 (2006)

20. Zhang, Y., Rockett, P.I.: A generic optimising feature extraction method using multiobjective genetic programming. Appl. Soft Comput. **11**(1), 1087–1097 (2011)

21. Lin, J., Ke, H., Chien, B., Yang, W.: Designing a classifier by a layered multi-population genetic programming approach. Pattern Recogn. **40**(8), 2211–2225 (2007)

22. Tran, B., Xue, B., Zhang, M.: Genetic programming for feature construction and selection in classification on high-dimensional data. Memet. Comput. **8**(1), 3–15 (2016)

23. Tran, B., Xue, B., Zhang, M.: Using feature clustering for GP-based feature construction on high-dimensional data. In: McDermott, J., Castelli, M., Sekanina, L., Haasdijk, E., García-Sánchez, P. (eds.) EuroGP 2017. LNCS, vol. 10196, pp. 210–226. Springer, Cham (2017). https://doi.org/10.1007/978-3-319-55696-3_14

24. Parkins, A., Nandi, A.: Genetic programming techniques for hand written digit recognition. Signal Process. **84**(12), 2345–2365 (2004)
25. LeCun, Y.: Mnist database of handwritten digits (1998). http://yann.lecun.com/exdb/mnist/
26. Sanderson, C.: LFWcrop face dataset (2014)
27. Samaria, F.S., Harter, A.C.: Parameterisation of a stochastic model for human face identification. In: 1994 Proceedings of the Second IEEE Workshop on Applications of Computer Vision, pp. 138–142. IEEE (1994)
28. Abadi, M., Agarwal, A., et al.: Tensorflow: Large-scale machine learning on heterogeneous distributed systems. arXiv preprint arXiv:1603.04467 (2016)

Comparing Rule Evaluation Metrics for the Evolutionary Discovery of Multi-relational Association Rules in the Semantic Web

Minh Duc Tran[1], Claudia d'Amato[2], Binh Thanh Nguyen[3],
and Andrea G. B. Tettamanzi[1(✉)]

[1] Université Côte d'Azur, CNRS, Inria, I3S, Sophia Antipolis, France
tdminh2110@yahoo.com, andrea.tettamanzi@unice.fr
[2] University of Bari, Bari, Italy
claudia.damato@uniba.it
[3] The University of Danang – University of Science and Technology,
Da Nang, Vietnam
ntbinh@dut.udn.vn

Abstract. We carry out a comparison of popular asymmetric metrics, originally proposed for scoring association rules, as building blocks for a fitness function for evolutionary inductive programming. In particular, we use them to score candidate multi-relational association rules in an evolutionary approach to the enrichment of populated knowledge bases in the context of the Semantic Web. The evolutionary algorithm searches for hidden knowledge patterns, in the form of SWRL rules, in assertional data, while exploiting the deductive capabilities of ontologies.

Our methodology is to compare the number of generated rules and total predictions when the metrics are used to compute the fitness function of the evolutionary algorithm. This comparison, which has been carried out on three publicly available ontologies, is a crucial step towards the selection of suitable metrics to score multi-relational association rules that are generated from ontologies.

Keywords: Evolutionary inductive programming
Description logics · Semantic Web

1 Introduction

Originally developed to enable a semantic and therefore intelligent retrieval of digital information resources or an intelligent navigation among them, the Semantic Web (SW) [3] has evolved into a vision of the Web of linked data aimed at enabling people to create and publish data stores on the Web, build vocabularies, write rules for handling data, and annotate resources semantically with metadata referring to ontologies, which are formal conceptualizations of

© Springer International Publishing AG, part of Springer Nature 2018
M. Castelli et al. (Eds.): EuroGP 2018, LNCS 10781, pp. 289–305, 2018.
https://doi.org/10.1007/978-3-319-77553-1_18

domains of interest acting as shared vocabularies where the meaning of the annotations is formally defined.

Data and annotated resources represent the assertional knowledge given the intensional definitions provided by ontologies. We will refer to their combination as an *ontological knowledge base*.

The description of data/resources in terms of ontologies is a key aspect of the SW. Interestingly, ontologies are also equipped with powerful deductive reasoning capabilities. However, due to the heterogeneous and distributed nature of the SW, ontological knowledge bases (KBs) may turn out to be incomplete and noisy wrt the domain of interest. Specifically, an ontology is incomplete when it is logically consistent (i.e., it contains no contradiction) but it lacks information (e.g., assertions, disjointness axioms, etc.) wrt the reference domain; while it is noisy when it is logically consistent but it contains invalid information wrt the reference domain. These situations may prevent the inference of relevant information or cause incorrect information to be derived.

Data mining techniques can be used to discover hidden knowledge patterns from ontological KBs, to be used for enriching an ontology both at terminological (schema) and assertional (facts) level, even in presence of incompleteness and/or noise. This is the goal of level-wise generate and test methods proposed in the inductive logic programming (ILP) [9,17,18], and in the SW community [12,14,15,22], which exploit just the assertional evidence of ontological KBs and, more recently, of approaches that exploit also the reasoning capabilities of the SW, like [7]. Even more recently, approaches that take advantage of the exploration capabilities of evolutionary algorithms jointly with the reasoning capabilities of ontologies have been proposed: this is the case of EDMAR [8,21] an evolutionary inductive programming approach capable of discovering hidden knowledge patterns in the form of multi-relational association rules (ARs) coded in SWRL [13], which can be added to the ontology, thus enriching its expressive power and increasing the assertional knowledge that can be derived from it. Additionally, discovered rules may suggest new axioms to be added to the ontology, such as transitivity and symmetry of a role, as well as concept/role inclusion.

The EDMAR algorithm uses a linear combination of the head coverage and confidence of a rule as its fitness. However, much work has been devoted in the data mining domain to devising metrics for evaluating the merit of association rules. In this paper, we focus on the question of what evaluation metrics can be most beneficial as a fitness function for the evolutionary discovery of multi-relational association rules from ontological KBs. To answer this question, we compare a number of popular asymmetric metrics by using them as fitness in EDMAR. In particular, we base the comparison on the number of the generated rules and of the total number of predictions achieved (the correct predictions) by EDMAR using each metric when applied to three publicly available ontological KBs.

The next section provides basics and definitions. Background information on the EDMAR algorithm is given in Sect. 3. Section 4 reports the details of the empirical comparison and a critical analysis of the results. Section 5 concludes.

2   Basics

We refer to ontological KBs described in Description Logics (DLs - a family of formal knowledge representation languages which are decidable fragments of first order logic) [2] (in practice some profile or subset of OWL,[1] the standard representation language in the SW), without restricting ourselves to any specific DL. As usual in DLs, we refer to a KB $\mathcal{K} = \langle \mathcal{T}, \mathcal{A} \rangle$ consisting of a TBox $\mathcal{T}$ containing the terminological axioms and an ABox $\mathcal{A}$ containing the assertional axioms. It should be recalled that DLs adopt the *open-world assumption* (OWA - what is known to be true is unknown). For more details concerning DLs see [2].

We address the problem of discovering relational ARs from ontological KBs.

Definition 1 (Relational Association Rule). *Given a populated ontological KB $\mathcal{K} = (\mathcal{T}, \mathcal{A})$, a relational association rule r for $\mathcal{K}$ is a Horn-like clause of the form: body $\rightarrow$ head, where (a) body is a generalization of a set of assertions in $\mathcal{K}$ co-occurring together; (b) head is a consequent that is induced from $\mathcal{K}$ and body.*

Problem Definition. Given $\mathcal{K} = (\mathcal{T}, \mathcal{A})$, a minimum "frequency threshold", θ_f, and a minimum "fitness threshold", θ_{fit}, we wish to discover as many frequent and fit hidden patterns (w.r.t θ_f and θ_{fit}) as possible, in the form of relational ARs, that may induce new assertions for $\mathcal{K}$.

Intuitively, a *frequent hidden pattern* is a generalization of a set of concept/role assertions co-occurring reasonably often (wrt a fixed frequency threshold) together, showing an underlying form of correlation that may be exploited for obtaining new assertions.

The rules to be discovered are represented in the Semantic Web Rule Language (SWRL) [13], which extends the set of OWL axioms with Horn-like rules.[2]

Definition 2 (SWRL Rule). *Given a KB $\mathcal{K}$, a SWRL rule is an implication between an antecedent (body) and a consequent (head) of the form: $B_1 \wedge B_2 \wedge \cdots B_n \rightarrow H_1 \wedge \cdots \wedge H_m$, where $B_1 \wedge \cdots \wedge B_n$ is the rule body and $H_1 \wedge \cdots \wedge H_m$ is the rule head. Each $B_1, \ldots, B_n, H_1, \ldots H_m$ is an atom.*

An atom is a unary or binary predicate of the form $C(s)$, $R(s_1, s_2)$, where the predicate symbol C is a concept name in $\mathcal{K}$, R is a role name in $\mathcal{K}$, s, s_1, s_2 are terms. A term is either a variable (denoted by x, y, z) or a constant (denoted by a, b, c) standing for an individual name or data value.

The discovered rules can be generally called *multi-relational* rules since multiple binary predicates $R(s_1, s_2)$ with different role names of $\mathcal{K}$ could appear in a rule. The intended meaning of a rule is: whenever the conditions in the antecedent hold, the conditions in the consequent must also hold. A rule having more than one atom in the head can be equivalently transformed, due to the

---

[1] https://www.w3.org/OWL/.

[2] The result is a KB with an enriched expressive power. More complex relationships than subsumption can be expressed.

safety condition (see Definition 3), into multiple rules, each one having the same body and a single atom in the head. Therefore, we will consider, w.l.o.g., only SWRL rules (hereafter just "rules") with one atom in the head.

Example 1 (SWRL rule). Given the rule fatherOf$(y, x) \wedge$ Male$(x) \rightarrow$ sonOf(x, y), sonOf(x, y) is its head, fatherOf$(y, x) \wedge$ Male(x) its body; fatherOf(y, x), Male(x), and sonOf(x, y) are atoms, and x, y are variables.

2.1 Language Bias

A *language bias* is a set of constraints giving a tight specification of the patterns worth considering, thus allowing to reduce the search space. Following [21], we are interested in rules having only atomic concepts and/or role names of $\mathcal{K}$ as predicate symbols, and individual names as constants. Only *connected* [12] and *non-redundant* [14] rules satisfying the *safety condition* [13] are considered.[3]

Given an atom A, let $T(A)$ denote the set of all the terms occurring in A and let $V(A) \subseteq T(A)$ denote the set of all the variables occurring in A e.g. $V(C(x)) = \{x\}$ and $V(R(x, y)) = \{x, y\}$. Such notation may be extended to rules straightforwardly.

Definition 3 (Safety Condition). *Given a KB $\mathcal{K}$ and a rule $r = B_1 \wedge B_2 \wedge \ldots B_n \rightarrow H$, r satisfies the* safety condition *if all variables appearing in the rule head also appear in the rule body; formally if: $V(H) \subseteq \bigcup_{i=1}^{n} V(B_i)$,*

Definition 4 (Connected Rule). *Given a KB $\mathcal{K}$ and a rule $r = B_1 \wedge B_2 \wedge \ldots B_n \rightarrow H$, r is connected if and only if every atom in r is transitively connected to every other atom in r.*

Two atoms B_i and B_j in r, with $i \neq j$, are connected if they share at least a variable or a constant i.e. if $T(B_i) \cap T(B_j) \neq \emptyset$. Two atoms B_1 and B_k in r are transitively connected if there exist in r, atoms $B_2, \ldots, B_{k-1}$, with $k \leq n$, such that, for all $i, j \in \{1, \ldots, k\}$ with $i \neq j$, $T(B_i) \cap T(B_j) \neq \emptyset$.

Example 2 (Disconnected rule). The rule wifeOf$(y, x) \wedge$ siblingOf$(z, w) \rightarrow$ spouseOf(x, y) is disconnected, since the atom siblingOf(z, w) does not share any variable with the other atoms.

Definition 5 (Closed Rule). *Given a KB $\mathcal{K}$ and a rule $r = B_1 \wedge B_2 \wedge \ldots B_n \rightarrow H$, r is closed if and only if every variable in r is closed.*

Each variable $v_j \in \bigcup_{i=1}^{n} V(B_i), j \in \{1, \ldots, k\}$, with $k \leq n$, is closed if it appears at least twice in r.

Example 3 (Open rule). Rule sonOf$(z, x) \rightarrow$ spouseOf(x, y) is not closed, since variables z and y are not closed.

---

[3] To guarantee decidability, only *DL-safe rules* are sought for [16], i.e., rules interpreted under the DL-safety condition, whose variables are bound only to explicitly named individuals in $\mathcal{K}$. When added to an ontology, DL-safe rules are decidable and generate sound, but not necessarily complete, results.

Definition 6 (Redundant Rule). *Given a KB $\mathcal{K}$ and a rule $r = B_1 \wedge B_2 \wedge \ldots B_n \rightarrow H$, r is a redundant rule if at least one atom in r is entailed by another atom in r with respect to $\mathcal{K}$, i.e., if, $\exists i \in \{0, 1, \ldots, n\}$, $\exists j \in \{0, 1, \ldots, n\}$, with $B_0 = H$, results: $B_j \models_{\mathcal{K}} B_i, i \neq j$*

Example 4 (Redundant Rule). Given $\mathcal{K}$ with $\mathcal{T} = \{\mathsf{Father} \sqsubseteq \mathsf{Parent}\}$ and the rule $r = \mathsf{Father}(x) \wedge \mathsf{Parent}(x) \rightarrow \mathsf{Human}(x)$ where Human is a primitive concept, r is redundant since the atom $\mathsf{Parent}(x)$ is entailed by the atom $\mathsf{Father}(x)$ with respect to $\mathcal{K}$.

2.2 Metrics for Rules Evaluation

For determining the rules of interest for discovery, metrics for assessing the quality of a rule are necessary. We summarize now the metrics we have considered.

Given a rule $r = B_1 \wedge \ldots \wedge B_n \rightarrow H$, let us denote:

- $\Sigma_H(r)$ the set of distinct bindings of the variables occurring in the head of r, formally: $\Sigma_H(r) = \{binding\ V(H)\}$
- $E_H(r)$ the set of distinct bindings of the variables occurring in the head of r provided the body and the head of r are satisfied, formally: $E_H(r) = \{binding\ V(H) \mid \exists\ binding\ V(B_1 \wedge \cdots \wedge B_n) : B_1 \wedge \cdots \wedge B_n \wedge H\}$. Since rules are connected and closed, $V(H) \subseteq V(B_1 \wedge \cdots \wedge B_n)$
- $M_H(r)$ the set of distinct bindings of the variables occurring in the head of r also appearing as binding for the variables occurring in the body of r, formally: $M_H(r) = \{binding\ V(H) \mid \exists\ binding\ V(B_1 \wedge \cdots \wedge B_n) : B_1 \wedge \cdots \wedge B_n\}$
- $P_H(r)$ the set of distinct bindings of the variables occurring in the head of r provided that the body and the head of r are satisfied. Particularly, this applies when a role atom is in the head of the considered rule. Formally: $P_H(r) = \{binding\ V(H) \mid \exists\ binding\ V(B_1 \wedge \cdots \wedge B_n) \cup v_{rng}(H') : B_1 \wedge \cdots \wedge B_n \wedge H'\}$ where
  - H and H' are role atoms with the same the predicate symbol R;
  - $V(H) \subseteq V(B_1 \wedge \cdots \wedge B_n)$ since rules are connected and closed
  - $v_{dom}(H) = v_{dom}(H')$ and $v_{rng}(H) \neq v_{rng}(H')$;

    with v_{dom} and v_{rng} standing for the domain and range variables respectively of the predicate symbol R
  - $v_{rng}(H') \notin V(B_1 \wedge \cdots \wedge B_n)$;
- Σ_i total number of individuals inside a KB.
- Given the rule $r = B_1 \wedge \ldots \wedge B_n \rightarrow H$, N is a number defined as follows:
$$N = \begin{cases} \Sigma_i, & \text{if } H \text{ is a concept atom;} \\ P^2_{\Sigma_i} = (\Sigma_i)!/(\Sigma_i - 2)!, & \text{if } H \text{ is a role atom.} \end{cases}$$

Like in [7, 12], the classical definitions (as used in [1]) are modified to ensure monotonicity, as summarized below. The range for these metrics is shown in Table 2.

Definition 7 (Rule Support). *Given a rule* $r = B_1 \wedge \ldots \wedge B_n \rightarrow H$, *its support is the number of distinct bindings of the variables in the head, provided the body and the head of r are satisfied jointly, formally:*

$$\text{supp}(r) = |E_H(r)|. \tag{1}$$

Example 5 (Computation of Rule Support). Given the rule $r = \text{feed}(x, y) \rightarrow \text{love}(x, y)$ and assuming the following bindings $\{\text{feed}(Anna, Dog), \text{feed}(Anna, Cat), \text{feed}(Peter, Pig), \text{love}(Anna, Dog), \text{love}(George, Cat)\}$ exist then $\text{supp}(r) = 1$, as there is just one binding for the rule head ($\text{feed}(Anna, Dog)$) allowing the head $\text{love}(Anna, Dog)$ and the body $\text{feed}(Anna, Dog)$ to be jointly satisfied.

Definition 8 (Head Coverage for a Rule). *Given the rule* $r = B_1 \wedge \ldots \wedge B_n \rightarrow H$, *its head coverage is the ration between the rule support and the distinct variable bindings from the head of the rule*

$$\text{headCoverage}(r) = |E_H(r)|/|\Sigma_H(r)|. \tag{2}$$

Example 6 (Computation of Head Coverage). Given the rule r in Example 5 and the corresponding bindings, $\text{headCoverage}(r) = \frac{1}{2}$ since there are two bindings for the head of r: $\{\text{love}(Anna, Dog), \text{love}(George, Cat)\}$.

Definition 9 (Rule Confidence). *Given a rule* $r = B_1 \wedge \ldots \wedge B_n \rightarrow H$, *its confidence is defined as the ratio of the number of the rule support and the number of bindings in the rule body:*

$$\text{conf}(r) = |E_H(r)|/|M_H(r)|. \tag{3}$$

Example 7 (Computation of Rule Confidence). Given the rule r in Example 5 and the corresponding bindings, $\text{conf}(r) = \frac{1}{3}$, since there are three bindings for the body of r: $\{\text{feed}(Anna, Dog), \text{feed}(Anna, Cat), \text{feed}(Peter, Pig)\}$.

An issue with Definition 9, is that an implicit closed-world assumption is made, since no distinction between *incorrect* predictions, i.e., bindings σ matching r such that $\mathcal{K} \models \neg H\sigma$, and *unknown* predictions, i.e., bindings σ matching r such that both $\mathcal{K} \models H\sigma$ and $\mathcal{K} \models \neg H\sigma$, is made. Reasoning on ontologies is grounded on the OWA and our goal is to maximize correct predictions, not just describe the data. Hence, following [12] we use *PCA Confidence* instead, which takes the OWA into account.

Definition 10 (Rule PCA-Confidence). *Given the rule* $r = B_1 \wedge \ldots \wedge B_n \rightarrow H$, *its PCA (Partial Completeness Assumption) confidence is defined as follows:*

$$\text{pcaconf}(r) = \begin{cases} |E_H(r)|/|M_H(r)|, & \text{if } H \text{ is a concept atom;} \\ |E_H(r)|/|P_H(r)|, & \text{if } H \text{ is a role atom.} \end{cases} \tag{4}$$

For Example 5, $\text{pcaconf}(r) = \frac{1}{2}$.

Definition 11 (Laplace for a Rule). *Given a rule* $r = B_1 \wedge \ldots \wedge B_n \to H$, *its Laplace* [6], *often used to grade rules for classification goals, is defined as*

$$\text{Laplace}(r) = \frac{|E_H(r)| + 1}{|M_H(r)| + 2} \tag{5}$$

For Example 5, Laplace$(r) = \frac{2}{5}$.

Definition 12 (Conviction for a Rule). *Given a rule* $r = B_1 \wedge \ldots \wedge B_n \to H$, *its Conviction* [5], *measuring the intensity of implication of a rule, is defined using the confidence metrics in the denominator:*

$$\text{conviction}(r) = \frac{N - |\Sigma_H(r)|}{N(1 - |\text{conf}(r)|)} \tag{6}$$

For Example 5, conviction$(r) = \frac{7}{5}$.

Definition 13 (Certainty factor for a Rule). *Given a rule* $r = B_1 \wedge \ldots \wedge B_n \to H$, *its Certainty Factor* [11] *represents uncertainty in the rule and is defined as follows:*

$$cf(r) = \begin{cases} \frac{\text{conf}(r) - \frac{|\Sigma_H(r)|}{N}}{1 - \frac{|\Sigma_H(r)|}{N}}, & \text{if } \text{conf}(r) > \frac{|\Sigma_H(r)|}{N} \ ; \\ \frac{\text{conf}(r) - \frac{|\Sigma_H(r)|}{N}}{\frac{|\Sigma_H(r)|}{N}}, & \text{if } \text{conf}(r) < \frac{|\Sigma_H(r)|}{N} \ ; \\ 0, & \text{if } \text{conf}(r) = \frac{|\Sigma_H(r)|}{N} \ . \end{cases} \tag{7}$$

For Example 5, cf$(r) = 0.286$.

Definition 14 (Added value for a Rule). *Given a rule* $r = B_1 \wedge \ldots \wedge B_n \to H$, *Added Value* [19] *for the rule r is defined as:*

$$\text{av}(r) = \text{conf}(r) - \frac{|\Sigma_H(r)|}{N} \tag{8}$$

This metric is more meaningful when the amount of evidence is large, for it relies on probabilities. For Example 5, av$(r) = 0.267$.

Definition 15 (J-Measure for a Rule). *Given a rule* $r = B_1 \wedge \ldots \wedge B_n \to H$, *its J-Measure* [20] *is defined according to the probability distribution of individuals as follows:*

$$J(r) = \frac{|E_H(r)|}{N} \log_2 \frac{N|E_H(r)|}{|M_H(r)||\Sigma_H(r)|} + \frac{|M_H(r)| - |E_H(r)|}{N} \log_2 \frac{N(|M_H(r)| - |E_H(r)|)}{|M_H(r)|(N - |\Sigma_H(r)|)}. \tag{9}$$

For Example 5, $J(r) = 0.045$.

Definition 16 (Gini index for a Rule). *Given a rule* $r = B_1 \wedge \ldots \wedge B_n \rightarrow H$, *its Gini Index* [4] *is defined according to the probability distribution of individuals from the sum of squared probabilities as follows:*

$$
\begin{aligned}
\text{gn}(r) = \frac{|M_H(r)|}{N} &\left[\left(\frac{|E_H(r)|}{|M_H(r)|} \right)^2 + \left(\frac{|M_H(r)| - |E_H(r)|}{|M_H(r)|} \right)^2 \right] - \left(\frac{|\Sigma_H(r)|}{N} \right)^2 \\
+ \frac{N - |M_H(r)|}{N} &\left[\left(\frac{|\Sigma_H(r)| - |E_H(r)|}{N - |M_H(r)|} \right)^2 + \left(\frac{(N - |M_H(r)|) - (|\Sigma_H(r)| - |E_H(r)|)}{N - |M_H(r)|} \right)^2 \right] \\
- \left(\frac{N - |\Sigma_H(r)|}{N} \right)^2 &.
\end{aligned}
$$

For Example 5, $\text{gn}(r) = 0.016$.

Definition 17 (Rule Precision). *Given the rule* $r = B_1 \wedge \ldots \wedge B_n \rightarrow H$, *its* precision *is the ratio of the number of correct predictions made by* r *and the total number of correct and incorrect predictions (predictions logically contradicting* $\mathcal{K}$), *leaving out the predictions with unknown truth value.*

This metric expresses the ability of a rule to perform correct predictions, but it is not able to take into account the induced knowledge, that is the *unknown* predictions. For this reason, the metrics proposed in [10] are also considered (for the evaluation in Sect. 4):

- *match rate*: number of predicted assertions in agreement with facts in the complete ontology, out of all predictions;
- *commission error rate*: number of predicted assertions contradicting facts in the full ontology, out of all predictions;
- *induction rate*: number of predicted assertions whose truth is unknown in the complete ontology, out of all predictions.

3   The EDMAR Algorithm

In this section, we provide background information about the EDMAR (Evolutionary Discovery of Multi-relational Association Rules) algorithm [21].

Given a populated ontological KB, EDMAR discovers frequent and accurate multi-relational association rules to be exploited for making predictions of new assertions in the KB. The rules discovered, besides complying with the chosen language bias (see Sect. 2.1), are bred to strike an optimal balance between generality and accuracy.

EDMAR, whose overall flow is shown by Algorithm 1, maintains a population of patterns (the individuals) and makes it evolve by iteratively applying a number of genetic operators. A pattern, represented as a list of atoms, of the form $C(x)$ or $R(x, y)$ and respecting the language bias, to be interpreted in conjunctive form, is the genotype of an individual and the corresponding rule is its phenotype, constructed using the first atom of the pattern as the head and the remaining atoms as the body.

Algorithm 1. The EDMAR algorithm.

Input: $\mathcal{K}$: ontological KB; θ_f: frequency threshold; n: the size of the population; p_{cross}: crossover
    probability; p_{mut}: mutation probability; τ: truncation proportion; θ_{fit}: fitness threshold;
Output: *pop*: set of frequent patterns discovered from $\mathcal{K}$
 1: Creating a list A_f of frequent atoms in $\mathcal{K}$
 2: Initialize a population *pop* of size n by using n times CREATENEWPATTERN() operator
 3: Compute fitness values for all of the patterns in *pop*
 4: Sort *pop* by decreasing fitness value
 5: Initialize *the number of generation* (equals to 0)
 6: **while** (*the number of generation* < MAX_GENERATIONS) **do**
 7:     **for** ($i = 0, 1, \ldots, \lfloor \tau n \rfloor$) **do**
 8:         CROSSOVER($pop[i]$, $pop[\lfloor \tau n \rfloor + i]$)
 9:         CROSSOVER($pop[i]$, $pop[2\lfloor \tau n \rfloor + i]$)
10:        Compute fitness value for all of offspring
11:        **for each** offspring **do**
12:            **with probability** p_{mut} **do** MUTATE(*offspring*)
13:        Add all of offspring to *pop*
14:     Sort *pop* by decreasing fitness value
15:     Remove patterns located at the end of *pop* so that the size of *pop* is exactly n
16:     Increase *the number of generation* by 1
17: Remove redundant and inconsistent rules from the final population *pop*
18: Remove rules where fitness value is less than θ_{fit} from the final population *pop*
19: **return** *pop*

EDMAR is steady-state: children are created by applying genetic operators on selected parents, and then the children are added back into the population to compete with individuals in the old population in order to allow transition into the new population at the next cycle. The selection operator chooses the best parents for reproduction. The genetic operators of initialization, crossover, and mutation, are designed to enforce the respect of the language bias.

The initial population is seeded with patterns consisting of atoms picked at random from a list of frequent atoms.

Before performing selection, patterns in the population are sorted by decreasing fitness value and a given parameter τ is used to assist in the selection of individuals. The selection operator is used before calling the crossover operator: the best $3\lfloor \tau n \rfloor$ individuals are selected for reproduction, and they are split into 3 groups of equal size and of decreasing fitness; each individual of group 1 (the best individuals) is selected twice to mate with each individual in group 2 and each individual in group 3, respectively.

The crossover operator produces two offspring patterns from two parent patterns by randomly rearranging the atoms of the parents. The operator proceeds by creating a set L including all the atoms in the two input patterns and choosing a target length for the two offspring; then, atoms are picked from L at random and added to either pattern until the target length is attained, possibly changing their variables to ensure the language bias is respected.

The mutation operator perturbs a pattern with a given probability p_{mut}, using two operations based on the idea of specialization and generalization in ILP: it applies the specialization operator, if the fitness of the pattern is above a given threshold θ_{mut}, or the generalization operator, if its fitness is below the threshold θ_{mut}, to the pattern undergoing it. The specialization operator appends a new atom to a pattern, while preserving the language bias. The generalization operator removes a random number of atoms located at the end of the body of the pattern. After removing atoms, the length of the body must remain at least one atom and preserve the language bias. In case a pattern is too long to undergo specialization or too short to undergo generalization, mutation will create a completely new body by picking atoms from the list of frequent atoms, while keeping the same head as the parent pattern and respecting the language bias.

The original EDMAR algorithm uses a fitness function defined as

$$\text{fitness}(r) = \text{headCoverage}(r) + \text{pcaconf}(r).$$

In this paper, however, we use a variety of metrics (see Sect. 2.2) to define the fitness of a pattern in view of comparing their respective performance.

Inconsistent rules, i.e., rules that are unsatisfiable when considered jointly with the ontology, are of no use for KB enrichment and have thus to be discarded. Since checking rules for consistency may be computationally very expensive, EDMAR does not check patterns for consistency during evolution. Instead, it defers this check and applies it to the final population only, by calling an off-the-shelf OWL reasoner.

Therefore, every rule r returned by EDMAR satisfies three conditions: (1) r is not a redundant rule (as per Definition 6); (2) $\mathcal{K} \cup r \not\models \bot$; (3) fitness$(r) \geq \theta_{\text{fit}}$ (θ_{fit} is chosen according to the effective range in Table 2).

We refer the reader to [21] for further details on the EDMAR algorithm.

4   Experiments and Results

To improve performance, we compare some popular asymmetric metrics used to assess the rules based on the ability to generate rules, the number of predictions and the number of unknown facts. Through this comparison, we might also select metrics that are suitable with data semantics. The best metrics could be considered and used in the next researches.

We applied our evolutionary algorithm to the same populated ontological KBs used in [8]: Financial,[4] describing the banking domain; Biological Pathways Exchange (BioPAX)[5] Level 2 Ontology, describing biological pathway data; and New Testament Names Ontology (NTNMerged),[6] describing named entities (people, places, and other classes) in the New Testament, as well as their attributes and relationships. Details on these ontologies are reported in Table 1.

To test the capability of the discovered rules to predict new assertional knowledge for each examined ontological KB, stratified versions of each ontology have been constructed (as described in [8]) by randomly removing, respectively, 20%, 30%, and 40% of the concept assertions, while the full ontology versions have been used as a testbed.

---

[4] http://www.cs.put.poznan.pl/alawrynowicz/financial.owl.
[5] http://www.biopax.org/release/biopax-level2.owl.
[6] http://www.semanticbible.com/ntn/ntn-view.html.

Table 1. Key facts about the ontological KBs used.

Ontology	# Concepts	# Roles	# Indiv	# Declared Assertions	# Decl.+Derived Assertions
Financial	59	16	1000	3359	3814
BioPAX	40	33	323	904	1671
NTNMerged	47	27	695	4161	6863

We performed 30 runs of the EA described in Sect. 3 for each stratified version and for each choice of fitness function using the following parameter setting:

$$n = 5,000; \quad p_{\text{mut}} = 5\%; \quad ,$$
$$\text{MAX\_GENERATIONS} = 200; \quad \theta_{\text{mut}} = 0.2;$$
$$\text{MAX\_RULE\_LENGTH} = 10; \quad \tau = \tfrac{1}{5}$$
$$\theta_{\text{fit}} = 0 \quad \theta_f = 1.$$
$$(\text{Conviction: } \theta_{\text{fit}} = 1)$$

Our experiments aimed at comparing of the results obtained by the EA using different rule evaluation metrics as fitness based on the three following criteria:

1. The *number of the rules discovered* by the EA.
2. The *induction rate*: if it is positive, this means assertions are predicted that could not be inferred from the stratified version. The higher the induction rate, the more novel predictions (unknown facts) are induced for the KB.
3. The *number of correct predictions* = *number of predictions* × *precision*, where the *number of predictions* is the number of predicted assertions and *precision* is defined in Definition 17.

Table 2. Symbols and range of metrics (the effective range is used to assist in the choice of θ_{fit}.)

Symbol	Metric	Range	Effective range
H	Head Coverage	[0, 1]	(0, 1]
C	Confidence	[0, 1]	(0, 1]
P	PCA-Confidence	[0, 1]	(0, 1]
L	Laplace	[0, 1]	(0, 1]
CV	Conviction	[0.5, +∞)	(1, +∞)
CF	Certainty Factor	[−1, 1]	(0, 1]
A	Added Value	[−0.5 , 1]	(0, 1]
J	J-Measure	[0, 1]	(0, 1]
G	Gini Index	[0, 1]	(0, 1]

Table 3 shows a comparison of the metrics (identified with the acronyms defined in Table 2) according to the first criterion. The second and third criteria are used to compare the predictive power of the discovered rules. In order to compare the metrics according to these criteria, we applied the rules discovered from the stratified versions to the full ontology versions and collected all predictions, i.e., the head atoms of the instantiated rules. Given the collected predictions, those already contained in the stratified ontology versions were discarded, while the remaining predicted facts were considered. A prediction is evaluated as *correct* if it is contained/entailed by the full ontology version and as *incorrect* if it is inconsistent with the full ontology version. All the results (see Tables 4 and 5) have been computed using the rules discovered by each metrics (see Table 3) based on 30 runs with the above parameter setting and have been measured in terms of *precision* (see Definition 17), *match*, *commission*, and *induction rate* (see Sect. 2.2). The statistic significance of all pairwise comparisons between metrics have been assessed using 1-tailed Welch's t-test.

Table 3. Comparison of the metrics by the number of discovered rules.

Ontology	Samp.	Total number of discovered rules by metric ± stdev								
		H	C	P	L	CV	CF	A	J	G
Financial	20%	26 ± 4	25 ± 4	25 ± 3	3,254 ± 30	4 ± 1	25 ± 3	26 ± 3	3 ± 1	487 ± 12
	30%	25 ± 3	25 ± 4	25 ± 4	3,301 ± 31	4 ± 1	26 ± 3	24 ± 4	4 ± 1	485 ± 9
	40%	23 ± 3	23 ± 3	22 ± 4	3,296 ± 31	3 ± 1	23 ± 4	21 ± 3	3 ± 1	479 ± 11
Biopax	20%	129 ± 13	122 ± 12	130 ± 10	4,293 ± 24	35 ± 5	118 ± 9	119 ± 9	58 ± 5	3,486 ± 182
	30%	128 ± 9	130 ± 13	130 ± 9	4,384 ± 22	33 ± 5	117 ± 8	110 ± 9	55 ± 5	3,658 ± 139
	40%	129 ± 11	136 ± 11	133 ± 8	4,530 ± 23	36 ± 5	124 ± 9	122 ± 7	59 ± 6	3,560 ± 157
NTNMerged	20%	1,157 ± 168	1,345 ± 423	1,418 ± 492	4,563 ± 53	382 ± 31	671 ± 36	656 ± 34	504 ± 22	2,040 ± 690
	30%	1,052 ± 353	947 ± 238	1,017 ± 370	4,805 ± 13	509 ± 39	743 ± 45	728 ± 48	460 ± 21	457 ± 90
	40%	1,088 ± 181	1,223 ± 177	1,295 ± 357	4,797 ± 22	397 ± 26	687 ± 38	664 ± 34	500 ± 26	1,506 ± 61

The EA achieves *precision = match rate + commission rate + induction rate = 1* and *commission error rate = 0* on all versions of all considered ontologies; this confirms its ability to discover accurate rules; as a consequence, the *number of correct predictions* coincides with the number of discovered predictions.

From the observations in Tables 3, 4 and 5, we can draw a few remarks:

1. *Laplace* has the highest number of discovered rules. However, it hardly produces any new knowledge (induction rate ≈ 0).

2. *Gini Index* scores the second highest number of discovered rules. However, this measure looks less robust when compared to other metrics, since large deviations among discovered rules show up for different stratified samples of the same ontology (see Table 3); sometimes, it produces much new knowledge, sometimes little or none (induction rate is not stable—see Table 5). In addition, the number of predictions is medium or low compared to other metrics (see Table 5).

3. Five metrics (*HeadCoverage, Confidence, PCA-Confidence, Certainty Factor,* and *Added Value*) allow the EA to generate the largest number of rules (see Table 3) and, which is even more relevant, to come up with rules that induce a large number of previously unknown facts (induction rate > 0), with a very large absolute number of correct predictions (see Tables 4 and 5).

4. Two metrics (*Conviction* and *J-Measure*) produce the smallest number of rules. Although both the induction rate and the number of predictions are acceptable, the low number of discovered rules may mean valuable rules are missed out.

From the above remarks, we may conclude that *HeadCoverage, Confidence, PCA-Confidence, Certainty Factor,* and *Added Value* are the best choices as an optimization criterion (i.e., fitness function) for EDMAR.

We also compared the experimental performance of EDMAR + the five best metrics to state-of-the-art methods which are closest to it in purpose, namely the original EDMAR algorithm (with *HeadCoverage + PCAConfidence* as fitness [21]) and the two level-wise generate-and-test algorithms RARD [7] and AMIE [12]. Table 6 reports the number of rules discovered by each system given each KB sample. We can remark the following:

1. The top-5 metrics discover more rules than RARD from NTNMerged, but fewer from the Financial and Biopax KBs. One reason is that RARD can discover also *open* rules, which are barred by EDMAR's language bias (see Definition 5); furthermore, the maximum length of a rule is 10 atoms. Another reason is that the number of individuals in Financial and Biopax is less than that of NTNMerged (see Table 1, last column). If one factor these differences out, the top-5 metrics are superior to RARD.

Table 4. Avg (± st.dev.) performance on each ontology of HeadCoverage (H), Confidence (C), PCA-Confidence (P), Laplace(L), Conviction(CV), Certainty Factor(CF) *precision = match rate + commission rate + induction rate*

	Ont.	Samp.	Match Rate	Com. Rate	Ind. Rate	Total # Predictions		Ont.	Samp.	Match Rate	Com. Rate	Ind. Rate	Total # Predictions
H	Financial	20%	0.855 ± 0.033	0	0.145 ± 0.033	47,232 ± 36,777	**L**	Financial	20%	1.0	0	0	122,432 ± 1,704
	Financial	30%	0.864 ± 0.044	0	0.136 ± 0.044	25,456 ± 34,174		Financial	30%	1.0	0	0	180,231 ± 2,801
	Financial	40%	0.861 ± 0.044	0	0.139 ± 0.044	23,207 ± 30,133		Financial	40%	1.0	0	0	230,736 ± 3,484
	BioPAX	20%	0.567 ± 0.031	0	0.433 ± 0.031	84,035 ± 15,018		BioPAX	20%	1.0	0	0	51,060 ± 866
	BioPAX	30%	0.591 ± 0.03	0	0.409 ± 0.03	85,499 ± 11,660		BioPAX	30%	1.0	0	0	78,488 ± 1,527
	BioPAX	40%	0.58 ± 0.027	0	0.42 ± 0.027	90,856 ± 14,048		BioPAX	40%	1.0	0	0	100,699 ± 1,600
	NTNMerged	20%	0.572 ± 0.026	0	0.428 ± 0.026	2,311,624 ± 287,858		NTNMerged	20%	0.994 ± 0.001	0	0.006 ± 0.001	197,374 ± 6,116
	NTNMerged	30%	0.564 ± 0.039	0	0.436 ± 0.039	2,314,346 ± 458,522		NTNMerged	30%	0.995	0	0.005	284,065 ± 5,806
	NTNMerged	40%	0.621 ± 0.027	0	0.379 ± 0.027	2,345,588 ± 357,565		NTNMerged	40%	0.996	0	0.004	323,205 ± 6,359
C	Financial	20%	0.848 ± 0.045	0	0.152 ± 0.045	43,151 ± 44,254	**CV**	Financial	20%	0 ± 0.001	0	1.0 ± 0.001	48,661 ± 41,318
	Financial	30%	0.860 ± 0.038	0	0.140 ± 0.038	27,589 ± 41,184		Financial	30%	0.001 ± 0.001	0	0.999 ± 0.001	43,078 ± 39,328
	Financial	40%	0.858 ± 0.051	0	0.142 ± 0.051	33,795 ± 41,880		Financial	40%	0.001 ± 0.002	0	0.999 ± 0.002	26,268 ± 33,679
	BioPAX	20%	0.574 ± 0.036	0	0.426 ± 0.036	79,454 ± 14,019		BioPAX	20%	0.08 ± 0.018	0	0.92 ± 0.018	44,971 ± 10,928
	BioPAX	30%	0.584 ± 0.027	0	0.416 ± 0.027	88,879 ± 12,890		BioPAX	30%	0.11 ± 0.017	0	0.89 ± 0.017	44,451 ± 10,557
	BioPAX	40%	0.582 ± 0.023	0	0.418 ± 0.023	96,884 ± 13,782		BioPAX	40%	0.102 ± 0.018	0	0.898 ± 0.018	50,457 ± 12,368
	NTNMerged	20%	0.618 ± 0.042	0	0.382 ± 0.042	1,437,868 ± 253,206		NTNMerged	20%	0.32 ± 0.019	0	0.68 ± 0.019	831,416 ± 183,095
	NTNMerged	30%	0.581 ± 0.036	0	0.419 ± 0.036	1,164,306 ± 167,173		NTNMerged	30%	0.344 ± 0.013	0	0.656 ± 0.013	1,123,266 ± 208,471
	NTNMerged	40%	0.670 ± 0.030	0	0.330 ± 0.030	1,557,516 ± 280,666		NTNMerged	40%	0.361 ± 0.015	0	0.639 ± 0.015	868,467 ± 174,865
P	Financial	20%	0.859 ± 0.055	0	0.141 ± 0.055	41,350 ± 46,196	**CF**	Financial	20%	0.877 ± 0.038	0	0.123 ± 0.038	31,656 ± 45,045
	Financial	30%	0.850 ± 0.055	0	0.150 ± 0.055	32,812 ± 41,501		Financial	30%	0.852 ± 0.057	0	0.148 ± 0.057	48,568 ± 45,051
	Financial	40%	0.859 ± 0.043	0	0.141 ± 0.043	29,762 ± 35,582		Financial	40%	0.857 ± 0.039	0	0.143 ± 0.039	31,068 ± 36,044
	BioPAX	20%	0.571 ± 0.028	0	0.429 ± 0.028	89,486 ± 11,303		BioPAX	20%	0.556 ± 0.026	0	0.444 ± 0.026	80,361 ± 8,700
	BioPAX	30%	0.584 ± 0.023	0	0.416 ± 0.023	92,392 ± 13,878		BioPAX	30%	0.581 ± 0.023	0	0.419 ± 0.023	78,933 ± 11,147
	BioPAX	40%	0.587 ± 0.027	0	0.413 ± 0.027	91,849 ± 11,960		BioPAX	40%	0.564 ± 0.035	0	0.436 ± 0.035	84,476 ± 12,647
	NTNMerged	20%	0.609 ± 0.046	0	0.391 ± 0.046	2,130,947 ± 380,546		NTNMerged	20%	0.565 ± 0.01	0	0.435 ± 0.01	1,039,112 ± 179,322
	NTNMerged	30%	0.588 ± 0.043	0	0.412 ± 0.043	1,409,235 ± 286,439		NTNMerged	30%	0.535 ± 0.014	0	0.465 ± 0.014	1,424,334 ± 180,205
	NTNMerged	40%	0.670 ± 0.042	0	0.330 ± 0.042	1,727,343 ± 262,891		NTNMerged	40%	0.557 ± 0.018	0	0.443 ± 0.018	2,110,928 ± 423,539

Table 5. Avg (± st.dev.) performance on each ontology of Added value (A), J-Measure (J) and Gini factor (G) *precision = match rate + commission rate + induction rate*

	Ont.	Samp.	Match Rate	Com. Rate	Ind. Rate	Total # Predictions
A	Financial	20%	0.859 ± 0.041	0	0.141 ± 0.041	33,358 ± 31,445
	Financial	30%	0.858 ± 0.041	0	0.142 ± 0.041	29,866 ± 31,123
	Financial	40%	0.859 ± 0.041	0	0.141 ± 0.041	29,870 ± 44,276
	BioPAX	20%	0.549 ± 0.032	0	0.451 ± 0.032	83,666 ± 11,663
	BioPAX	30%	0.578 ± 0.029	0	0.422 ± 0.029	78,059 ± 9,368
	BioPAX	40%	0.579 ± 0.02	0	0.421 ± 0.02	84,483 ± 10,376
	NTNMerged	20%	0.563 ± 0.012	0	0.437 ± 0.012	966,840 ± 204,430
	NTNMerged	30%	0.541 ± 0.014	0	0.459 ± 0.014	1,324,518 ± 282,410
	NTNMerged	40%	0.566 ± 0.014	0	0.434 ± 0.014	1,632,633 ± 218,033
J	Financial	20%	0 ± 0.001	0	1.0 ± 0.001	10,148 ± 13,149
	Financial	30%	0.001 ± 0.001	0	0.999 ± 0.001	32,052 ± 39,154
	Financial	40%	0	0	1.0	36,204 ± 40,910
	BioPAX	20%	0.083 ± 0.011	0	0.917 ± 0.011	82,799 ± 11,596
	BioPAX	30%	0.108 ± 0.013	0	0.892 ± 0.013	80,797 ± 13,564
	BioPAX	40%	0.11 ± 0.013	0	0.89 ± 0.013	90,480 ± 12,579
	NTNMerged	20%	0.294 ± 0.008	0	0.706 ± 0.008	1,317,526 ± 207,005
	NTNMerged	30%	0.301 ± 0.011	0	0.699 ± 0.011	1,765,003 ± 242,269
	NTNMerged	40%	0.319 ± 0.011	0	0.681 ± 0.011	2,387,450 ± 698,911

	Ont.	Samp.	Match Rate	Com. Rate	Ind. Rate	Total # Predictions
G	Financial	20%	0.182 ± 0.007	0	0.818 ± 0.007	20,321 ± 22,967
	Financial	30%	0.181 ± 0.01	0	0.819 ± 0.01	49,443 ± 42,556
	Financial	40%	0.186 ± 0.009	0	0.814 ± 0.009	20,645 ± 18,367
	BioPAX	20%	1.0	0	0	30,839 ± 1,632
	BioPAX	30%	1.0	0	0	45,063 ± 1,727
	BioPAX	40%	1.0	0	0	62,941 ± 2,781
	NTNMerged	20%	0.768 ± 0.054	0	0.232 ± 0.054	199,745 ± 50,410
	NTNMerged	30%	0.725 ± 0.023	0	0.275 ± 0.023	82,059 ± 13,066
	NTNMerged	40%	0.785 ± 0.007	0	0.215 ± 0.007	258,454 ± 10,162

2. The top-5 metrics discover more rules than AMIE from Financial and Biopax and a comparable number from the NTNMerged KB. One limitation of deterministic level-wise generate-and-test methods like AMIE and RARD is that they cannot scale up to rules longer than 3 atoms, while EDMAR (with any metrics) can easily discover rules of 10 atoms (and possibly more).
3. EDMAR's original fitness function outperforms each of the top-5 metrics; however, it is a combination of two of them. This suggests a new promising direction of research, that is to try to find an optimal fitness function for EDMAR by combining the individual metrics studies in this paper.

Table 6. Comparison of the number of discovered rules.

Ontology	Samp.	# The total number of rules discovered							
		H	C	P	CF	A	EDMAR	RARD	AMIE
Financial	20%	26 ± 4	25 ± 4	25 ± 3	25 ± 3	26 ± 3	27 ± 3	177	2
	30%	25 ± 3	25 ± 4	25 ± 4	26 ± 3	24 ± 4	26 ± 3	181	2
	40%	23 ± 3	23 ± 3	22 ± 4	23 ± 4	21 ± 3	24 ± 4	180	2
Biopax	20%	129 ± 13	122 ± 12	130 ± 10	118 ± 9	119 ± 9	132 ± 10	298	8
	30%	128 ± 9	130 ± 13	130 ± 9	117 ± 8	110 ± 9	118 ± 12	283	8
	40%	129 ± 11	136 ± 11	133 ± 8	124 ± 9	122 ± 7	137 ± 12	272	0
NTNMerged	20%	1,157 ± 168	1,345 ± 423	1,418 ± 492	671 ± 36	656 ± 34	1,834 ± 782	243	1,129
	30%	1,052 ± 353	947 ± 238	1,017 ± 370	743 ± 45	728 ± 48	1,235 ± 495	225	1,022
	40%	1,088 ± 181	1,223 ± 177	1,295 ± 357	687 ± 38	664 ± 34	1,810 ± 733	239	1,063

5   Conclusions

The results of the empirical comparison of a number of popular asymmetric metrics to be used as fitness functions for evolutionary inductive programming allow us to identify five metrics as the most promising candidates for further exploration.

Future work might focus on four main aspects: (i) exploration of various possible combination of the promising metrics; (ii) development of other metrics suited for scenarios based on the OWA; (iii) scalability, by considering large datasets from the Linked Data Cloud; (iv) parallelization according to programming models such MapReduce to take advantage of frameworks like Hadoop, in order to be able to perform big data analytics.

References

1. Agrawal, R., Imielinski, T., Swami, A.N.: Mining association rules between sets of items in large databases. In: Proceedings of the International Conference on Management of Data, pp. 207–216. ACM Press (1993)
2. Baader, F., Calvanese, D., McGuinness, D.L., Nardi, D., Patel-Schneider, P.F. (eds.): The Description Logic Handbook: Theory, Implementation, and Applications. Cambridge University Press, New York (2003)
3. Berners-Lee, T., Hendler, J., Lassila, O.: The semantic web. Scientific American (2001)

4. Breiman, L., Friedman, J., Olshen, R., Stone, C.: Classification and regression trees, New York, USA (1984)
5. Brin, S., Motwani, R., Ullman, J.D., Tsur, S.: Dynamic itemset counting and implication rules for market basket data. In: Proceedings of 1997 ACM SIGMOD International Conference on Management of Data, pp. 255–264 (1997)
6. Clark, P., Boswell, R.: Rule induction with CN2: some recent improvements. In: Proceedings of the Fifth European Conference, pp. 151–163 (1991)
7. d'Amato, C., Staab, S., Tettamanzi, A., Tran, D.M., Gandon, F.: Ontology enrichment by discovering multi-relational association rules from ontological knowledge bases. In: Proceedings of SAC 2016. ACM (2016)
8. d'Amato, C., Tettamanzi, A.G.B., Minh, T.D.: Evolutionary discovery of multi-relational association rules from ontological knowledge bases. In: Blomqvist, E., Ciancarini, P., Poggi, F., Vitali, F. (eds.) EKAW 2016. LNCS (LNAI), vol. 10024, pp. 113–128. Springer, Cham (2016). https://doi.org/10.1007/978-3-319-49004-5_8
9. Divina, F.: Genetic Relational Search for Inductive Concept Learning: A Memetic Algorithm for ILP. LAP LAMBERT Academic Publishing (2010)
10. Fanizzi, N., d'Amato, C., Esposito, F.: Learning with kernels in description logics. In: Železný, F., Lavrač, N. (eds.) ILP 2008. LNCS (LNAI), vol. 5194, pp. 210–225. Springer, Heidelberg (2008). https://doi.org/10.1007/978-3-540-85928-4_18
11. Fu, L.M., Shortliffe, E.H.: The application of certainty factors to neural computing for rule discovery. IEEE Trans. Neural Netw. **11**, 647–657 (2000)
12. Galárraga, L., Teflioudi, C., Hose, K., Suchanek, F.: AMIE: association rule mining under incomplete evidence in ontological knowledge bases. In: WWW 2013, pp. 413–422. ACM (2013)
13. Horrocks, I., Patel-Schneider, P.F., Boley, H., Tabet, S., Grosof, B., Dean, M.: SWRL: a semantic web rule language combining OWL and RuleML (2004). http://www.w3.org/Submission/2004/SUBM-SWRL-20040521/
14. Józefowska, J., Lawrynowicz, A., Lukaszewski, T.: The role of semantics in mining frequent patterns from knowledge bases in description logics with rules. Theory Pract. Logic Program. **10**(3), 251–289 (2010)
15. Lisi, F.A.: AL-QuIn: an onto-relational learning system for semantic web mining. Int. J. Semant. Web Inf. Syst. 7(3), 1–22 (2011)
16. Motik, B., Sattler, U., Studer, R.: Query answering for OWL-DL with rules. Web Semantics **3**(1), 41–60 (2005)
17. Muggleton, S., Tamaddoni-Nezhad, A.: QG/GA: a stochastic search for progol. Mach. Learn. **70**(2–3), 121–133 (2008)
18. Reiser, P., Riddle, P.: Scaling up inductive logic programming: an evolutionary wrapper approach. Appl. Intell. **15**(3), 181–197 (2001)
19. Sahar, S., Mansour, Y.: An empirical evaluation of objective interestingness criteria. In: SPIE Conference on Data mining and Knowledge Discovery, pp. 63–74 (1999)
20. Smyth, P., Goodman, R.: Rule Induction Using Information Theory. MIT Press, Cambridge (1991)
21. Tran, M.D., d'Amato, C., Nguyen, B.T., Tettamanzi, A.G.B.: An evolutionary algorithm for discovering multi-relational association rules in the semantic web. In: GECCO, pp. 513–520. ACM (2017)
22. Völker, J., Niepert, M.: Statistical schema induction. In: Antoniou, G., Grobelnik, M., Simperl, E., Parsia, B., Plexousakis, D., De Leenheer, P., Pan, J. (eds.) ESWC 2011. LNCS, vol. 6643, pp. 124–138. Springer, Heidelberg (2011). https://doi.org/10.1007/978-3-642-21034-1_9

Genetic Programming Hyper-Heuristic with Cooperative Coevolution for Dynamic Flexible Job Shop Scheduling

Daniel Yska, Yi Mei$^{(\boxtimes)}$, and Mengjie Zhang

Victoria University of Wellington, Wellington, New Zealand
{daniel.yska,yi.mei,mengjie.zhang}@ecs.vuw.ac.nz

Abstract. Flexible Job Shop Scheduling (FJSS) problem has many real-world applications such as manufacturing and cloud computing, and thus is an important area of study. In real world, the environment is often dynamic, and unpredicted job orders can arrive in real time. Dynamic FJSS consists of challenges of both dynamic optimisation and the FJSS problem. In Dynamic FJSS, two kinds of decisions (so-called *routing* and *sequencing* decisions) are to be made in real time. Dispatching rules have been demonstrated to be effective for dynamic scheduling due to their low computational complexity and ability to make real-time decisions. However, it is time consuming and strenuous to design effective dispatching rules manually due to the complex interactions between job shop attributes. Genetic Programming Hyper-heuristic (GPHH) has shown success in automatically designing dispatching rules which are much better than the manually designed ones. Previous works only focused on standard job shop scheduling with only the sequencing decisions. For FJSS, the routing rule is set arbitrarily by intuition. In this paper, we explore the possibility of evolving both routing and sequencing rules together and propose a new GPHH algorithm with Cooperative Co-evolution. Our results show that co-evolving the two rules together can lead to much more promising results than evolving the sequencing rule only.

Keywords: Job Shop Scheduling · Genetic Programming
Hyper-heuristics · Cooperative Co-evolution

1 Introduction

In the modern industrial world, processing and manufacturing are global industries which are central to the economies of virtually every country. In a large factory setting, the efficient allocation of jobs to machines is therefore an extremely important concept that businesses must consider to increase throughput, decrease costs and increase profitability [15]. In a virtual setting, the idea of the efficient allocation of jobs to machines can also be applied to cloud resources. There are further applications to be found in timetabling, sports scheduling,

© Springer International Publishing AG, part of Springer Nature 2018
M. Castelli et al. (Eds.): EuroGP 2018, LNCS 10781, pp. 306–321, 2018.
https://doi.org/10.1007/978-3-319-77553-1_19

health care scheduling and crew scheduling. This study of the allocation of jobs to machines is therefore a hugely important and relevant area of study to create more efficient outcomes in the modern world, saving time and resources.

In the Job Shop Scheduling (JSS) problem, there is a set of jobs to be completed, and a set of machines which can process the jobs [27]. A solution to this problem is an ordered schedule of assignments of jobs to machines, so that all jobs are completed. These schedules are optimised relative to some objective, such as minimising the makespan and flowtime.

Flexible JSS (FJSS) problem is an extension of JSS. In JSS, each job operation only has one candidate machine to process it. In contrast, an operation of a job may have multiple candidate machines (*options*) in FJSS. As a result, FJSS involves allocating job operations to machines (i.e. *routing* problem) as well as selecting jobs from the queue of an idle machine to be processed next (i.e. *sequencing* problem). This makes FJSS more challenging than JSS.

FJSS is NP-hard since it has JSS as its special case (where all the operations have only one candidate machine). Thus, traditional optimisation methods such as branch-and-bound [18] is applicable when the problem size is not large. In this case, heuristic search methods such as simulated annealing [34], tabu search [26] and genetic algorithm [35] show promise in finding reasonably good solutions in a short time. However, in real world, the environment is usually *dynamic*, and unpredicted jobs can arrive at any time. The decisions made about which job to be processed next must be able to factor in the changing state of the system quickly and computationally cheaply. Therefore traditional optimisation techniques are infeasible for dynamic JSS due to their high computational complexity.

Dispatching Rules (DRs) have been used extensively in JSS (e.g. [3]) due to their computational efficiency. Whenever a machine becomes idle, a DR calculates a priority value for each job waiting in its queue and selects the most prior job to process next. Such computation is carried out at each decision point (e.g. when a machine becomes idle) and can be done efficiently. A variety of DRs have been designed manually to handle different scenarios. An overview of the manually designed DRs can be found in [30].

Manually designing DRs is time consuming and very demanding on domain expertise. The existing manually designed rules tend to be overly simplistic, with plenty of literature showing that many manually designed rules only perform well for certain objectives and in certain job shops [17,29,31]. Recently, Genetic Programming Hyper-heuristics (GPHH) has been successfully applied to automatically designing (evolving) DRs for scheduling [5,19,20,23], and the evolved DRs are much more effective than the manually designed DRs. However, the existing works mainly focused on evolving the *sequencing* rules, i.e. the rules selecting the operations from the queue of the idle machine to process next. The *routing* rule (i.e. the rule to select a candidate machine to process the given operation) is normally specified intuitive (e.g. selecting the machine with the least waiting time in [32]). Such simple routing rules are by no means the best and there is a potential to design routing rules that cooperates with the sequencing rules better in the given scheduling scenario. This motivates us to evolve the sequencing and routing rules simultaneously. To this end, we adopt the Cooperative Co-evolution (CC) [28] framework, which is a natural framework to evolve multiple components together. It has also been applied in JSS for co-evolving the DR and due date assignment rule [24].

1.1 Goals

In this paper, we aim to find more promising routing and sequencing rules for FJSS. Specifically, we aim to achieve the following research objectives.

1. Compare between different manually designed routing rules on different FJSS scenarios to understand which manually designed routing rule performs the best in general.
2. Propose a GP with Cooperative Co-evolution (called CCGP) for co-evolving the routing and sequencing rules simultaneously.
3. Compare CCGP with the GP that evolves sequencing rule with pre-specified routing rule (called SeqGP) to evaluate the performance of CCGP.
4. Conduct analysis on the characteristics of the rules evolved by CCGP to gain new knowledge about the structure of the effective routing rules for FJSS.

1.2 Organisation

The rest of the paper is organised as follows: Sect. 2 gives the problem description and related work. Then, the proposed CCGP is proposed in Sect. 3. Experiment studies are carried out in Sects. 4 and 5. Finally, Sect. 6 gives the conclusions and future work.

2 Background

2.1 Flexible Job Shop Scheduling

FJSS is to process a set of jobs $\mathcal{J} = \{J_1, \ldots, J_n\}$ with a set of machines $\mathcal{M} = \{M_1, \ldots, M_m\}$. Each job J_j has an arrival time $t_0(J_j)$ and a sequence of operations $O_{1,j}, \ldots, O_{l_j,j}$. Each operation $O_{i,j}$ has a set of candidate machines $\pi_{i,j} \subseteq \mathcal{M}$. It can be processed by any machine $\pi_{i,j,k} \in \pi_{i,j}$. The duration of processing operation $O_{i,j}$ with machine $\pi_{i,j,k}$ is $\delta_{i,j,k}$. One cannot start processing an operation until its preceding operations have been completed. Each machine can process at most one operation at a time, and each operation is processed by exactly one machine without interruption. The goal of FJSS is to find a feasible schedule to optimise some objective(s). The commonly considered JSS objectives include minimising the makespan ($C_{\max}$), total flowtime ($\sum C_j$), total weighted tardiness ($\sum w_j T_j$), number of tardy jobs, etc. [27].

JSS is a special case of FJSS, where for each operation $O_{i,j}$, $|\pi_{i,j}| = 1$. In other words, each operation can be processed by only one machine. In this case, no routing decision needs to be made.

2.2 Related Work

The FJSS problem was first identified by Brucker and Schlie [6] in 1990, where a solution of a polynomial algorithm was suggested to solve each of the routing and sequencing sub-problems for a two job system. Early studies focused on finding FJSS solutions using traditional optimisation approaches. Brandimarte [4] proposed using a hierarchical method to minimise the makespan for a FJSS system. In his work, he used a two-level tabu search algorithm in combination with the decomposition of FJSS into routing and job shop scheduling sub problems. In his work, Brandimarte also created a class of flexible job shops that would become used as a benchmark by future researchers [2]. Norman and Bean [25] developed a genetic algorithm with random key representation, elitist reproduction, immigration mutation as well as Bernoulli crossover to solve the FJSS problem with the objective of minimising total tardiness. In 2002, Kacem et al. [16] proposed a hybrid approach for solving the FJSS problem, using localisation for the routing component, and three manually designed dispatching rules for the sequencing component. An advanced genetic algorithm was proposed for evolving arrangements of jobs and machines.

In recent decades, hyper-heuristics [7] have attracted more and more research attention, as they can find heuristics (i.e. dispatching rules in JSS) rather than solutions, and thus are more flexible and scalable. More importantly, the evolved heuristics can handle dynamic environment much more effectively than traditional (re-)optimisation approaches. In 2001, Dimopoulos and Zalzala [9] used GP to evolve dispatching rules for JSS, for single machine scheduling with a terminal set of scheduling attributes (processing time, due date, number of jobs, release time, etc.) with a standard function set. These evolved dispatching rules performed well and were better than traditional manually designed rules even for unseen and large instances. Then in 2006, Geiger et al. [10] presented a learning system which combined GP with a simulation model for an industrial facility. This proposed GP method creates a rule assigned priority to jobs on a single machine in both static and dynamic environments. This paper quickly produced many dispatching rules which rivalled results produced by rules found in past decades. A method for evolving dispatching rules for multiple machines was proposed, which used modified genetic operators. Miyashita [22] in 2000 developed an automatic method of evolved customised dispatching rules for a JSS environment, using GP. In his work, he considered the JSS problem as being a multi-agent problem, where each agent represents a resource (machine or work station). This multi-agent model was explored using GP, and produced good results, however prior knowledge of the JSS environment was required. This limits the application of this work to only static environments.

In 2007, Tay and Ho [33] proposed a GP method to evolve dispatching rules for a FJSS environment which were optimised for multiple objectives. These multiple objectives were treated as a single objective by linearly combined their objective functions. The proposed GP method can be thought of as a priority function which calculated the priority of operations in the queue of a single machine, based on static and dynamic attributes in the job shop. The dispatching rules evolved outperformed other manually designed dispatching rules, although the use of machine attributes was not considered. This system was assessed later in 2010 by Hildebrandt et al. [12] which showed that in some dynamic JSS instances, the evolved rules by Tay and Ho [33] performed only slightly better than the earliest release date rule, and worse the than shortest processing time rule, which are very simplistic. Hildebrandt et al. [12] then used GP to evolve dispatching rules in four simulations (all with 10 machines, with a combination of two utilisation levels and two job types) for the single objective of mean flow time. Their evolved rules were robust, performing very well in both different environments (50 machines with varying processing time distributions) and the original training environments. In 2014, Nguyen et al. [24] used cooperative coevolution GP to evolve due date assignment rules and dispatching rules, for multi-objective JSS. In this work, Nguyen et al. showed that the evolved scheduling policies performed very well on unseen simulation scenarios, given different shop settings. In 2016, Mei et al. [21] used GP to evolve dispatching rules for JSS for a single objective. Feature selection was then performed on the terminal set of the dispatching rules, removing extraneous terminal attributes and reducing the problem search space. This led to significantly better dispatching rules evolved by GP on both training and test instances.

3   Genetic Programming with Cooperative Co-evolution

The pseudo-code of the proposed CCGP is described in Algorithm 1. In the proposed CCGP, there are two subpopulations $\mathbf{P}_r = \{P_{r,1}, P_{r,2}, \ldots\}$ and $\mathbf{P}_s = \{P_{s,1}, P_{s,2}, \ldots\}$, where $\mathbf{P}_r$ stands for the population of routing rules and $\mathbf{P}_s$ stands for the population of sequencing rules. In addition, a context vector $\mathbf{cv} = (cv_r, cv_s)$ is maintained for fitness evaluation. At first, the two populations are randomly initialised by ramp half-and-half method, and the context vector is randomly initialised from the populations. Then, at each generation, the routing rules and sequencing rules are evolved separately using the crossover/mutation/reproduction operator of GP. Then, each newly generated rule is evaluated by the evaluate(·) method. Finally, the context vector is updated by replacing the routing and sequencing components with the best individuals in the corresponding population, if they have better fitness values. In the minimisation case in FJSS (e.g. makespan and flowtime are to be minimised), a smaller fitness value is better.

The fitness evaluation procedure is given in Algorithm 2. It takes a routing rule p_r, a sequencing rule p_s, and a set of FJSS instances $\mathcal{I}_{\text{train}}$ (i.e. *training set*), and returns a fitness value. For each training instance, it constructs a discrete event simulation based on the instance, the routing and sequencing rules, and run the simulation to generate a schedule.

At the beginning of the simulation, all machines are idle, and there may be some initial jobs ready to be processed (ready time 0). In the dynamic FJSS scenarios, unpredicted job arrival events are generated randomly as well. Then online decisions are made as follows until all the jobs have been completed.

Algorithm 1. Pseudo-code of the proposed CCGP

---

1   Randomly initialise $\mathbf{P}_r$ and $\mathbf{P}_s$ by ramp half-and-half;
2   $cv_r \leftarrow P_{r,1}$, $cv_s \leftarrow P_{s,1}$;   // arbitrarily initialise context vector
3   **while** *Stopping criteria not met* **do**
       // Evolve the routing rules
4      $\mathbf{P}'_r \leftarrow \mathtt{elite}(\mathbf{P}_r)$;   // copy the elites to the new population
5   **while** $|\mathbf{P}'_r| < \mathrm{popsize}$ **do**
6   Generate offspring(s) by applying the crossover/mutation/reproduction operator to $\mathbf{P}_r$;
7   Add the generated offspring(s) to $\mathbf{P}'_r$;
8   **end**
       // Evolve the sequencing rules
9      $\mathbf{P}'_s \leftarrow \mathtt{elite}(\mathbf{P}_s)$;   // copy the elites to the new population
10   **while** $|\mathbf{P}'_s| < \mathrm{popsize}$ **do**
11   Generate offspring(s) by applying the crossover/mutation/reproduction operator to $\mathbf{P}_s$;
12   Add the generated offspring(s) to $\mathbf{P}'_s$;
13   **end**
       // fitness evaluation
14   **foreach** $p \in \mathbf{P}'_r$ **do** $\mathtt{fit}(p) \leftarrow \mathtt{evaluate}(p, cv_s, \mathcal{I}_{\mathrm{train}})$;
15   **foreach** $p \in \mathbf{P}'_s$ **do** $\mathtt{fit}(p) \leftarrow \mathtt{evaluate}(cv_r, p, \mathcal{I}_{\mathrm{train}})$;
16     $\mathbf{P}_r \leftarrow \mathbf{P}'_r$, $\mathbf{P}_s \leftarrow \mathbf{P}'_s$;   // update subpopulations
       // update context vector
17     $cv'_r \leftarrow \arg\min_{p \in \mathbf{P}_r} \mathtt{fit}(p)$, $cv'_s \leftarrow \arg\min_{p \in \mathbf{P}_s} \mathtt{fit}(p)$;
18   **if** $\mathtt{fit}(cv'_r) < \mathtt{fit}(cv_r)$ **then** $cv_r \leftarrow cv'_r$;
19   **if** $\mathtt{fit}(cv'_s) < \mathtt{fit}(cv_s)$ **then** $cv_s \leftarrow cv'_s$;
20 **end**
21 **return** $\mathbf{cv} = (cv_r, cv_s)$;

- Whenever a job becomes ready to be processed, if its next operation has only one candidate machine, then place the job into the queue of the candidate machine. Otherwise, apply the *routing rule* to select the machine to process the job, and place the job to the queue of the selected machine.
- Whenever a machine is idle and its queue is not empty, apply the *sequencing rule* to select the next job from the queue, and start processing the next job.

A simulation essentially generates a schedule (with starting and finishing time of each job). Then, we can calculate the *normalised* objective value (e.g. makespan and flowtime) of the schedule. Finally, the fitness value is set to the average value of all the normalised objective values (line 7). Here, the normalisation (line 5) is with respect to a reference value $\mathtt{obj}^*(I)$, which can be set to either the best known (lower bound of) objective value of the instance, or the objective value obtained by applying benchmark routing and sequencing rules.

As shown in Algorithm 1 (lines 14 and 15), for evaluating a routing (sequencing) rule, it is combined with the sequencing (routing) component of the context vector so that the discrete event simulation in Algorithm 2 can be constructed.

Algorithm 2. evaluate($p_r, p_s, \mathcal{I}_{train}$)

Input: A routing rule p_r, a sequencing rule p_s, a set of FJSS instances $\mathcal{I}_{train}$
Output: A fitness value
1 $f \leftarrow 0$;
2 **foreach** $I \in \mathcal{I}_{train}$ **do**
3 | Construct a discrete event simulation based on p_r, p_s and I;
4 | Generate a schedule $S(p_r, p_s, I)$ by running the discrete event simulation;
5 | $f \leftarrow f + \frac{\text{obj}(S(p_r, p_s, I))}{\text{obj}^*(I)}$;   // normalisation cross instances
6 **end**
7 **return** $f/|\mathcal{I}_{train}|$;   // average over the training set

---

4   Experiment Settings

To evaluate the proposed CCGP, we conducted experiments on both static and dynamic FJSS datasets. The static instances are commonly used in the evaluation of FJSS methods [2], and their lower and upper bounds of makespan are known. Specifically, there are 4 static FJSS datasets, namely the Barnes dataset [1], Brandimarte dataset [4], Dauzere dataset [8] and Hurink dataset [14]. The Barnes dataset consists of 21 instances with 10 or 15 jobs. Each job has 11 to 18 operations, and each operation has 1.07–1.3 candidate machines. Thus, the Barnes dataset is small and has relatively low flexibility. The Brandimarte dataset has 10 small sized instances (no more than 20 jobs and 15 machines, each job has 5–15 operations) and medium flexibility (each operation has 2–6 machine options). The Dauzere dataset consists of 18 instances with similar size and flexibility as the Brandimarte dataset. There are 66 instances in the Hurink dataset, which can be divided into 4 subsets with increasing flexibility, namely sdata, edata, rdata and vdata. The sdata instances are essentially JSS instances, as no operation can be processed by more than one machine. In the most flexible vdata instances, all the operations can be processed by multiple machines.

For dynamic simulation, the configuration is given in Table 1, which has been used in previous studies (e.g. [11,20]).

The parameter setting of CCGP is standard, as given in Table 2. The terminal set of CCGP is described in Table 3. The terminals are adapted from the JSS terminals proposed in [20]. The terminals involving the future operations (e.g. NPT and WKR) are modified to take into account the machine-dependent processing times. For each future operation, the processing time is set to the median processing time of all the options.

The function set of CCGP is set to $\{+, -, *, /, \min, \max\}$, where "/" is the protected division that returns 1 if divided by 0. The "min" and "max" operators take two arguments, and return the minimal (maximal) value between them.

In the experiment, we will compare CCGP with the GP counterpart with routing rule fixed to the Least Work in Queue (LWQ) rule, and evolving the sequencing rule only. For the sake of convenience, the counterpart will be denoted as SeqGP hereafter. For fair comparison, the population size of SeqGP is set to

1024 so that the number of fitness evaluations per generation is the same as CCGP. All the other parameters are the same for SeqGP and CCGP.

Table 1. The dynamic JSS simulation system configuration.

Parameter	Value
#machines	10
#jobs (#warmup jobs)	5000 (1000)
#operations per job	Uniform discrete distribution between 1 and 10
#Machines per operation	Uniform discrete distribution between 1 and 10
Job arrival process	Poisson process
Utilisation level	{0.85, 0.95}
Processing time	Uniform discrete distribution between 1 and 99
Job weights	20% with weight 1, 60% with weight 2, 20% with weight 4

Table 2. The parameter setting of CCGP.

Parameter	Value
Number of subpopulations	2
Subpulation size (**popsize**)	512
Maximal depth	8
Crossover/Mutation/Reproduction rates	80%/15%/5%
Parent selection	Tournament selection with size 7
Elitism	2 best individuals
Number of generations	51

Table 3. The terminal set of CCGP.

Notation	Description
NIQ	Number of Operations in a Machine's Queue
WIQ	Work In a Machine's Queue
MWT	Waiting Time of a Machine
PT	Processing Time of an Operation on a given Machine
NPT	**Median** Processing Time for the Next Operation on Machine options
OWT	The Waiting Time of an Operation
WKR	**Median** Amount of Work Remaining for a Job
NOR	The Number of Operations Remaining in a Job
W	Weight of a Job
TIS	Time In System

5   Results and Discussions

5.1   Comparing Manually Designed Routing Rules for SeqGP

SeqGP requires a pre-specified routing rule for evaluating the evolved sequencing rules. In existing studies, only the *least waiting time assignment* routing rule was considered [13, 32] without investigating whether it is the best routing rule. In this paper, we first compare a set of commonly used manually designed routing rules on the static FJSS instances to identify the best routing rule for SeqGP.

Specifically, four manually designed routing rules are taken into account in the comparison. They are described as follows:

1. *Least Work in Queue* (LWQ): select the machine with the least work (total processing time) in its queue;
2. *Least Queue Size* (LQS): select the machine with the least queue size (number of operations in the queue);
3. *Earliest Ready Time* (ERT): select the machine that will become ready (idle) the earliest;
4. *Shortest Busy Time* (SBT): select the machine with the shortest busy time so far.

Among the above routing rules, the ERT is essentially the same as the least waiting time rule used in previous studies (e.g. [13, 32]).

For each routing rule, the SeqGP with that routing rule was run on each static instance for 30 times (except the 66 Hurink-sdata instances, which are essentially JSS instances). Then, a routing rule is considered as a *"winner"* of an instance if it achieved the best mean makespan over the 30 runs (there may be multiple winners). Then, we compare the number of instances where each routing rule was a winner.

Table 4 shows the number of instances in each static dataset where each routing rule was a winner. It can be seen that LWQ was a winner for most instances (127 out of 247), followed by ERT. More specifically, the advantage of ERT over LWQ was only on the Barnes dataset, which was the very inflexible. As the flexibility increases, the advantage of LWQ becomes more obvious.

The findings in this subsection is interesting as it identifies LWQ as a better routing rule than ERT, which has been used in previous studies, for static FJSS. In subsequent experiments, we set LWQ as the fixed routing rule for SeqGP.

5.2   Optimisation Performance on Static Instances

The first set of experiments aims to verify the optimisation performance of SeqGP and CCGP on the static FJSS instances, without a training and test (generalisation) process. This way, one can investigate the effectiveness of (co-)evolving dispatching rules as compared to directly optimising FJSS solutions.

For the static instances, the objective is to minimise the makespan. For each static instance, CCGP and SeqGP were run 30 times independently, and the

Table 4. The number of instances in each static dataset where each compared routing rule was a winner.

Dataset	#Instances	LWQ	LQS	ERT	SBT
Barnes	21	0	0	13	8
Brandimarte	10	8	0	2	2
Dauzere	18	16	0	1	1
Hurink-edata	66	31	3	27	9
Hurink-rdata	66	39	0	27	0
Hurink-vdata	66	33	0	33	0
Total	247	127	3	103	20

normalised makespans (makespan over the known lower bound) of the best rules were recorded. In addition, two manually designed sequencing rules, i.e. First-Come-First-Serve (FCFS) and Shortest Processing Time (SPT), are also taken into comparison.

Table 5 shows the summary of the compared algorithms over 30 independent runs for the static datasets. FCFS and SPT are deterministic rules. Therefore, for each dataset, the average normalised makespan value cross all the instances of that dataset is shown. SeqGP and CCGP are stochastic algorithms. Therefore, for each dataset, the mean and standard deviation over the 30 runs are given. In addition, for each instance, Wilcoxon rank sum test with significance level of 0.05 was conducted between the 30 results obtained by CCGP and SeqGP. Then, for each dataset, the numbers of instances that CCGP performed significantly better than SeqGP ("W"), comparable with SeqGP ("D"), and significantly worse than SeqGP ("L") are given.

Table 5. The normalised makespan (MK/LB) with respect to lower bound of the compared algorithms over 30 independent runs for the static datasets.

Dataset	#Instances	FCFS	SPT	SeqGP	CCGP	W-D-L
Barnes	21	1.270	1.238	1.079(0.0021)	1.065(0.0026)	15-6-0
Brandimarte	10	1.431	1.501	1.229(0.0042)	1.062(0.0045)	8-2-0
Dauzere	18	1.244	1.227	1.086(0.0012)	1.061(0.0019)	16-2-0
Hurink-edata	66	1.247	1.241	1.070(0.0009)	1.048(0.0014)	50-15-1
Hurink-rdata	66	1.271	1.292	1.123(0.0012)	1.062(0.0014)	64-2-0
Hurink-vdata	66	1.312	1.324	1.215(0.0005)	1.019(0.0008)	64-2-0

From Table 5, it is obvious that both SeqGP and CCGP dramatically outperformed the manually designed rules (FCFS and SPT). In addition, CCGP performed much better than SeqGP. Overall, FCFS and SPT obtained solutions which are 25%–50% worse than the lower bound. SeqGP obtained solutions that are 7%–23% worse than the lower bound. All the solutions obtained by CCGP are less than 7% worse than the lower bound. The most obvious advantage of CCGP over SeqGP occurred on the Brandimarte and Hurink-vdata datasets, which have reasonable large problem sizes and flexibility.

More specifically, CCGP statistically significantly outperformed SeqGP on most static instances (e.g. 64 out of 66 of the Hurink-rdata and Hurink-vdata instances). CCGP was defeated by SeqGP on only one Hurink-edata instance out of the total 247 static instances. This clearly demonstrates the advantage of CCGP over SeqGP on solving static FJSS instances.

Figure 1 shows the convergence curves of SeqGP and CCGP on three representative instances (the ribbon is the standard deviation over the 30 runs), on which CCGP performed significantly better than, worse than, and comparable with SeqGP. All the other instances showed similar patterns. From the figure, it is clear that CCGP started from a much higher makespan due to the random initial routing rule. Then, it converged very fast, and achieved local optima within 20 generations.

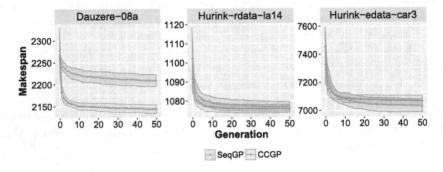

Fig. 1. The convergence curves of the makespan of the 30 runs of SeqGP and CCGP.

Finally, CCGP can obtain solutions that are less than 7% worse than the lower bound, which can be seen as a promising optimisation performance for static FJSS instances.

5.3   Generalisation Performance on Dynamic Instances

The experiments in the dynamic environment is to examine the generalisation performance of SeqGP and CCGP. We consider 2 utilisation levels (0.85 and 0.95) and 3 objectives in the dynamic environment. Specifically, we consider minimising (1) mean flowtime (Fmean), (2) max flowtime (Fmax) and (3) mean weighted flowtime (MWF). This results in $3 \times 2 = 6$ scenarios. For each scenario, SeqGP and CCGP were run 30 times independently over a training set. The training set consists of a single dynamic FJSS simulation. To improve generalisation, the random seed for generating the training simulation changes per generation. After the training process, the best rule of the last generation is then tested on an unseen test set to evaluate its test performance. The test set consists of 50 dynamic simulations using the same configurations as the training set, but different random seeds.

For the dynamic simulations, the lower bound objective values are unknown. Therefore, the normalisation is with respect to the objective value obtained by a benchmark dispatching rule (routing plus sequencing rules). Here, the benchmark routing rule is fixed to LWQ for all the scenarios. The benchmark sequencing rule is specified depending on the scenario. Based on our preliminary work [20], we set the benchmark sequencing rule to FCFS for the scenarios minimising Fmax, to SPT for the scenarios minimising Fmean, and to WSPT for the scenarios minimising MWF.

Figure 2 shows the convergence curves of the test fitness obtained by SeqGP and CCGP over the 6 dynamic scenarios. From the figure, it is obvious that CCGP significantly outperformed SeqGP in all the 6 scenarios. The Wilcoxon rank sum test with significance level of 0.05 also confirmed the significance. The convergence curves of CCGP are almost always below the curves of SeqGP. For the scenarios minimising Fmean and MWF, CCGP successfully initialised much more effective routing rules even from the first generation.

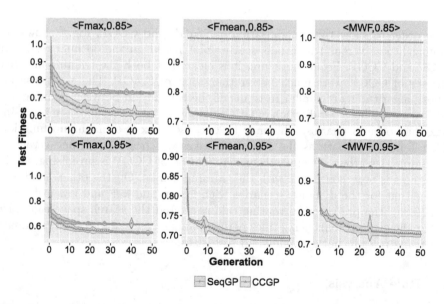

Fig. 2. The convergence curves of the test fitness obtained by SeqGP and CCGP.

Figure 3 shows the convergence curves of the size of the sequencing rules obtained by SeqGP and CCGP. It can be seen that the two algorithms have similar convergence curves in terms of sequencing rule size, i.e. evolving routing rules does not seem to make the sequencing rule simpler or more complex.

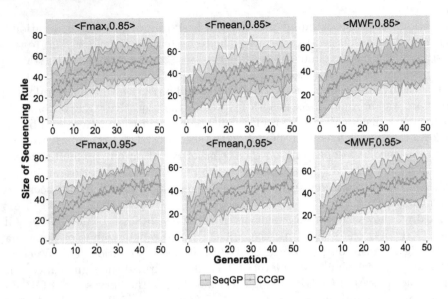

Fig. 3. The convergence curves of the **sequencing** rule size obtained by SeqGP and CCGP.

In order to show the generalisation of SeqGP and CCGP, Fig. 4 shows the training fitness versus test fitness scatter plot based on the 30 final results of SeqGP and CCGP. From the figure, it is clear that both the training and test fitnesses of CCGP were much better than that of SeqGP. The generalisation of both algorithms are similar in terms of the correlation between training and test fitnesses. The generalisation of CCGP is poorer for the scenarios minimising Fmax than other scenarios. This may be because Fmax is a maximum function, which is not so smooth as the other objectives which are based on average as the sample size grows. Overall, the generalisation of CCGP is promising, as the test fitness is very consistent with the training fitness. On the other hand, one can see that for the dynamic scenarios with Fmean and MWF and low utilisation level (0.85), the pre-specified routing rule restricted the search space too much so that the evolved sequencing rules perform almost the same as the benchmark sequencing rules in both training and test instances.

5.4   Rule Analysis

Equation 1 shows an example routing rule evolved by CCGP for the scenario $\langle$MWF, 0.95$\rangle$.

$$\min\{NIQ \times PT, WIQ\} + \frac{W}{MWT \times PT} - \min\{MWT \times W, NIQ \times NOR\}. \quad (1)$$

It mainly consists of three components. The first component $\min\{NIQ \times PT, WIQ\}$ is similar to WIQ, i.e. the number of operations in queue times

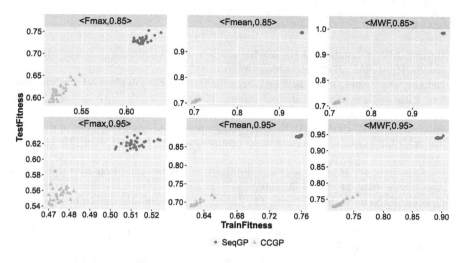

Fig. 4. The training fitness versus test fitness scatter plot based on the 30 final results of SeqGP and CCGP.

the processing time of an operation is similar to the total processing time in queue. The second and third terms show that the routing rule prefers machines with larger MWT, i.e. the earliest available machine (MWT = current time − machine ready time). This preference is more obvious if the current job has a larger weight. That is, the routing rule tries to finish the more important jobs as early as possible. In summary, CCGP can automatically evolve routing rules that contain sensible patterns consistent with intuition for making routing decisions.

6   Conclusions and Future Work

This paper is the very first piece of work to co-evolving routing and sequencing rules simultaneously for dynamic FJSS, and significantly extends the previous work on both static and dynamic FJSS. Through comprehensive experiments, we had several interesting findings. First, we found that the commonly used pre-specified routing rule is not the best one for static FJSS. We found a better routing rule, which is LWQ (least work in queue). Then, we developed the GPHH with the routing rule fixed to LWQ (named SeqGP), and the Cooperative Co-evolution GP (CCGP) that co-evolves the routing and sequencing rules simultaneously. The results show that CCGP performed much better than SeqGP in both static and dynamic scenarios. This demonstrates that the routing rules evolved by CCGP are much better than the rules that are manually designed and fixed in SeqGP. In other words, there is a great potential to find much more effective routing rules for FJSS, especially in the dynamic environment.

In the future, we will focus on further improving the effectiveness of CCGP. In this paper, only a baseline CC framework is adopted. We will consider incorporating other domain specific strategies such as feature selection and construction to improve the effectiveness and efficiency of the GP search.

References

1. Barnes, J.W., Chambers, J.B.: Solving the job shop scheduling problem with tabu search. IIE Trans. **27**(2), 257–263 (1995)
2. Behnke, D., Geiger, M.J.: Test instances for the flexible job shop scheduling problem with work centers. Technical report, Helmut Schmidt Universitat, January 2012
3. Blackstone, J.H., Phillips, D.T., Hogg, G.L.: A state-of-the-art survey of dispatching rules for manufacturing job shop operations. Int. J. Prod. Res. **20**(1), 27–45 (1982)
4. Brandimarte, P.: Routing and scheduling in a flexible job shop by tabu search. Ann. Oper. Res. **41**, 157–183 (1993)
5. Branke, J., Nguyen, S., Pickardt, C., Zhang, M.: Automated design of production scheduling heuristics: a review. IEEE Trans. Evol. Comput. **20**(1), 110–124 (2016)
6. Brucker, P., Schlie, R.: Job-shop scheduling with multi-purpose machines. Computing **45**(4), 369–375 (1990)
7. Burke, E., Gendreau, M., Hyde, M., Kendall, G., Ochoa, G., Özcan, E., Qu, R.: Hyper-heuristics: a survey of the state of the art. J. Oper. Res. Soc. **64**(12), 1695–1724 (2013)
8. Dauzere-Peres, S., Paulli, J.: An integrated approach for modeling and solving the general multiprocessor job-shop scheduling problem using tabu search. Ann. Oper. Res. **70**, 281–306 (1997)
9. Dimopoulos, C., Zalzala, A.: Invesigating the use of genetic programming for a classic one-machine scheduling problem. Adv. Eng. Softw. **32**, 489–498 (2001)
10. Geiger, C.D., Uzsoy, R., Aytug, H.: Rapid modeling and discovery of priority dispatching rules: an autonomous learning approach. J. Sched. **9**, 7–34 (2006)
11. Hildebrandt, T., Branke, J.: On using surrogates with genetic programming. Evol. Comput. **23**(3), 343–367 (2015)
12. Hildebrandt, T., Heger, J., Scholz-Reiter, B.: Towards improved dispatching rules for complex shop floor scenarios–a genetic programming approach. In: Proceedings of the 12th Annual Conference on Genetic and Evolutionary Computation Conference (2010)
13. Ho, N.B., Tay, J.C.: Genace: an efficient cultural algorithm for solving the flexible job-shop problem. In: IEEE Congress on Evolutionary Computation, vol. 2, pp. 1759–1766. IEEE (2004)
14. Hurink, J., Jurisch, B., Thole, M.: Tabu search for the job-shop scheduling problem with multi-purpose machines. Oper. Res. Spektrum **15**(4), 205–215 (1994)
15. Jones, A., Rabelo, L.C., Sharawi, A.T.: Survey of job shop scheduling techniques. In: Wiley Encyclopedia of Electrical and Electronics Engineering. John Wiley & Sons, New York (1999)
16. Kacem, I., Hammadi, S., Borne, P.: Approach by localization and multiobjective evolutionary optimization for flexible job-shop scheduling problems. IEEE Trans. Syst. Man Cybern. **32**(1), 1–13 (2002)
17. Kiran, A.S.: Simulation studies in job shop scheduling - I a survey. Comput. Ind. Eng. **8**(2), 87–93 (1984)
18. Land, A.H., Doig, A.G.: An automatic method of solving discrete programming problems. Econometrica **28**(3), 497–520 (1960)
19. Mei, Y., Nguyen, S., Xue, B., Zhang, M.: An efficient feature selection algorithm for evolving job shop scheduling rules with genetic programming. IEEE Trans. Emerg. Top. Comput. Intell. **1**(5), 339–353 (2017)

20. Mei, Y., Nguyen, S., Zhang, M.: Evolving time-invariant dispatching rules in job shop scheduling with genetic programming. In: McDermott, J., Castelli, M., Sekanina, L., Haasdijk, E., García-Sánchez, P. (eds.) EuroGP 2017. LNCS, vol. 10196, pp. 147–163. Springer, Cham (2017). https://doi.org/10.1007/978-3-319-55696-3_10

21. Mei, Y., Zhang, M., Nyugen, S.: Feature selection in evolving job shop dispatching rules with genetic programming. In: GECCO (2016)

22. Miyashita, K.: Job-shop scheduling with GP. In: Genetic and Evolutionary Computation Conference (2000)

23. Nguyen, S., Mei, Y., Zhang, M.: Genetic programming for production scheduling: a survey with a unified framework. Complex Intell. Syst. **3**(1), 41–66 (2017)

24. Nguyen, S., Zhang, M., Johnston, M., Tan, K.C.: Automatic design of scheduling policies for dynamic multi-objective job shop scheduling via cooperative coevolution genetic programming. IEEE Trans. Evol. Comput. **18**(2), 193–208 (2014)

25. Norman, B.A., Bean, J.C.: A genetic algorithm methodology for complex scheduling problems. Nav. Res. Logist. **46**(2), 199–211 (1999)

26. Nowicki, E., Smutnicki, C.: A fast taboo search algorithm for the job shop problem. Manag. Sci. **42**(6), 797–813 (1996)

27. Pinedo, M.L.: Scheduling: Theory, Algorithms and Systems. Springer, New York (2012)

28. Potter, M.A., Jong, K.A.D.: Cooperative coevolution: an architecture for evolving coadapted subcomponents. Evol. Compt. **8**, 1–29 (2000)

29. Ramasesh, R.: Dynamic job shop scheduling: a survey of simulation research. Omega **18**(1), 43–57 (1990)

30. Sels, V., Gheysen, N., Vanhoucke, M.: A comparison of priority rules for the job shop scheduling problem under different flow time- and tardiness-related objective functions. Int. J. Prod. Res. **50**(15), 4255–4270 (2012)

31. Subramaniam, V., Ramesh, T., Lee, G.K., Wong, Y.S., Hong, G.S.: Job shop scheduling with dynamic fuzzy selection of dispatching rules. Int. J. Adv. Manuf. Technol. **16**, 759–764 (2000)

32. Tay, J., Ho, N.: Evolving dispatching rules using genetic programming for solving multi-objective flexible job-shop problems. Comput. Ind. Eng. **54**(3), 453–473 (2008)

33. Tay, J.C., Ho, N.B.: Evolving dispatching rules using genetic programming for solving multi-objective flexible job-shop problems. Technical report, Evolutionary and Complex Systems Program, School of Computer Engineering, Nanyang Technological University (2007)

34. van Laarhoven, P.J.M., Aarts, E.H.L., Lenstra, J.K.: Job shop scheduling by simulated annealing. Oper. Res. **40**(1), 113–125 (1992)

35. Zhang, G., Gao, L., Shi, Y.: An effective genetic algorithm for the flexible job-shop scheduling problem. Expert Syst. Appl. **38**(4), 3563–3573 (2011)

Author Index

Printed in the United States
By Bookmasters